Fields and Particles

H. Mitter W. Schweiger (Eds.)

Fields and Particles

Proceedings of the
XXIX Int. Universitätswochen für Kernphysik
Schladming, Austria, March 1990

With 23 Figures

Springer-Verlag
Berlin Heidelberg New York London
Paris Tokyo Hong Kong Barcelona

Professor Dr. Heinrich Mitter
Dr. Wolfgang Schweiger
Institut für Theoretische Physik, Karl-Franzens-Universität, Universitätsplatz 5,
A-8010 Graz, Austria

ISBN-13:978-3-642-76092-1 e-ISBN-13:978-3-642-76090-7
DOI: 10.1007/978-3-642-76090-7

Softcover reprint of the hardcover 1st edition 1990

The text was processed by the authors using the $\TeX$ macro package from Springer-Verlag.

2155/3140-543210 – Printed on acid-free paper

Preface

This volume contains the written versions of invited lectures presented at the 29th "Internationale Universitätswochen für Kernphysik" in Schladming, Austria, in March 1990.

The generous support of our sponsors, the Austrian Ministry of Science and Research, the Government of Styria, and others, made it possible to invite expert lecturers. In choosing the topics of the course we have tried to select some of the currently most fiercely debated aspects of quantum field theory. It is a pleasure for us to thank all the speakers for their excellent presentations and their efforts in preparing the lecture notes.

After the school the lecture notes were revised by the authors and partly rewritten in TEX. We are also indebted to Mrs. Neuhold for the careful typing of those notes which we did not receive in TEX.

Graz, Austria
July 1990

H. Mitter
W. Schweiger

Contents

An Introduction to Integrable Models and Conformal Field Theory

H. Grosse

Institut für Theoretische Physik, Universität Wien
Boltzmanngasse 5, A – 1090 Wien, Austria

Abstract: We review first the steps which lead to solutions of continuous integrable models in two dimensions. Next we discuss in more detail solvable models of statistical physics. The central role of the Yang Baxter equation is emphasized. The connection to Braids and Knots and certain algebras is mentioned.

The continuum limit of lattice models may yield a conformal invariant field theory. The Virasoro algebra is realized in a special manner. The central extension is related to the conformal anomaly. Unitary highest weight representations restrict both the coefficient in front of the anomaly and the conformal weights. The latter are directly related to the critical exponents. The operator product expansion as well as the classification of field operators is mentioned finally.

1. Introduction

Various subjects became more and more interrelated recently (Fig. 1):

1.1 Continuous Integrable Models

More than 150 years ago J. Scott-Russell observed solitons interacting with each other. 1898 Korteweg and de Vries wrote down an equation for $v(t,x)$:

$$v_t = 6vv_x - v_{xxx} \tag{1.1}$$

which was supposed to describe water waves in a channel. Fermi, Pasta and Ulam observed already by numerical experiments that certain modes of a dynamical system may dominate. But it was not before 1967 when Gardner, Green, Kruskal and Miura invented the inverse scattering method to *solve* the KdV equation. Especially the soliton solutions were obtained explicitly. Lax reformulated their scheme. Soon after these discoveries many hierarchies were obtained. Among them there are the Nonlinear-Schrödinger equation, the Sine-Gordon equation, the Toda lattices but more recently the Kadomtsev-Petviashvili type equations in three dimensions were also shown to be integrable. Especially during these more recent developments many more insights have been obtained. Special vertex operators "create" solitons. The Bose-Fermi

Fields and Particles Editors: H. Mitter · W. Schweiger

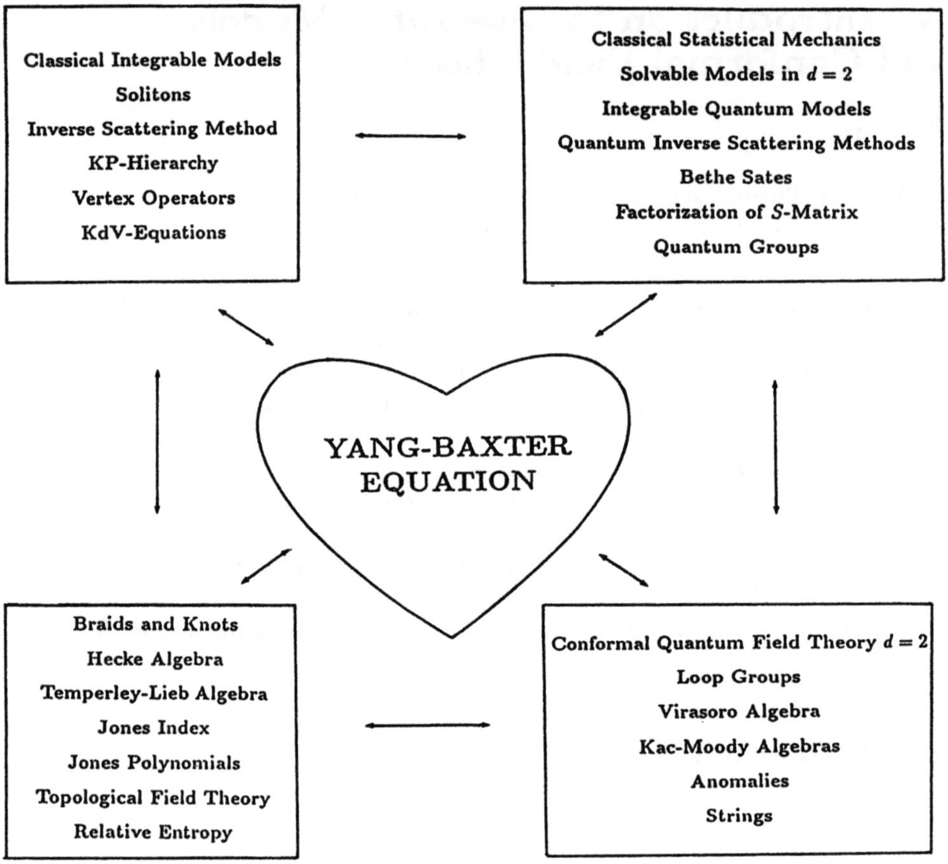

Fig. 1. Interrelations between various subjects

correspondence allows to identify orbits of infinite-dimensional groups with nonlinear equations. An interesting connection to the Riemann-Hilbert problem resulted.

1.2 "Solvable" Models of Statistical Physics

The subject started when Lenz asked his student Ising to study the thermodynamics of a one-dimensional spin system. The teacher became disappointed since no nontrivial phase transition was found. Heisenberg wrote down a more general quantum spin model with Hamiltonian

$$H = -J \sum_{j=1}^{N} \boldsymbol{\sigma}_j \boldsymbol{\sigma}_{j+1}, \qquad \sigma_j^k = \mathbf{1} \otimes \dots \underbrace{\sigma^k}_{j\text{-th place}} \otimes \mathbf{1} \dots \otimes \mathbf{1}, \tag{1.2}$$

which is supposed to describe ferromagnetism for $J > 0$. Physically J results from the exchange integral, but even its sign is hard to evaluate. σ^k are Pauli matrices. If we allow only for spin up and down degrees of freedom the Ising model results. The great step forward was done by Onsager who calculated the free energy of the two-dimensional Ising model. If we allow for an anisotropic interaction in spin space of (1.2) we connect to vertex models. A special case of the six-vertex model yields a two-dimensional description of ice.

The more general eight-vertex models were studied by Baxter and others. Among the many other models we mention especially the Potts models, since the algebra obtained from operators entering the transfer matrix, the Temperley-Lieb algebra, is related to the Hecke algebra and finally to braids and knots.

1.3 The Yang-Baxter Relation

Already in 1938 Bethe found an efficient method to obtain eigenfunctions of (1.2). The Bethe states are factorized in the same way as the S-matrices of certain two-dimensional models. From a quantized form of the inverse scattering method an algebraic scheme results. The algebras of operators entering this algebraic Bethe ansatz are closely connected to quantum groups.

It was realized that a large number of integrable models are obtained from solving consistency conditions, which are nowadays connected to the names of Yang and Baxter. The commutativity of the transfer matrices to different spectral values yields immediately an infinite number of conserved quantities and indicates the integrability of the model.

1.4 Braids and Knots

The algebraic steps by which we obtain the Yang-Baxter relations are closely connected to the braiding relations. It is therefore not surprising that invariant knot polynomials can be obtained from integrable lattice models. The constant entering the Temperley-Lieb algebra is related to the Jones index. The partition function of the Potts models connects to chromatic polynomials. The Beraha numbers play a special role. Jones polynomials can be obtained as correlation functions within a topological field theory. Jones obtained the special values for his index by studying embeddings of type II von Neumann algebras. Realizations of such algebras are given by conformal field theory.

1.5 Conformal Field Theory $d = 2$

The construction of correlation functions in the continuum limit consists of tedious steps and has been done explicitly only for the Ising model in two dimensions. Following standard wisdom we should follow renormalization group trajectories while taking the limit where the lattice constant goes to zero. The temperature should approach the critical one, and the correlation length should diverge. If we obtain a conformal invariant field theory, then the Fourier coefficients of the energy momentum tensor yield a representation of the Virasoro algebra. More precisely, a ray representation is obtained due to the occurrence of a central extension term. The coefficient in front of this conformal anomaly as well as the eigenvalue of one of the generators determine properties of the representation. The latter is called a conformal weight and enters the behaviour of correlation functions and determines critical exponents. For unitary representations both constants are severely restricted. This allows to classify possible two-dimensional conformal covariant models. Depending on the behaviour under conformal transformations we introduce the notion of primary and secondary fields. A further useful tool to study these models consists in using the operator product expansion. Conformal blocks form a closed algebra. Correlation functions are obtained as solutions of differential equations.

2. Integrable Continuum Models – Inverse Scattering Method – Solitons

2.1 A General Scheme for Solving (Linear) Problems

We are all familiar with integrable systems having a finite number of degrees of freedom. Most books and courses in mechanics deal with such systems; e.g. the oscillator or the Coulomb problem (Fig. 2). In these examples we can find enough globally defined conserved quantities. More precisely consider a $2n$-dimensional phase space with local coordinates (q_i, p_i) and canonical two form $\omega = \sum_i dq_i \wedge dp_i$. Poisson brackets are given by $\{f, g\} = \omega(X_f, X_g)$ where X_f denotes the vector field generated by f and the equations of motion read $\dot{q}_i = \{q_i, H\}$, $\dot{p}_i = \{p_i, H\}$, with H being the Hamiltonian. A well-known theorem says:

Liouville-Arnold: If there exist n globally defined conserved quantities K_i in involution $\{K_i, K_j\} = 0$, then there exists a transformation to new variables φ_i, I_i (cyclic action-angle variables) such that the new time evolution is given by $\varphi_i(t) = \omega_i t + \varphi_i(0)$. I_i are constants and $\omega = \sum_i d\varphi_i \wedge dI_i$.

The scheme for solving the time evolution is therefore given by the diagram:

$$\begin{array}{ccc} q_i(0), p_i(0) & \overset{\text{“direct” step}}{\longrightarrow} & \varphi_i(0), I_i(0) \\ ?\downarrow & & \downarrow \ \text{free time evolution} \\ q_i(t), p_i(t) & \overset{\text{“inverse” step}}{\longleftarrow} & \varphi_i(t), I_i(t) \end{array}$$

As a further example, where a similar scheme applies, we solve the linear diffusion equation $\Phi_t = \Phi_{xx}$ with infinite number of degrees of freedom. From

Fig. 2. Examples of well-known integrable systems

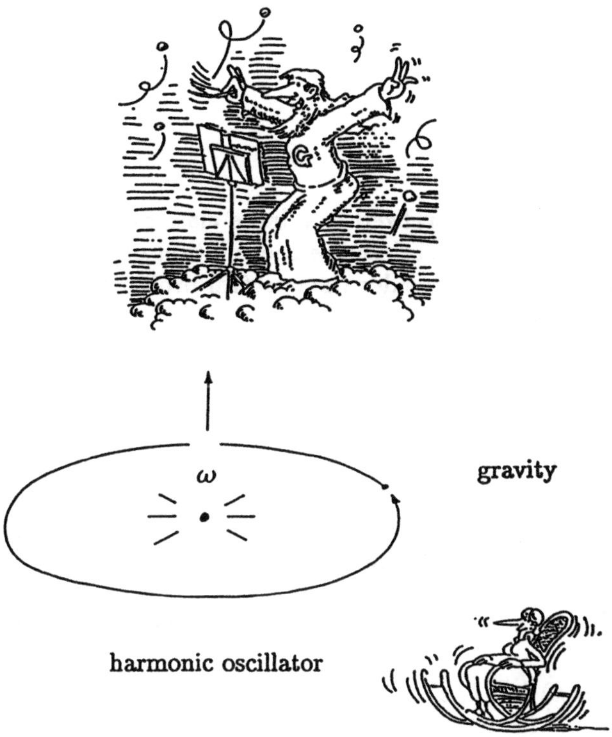

the dispersion law $\omega = -ik^2$ we obtain the solution of the initial value problem through Fourier transformation (F.T.):

$$\Phi(t,x) = \int_{-\infty}^{\infty} dk e^{ikx} \tilde{\Phi}(t,x), \qquad \tilde{\Phi}(t,x) = e^{-k^2 t} \tilde{\Phi}(0,k). \tag{2.1}$$

A rapid spreading of wave packets occurs. The scheme for solving looks similar as before

$$\begin{array}{ccc} \Phi(0,x) & \xrightarrow{\text{F.T., "direct" step}} & \tilde{\Phi}(0,k) \\ ?\downarrow & & \downarrow \text{ free time evolution} \\ \Phi(t,x) & \xleftarrow{(\text{F.T.})^{-1}, \text{"inverse" step}} & \tilde{\Phi}(t,k) \end{array}$$

We note that there exists an infinite number of local constants of motion. First of all we may integrate the equation of motion and realize that the spatial integral of $\Phi(t,x)$

$$\frac{\partial}{\partial t} \int_{-\infty}^{\infty} dx \Phi(t,x) = \int_{-\infty}^{\infty} dx \frac{\partial^2}{\partial x^2} \Phi(t,x) = 0 \tag{2.2}$$

is conserved. Moreover all the Fourier coefficients of $\Phi(t,x)$

$$e^{q^2t}\int_{-\infty}^{\infty}\frac{dx}{2\pi}e^{-iqx}\Phi(t,x)=\tilde{\Phi}(0,k) \tag{2.3}$$

provide us with constants of motion.

2.2 The Direct Step

The above mentioned scheme may be generalized to a large class of nonlinear evolution equations. As an example we shall describe the solution of the KdV equation (1.1). As for the direct step we consider the spectral problem for the one-dimensional Schrödinger operator

$$(-\frac{d^2}{dx^2}+v(x))\psi(x)=E\psi(x), \tag{2.4}$$

for real potential v, such that $(1+x^2)|v|\in L^1(\mathbb{R})$. We introduce Jost solutions f_1, f_2 with spatial behaviour at infinity

$$\lim_{x\to\infty}f_1(k,x)e^{-ikx}=1, \qquad \lim_{x\to-\infty}e^{ikx}f_2(k,x)=1. \tag{2.5}$$

Since $f_1(-k,x)$ solves Equ. (2.4) too, there exists a relation between these three functions

$$f_2(k,x)=a(k)f_1(-k,x)+b(k)f_1(k,x). \tag{2.6}$$

The physical solution $\psi(k,x)$ should describe scattering from one side and is connected to Jost solutions by

$$\psi(k,x)=T(k)f_2(k,x)=R(k)f_1(k,x)+f_1(-k,x), \tag{2.7}$$

where we introduced reflection and transmission coefficients of the one-dimensional scattering problem. We compare (2.6) to (2.7) and relate $b=R/T$ and $a=1/T$ to R and T.

From the Volterra integral equation obeyed by f_1 we deduce analyticity properties of that function and the fact that $|f_1(k,x)-\exp(ikx)|\in L^2((-\infty,\infty),dk)$. A theorem, similar to the Paley-Wiener one, due to Boas allows to deduce support properties of the Fourier tranform in k. This yields a representation for f_1 of the type

$$f_1(k,x)=e^{ikx}+\int_x^{\infty}dyK(x,y)e^{iky}. \tag{2.8}$$

Note that $K(x,y)$ is independent of the spectral parameter k.

2.3 The Inverse Step

The inverse problem for (2.4) consists in recovering the potential from scattering data. The two-dimensional drum problem became famous: M. Kac asked the question as to whether one can hear the shape of a drum. We consider the operator $(-\Delta)|_{\Omega}$ with Dirichlet boundary conditions on Ω. Does the knowledge of all frequencies determine the shape? We may expand the trace of the heat kernel operator

$$\mathrm{Tr}\ \exp(\beta(-\Delta)|_{\Omega}) \overset{\beta\searrow 0}{\cong} \frac{c_0|\Omega|}{\beta} + \frac{c_1 L}{\beta^{1/2}} + c_2(1-n) + \ldots \tag{2.9}$$

and observe that the area of the drum $|\Omega|$, the length of the boundary L and the number of holes n are determined together with the frequencies. Since $|\Omega| \leq L^2/4\pi$ is a standard isoperimetric inequality which becomes an equality iff the drum is a circle. We deduce uniqueness of the inverse problem for the circular case. The general problem is very complicated. A counterexample exists in 16 dimensions.

Coming back to the Schrödinger potential problem we first connect the kernel $K(x,y)$ to the potential. We apply the Schrödinger operator to $f_1 - e^{ikx}$ and Fourier transform in k:

$$(-\frac{\partial^2}{\partial x^2} + \frac{\partial^2}{\partial y^2} + v(x))K(x,y) = -v(x)\delta(x-y). \tag{2.10}$$

Introduce light cone coordinates $\xi = y + x$, $\eta = y - x$, integrate (2.10) from $-\varepsilon$ to ε, observe that $K(x,y)$ vanishes for $x > y$ and deduce that

$$-2\frac{d}{dx}K(x,x) = v(x). \tag{2.11}$$

The connection of $K(x,y)$ to the scattering data $\{R(k), \varepsilon_\ell, c_\ell\}$, where ε_ℓ denote the energy eigenvalues and c_ℓ bound state wave function normalization constants, is more tricky. We rewrite (2.7) so that all quantities have a Fourier transform

$$\begin{aligned}(T(k)-1)f_2(k,x) = R(k)(f_1(k,x) - e^{ikx}) + (f_1(-k,x) - e^{-ikx})\\ -(f_2(k,x) - e^{-ikx}) + R(k)e^{ikx}.\end{aligned} \tag{2.12}$$

We obtain contributions to the F.T. of the l.h.s. from bound states

$$\int_{-\infty}^{\infty} \frac{dk}{2\pi}(T(k)-1)e^{iky}f_2(k,x) = -\sum_{\ell=1}^{N} c_\ell^2 e^{-\kappa_\ell y} f_1(i\kappa_\ell, x), \tag{2.13}$$

where we used the proportionality of f_1 and f_2 at $k = i\kappa_\ell$. If we define a kernel G through the scattering data by

$$G(x,y) = \sum_{\ell=1}^{N} c_\ell^2 e^{-\kappa_\ell(x+y)} + \int_{-\infty}^{\infty} \frac{dk}{2\pi} e^{ik(x+y)} R(k), \tag{2.14}$$

the F.T. of (2.12) is rewritten and becomes the **Gelfand-Levitan-Marchenko** equation

$$K(x,y) + G(x,y) + \int_x^\infty dz K(x,z) G(z,y) = 0, \qquad x < y. \tag{2.15}$$

We note the meaning of c_ℓ: If $f_1(i\kappa_\ell, x) \overset{x\to\infty}{\simeq} e^{-\kappa_\ell x}$ then $c_\ell^{-2} = \int dx f_1^2(i\kappa_\ell, x)$.

2.4 Solutions of the GLM Equation for $R \equiv 0$

All one-dimensional totally reflectionless mirrors can be obtained from (2.15) easily. They form all soliton solutions of the KdV equation (Fig. 3). For $R(k) \equiv 0$ (2.15) becomes a Fredholm equation with a separable kernel. From the ansatz

$$K(x,y) = -\sum_\ell c_\ell \psi_\ell(x) e^{-\kappa_\ell y} \Longrightarrow (1 + C(x))\psi_\ell = e^{-\kappa_\ell x} c_\ell,$$

$$C_{\ell m}(x) = \frac{c_\ell c_m}{\kappa_\ell + \kappa_m} \exp(-(\kappa_\ell + \kappa_m)x) \tag{2.16}$$

we obtain

$$K(x,x) = \operatorname{Tr}(1+C)^{-1} \frac{d}{dx} C = \frac{d}{dx} \ln \det(1 + C) \tag{2.17}$$

and all reflectionless potentials are given by

$$V(x) = -2 \frac{d^2}{dx^2} \ln \det(1 + C(x)). \tag{2.18}$$

Among them are the standard examples $-n(n+1)/\cosh^2 x$ with $n \in \mathbb{N}$.

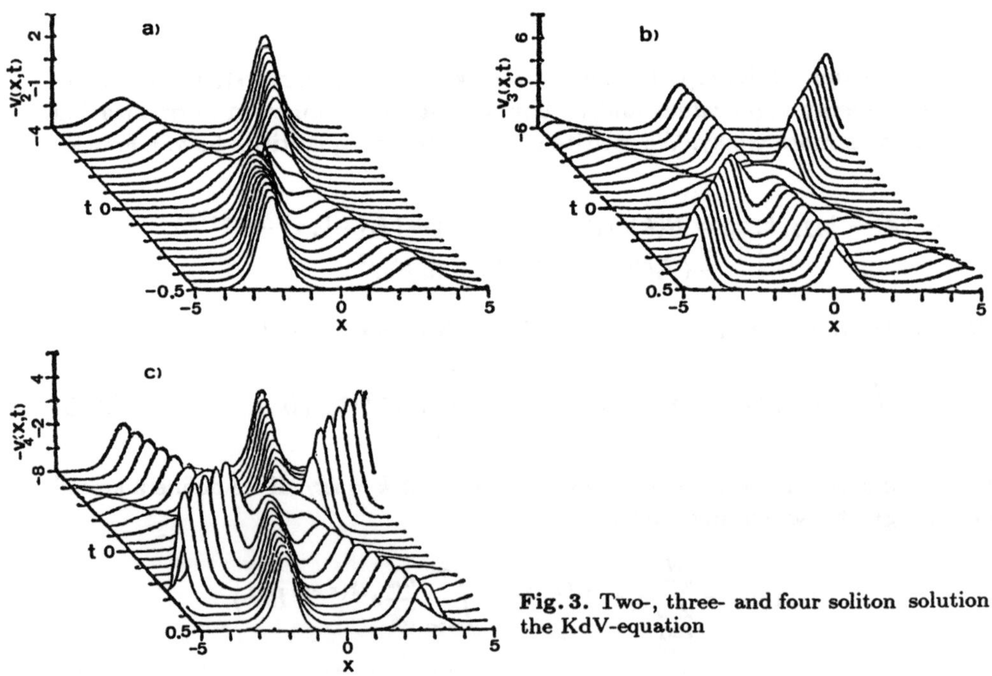

Fig. 3. Two-, three- and four soliton solution of the KdV-equation

2.5 Solving the KdV Equation

The scheme mentioned in (2.1) applies also here: We take $v(t,x)$ as a potential in a Schrödinger problem and transform to scattering data:

$$\begin{array}{ccc} v(0,x) & \xrightarrow{\text{Schrödinger Equ., "direct" step}} & \{R_0(k), \varepsilon_\ell(0), c_\ell(0)\} \\ ?\downarrow & & \downarrow \quad \text{"free" time evolution} \\ v(t,x) & \xleftarrow{\text{GLM Equ., "inverse" step}} & \{R_t(k), \varepsilon_\ell(t), c_\ell(t)\} \end{array}$$

In order to "solve" the initial value problem, we have to determine the time evolution of the scattering data of the auxiliary problem assuming that $v(t,x)$ evolves according to the KdV equation. As a first example we evaluate the change of ε_ℓ:

$$\delta\varepsilon_\ell(t) = \int_{-\infty}^{\infty} dx \varphi_\ell^2 (6vv_x - v_{xxx}) = 0. \tag{2.19}$$

By partial integrations we obtain the invariance of the spectrum of the Schrödinger operator under the KdV flow: $\varepsilon_\ell(t) = \varepsilon_\ell(0)$. This indicates the stability of solitons. In a similar way we obtain the time evolution of the other data

$$R_t(k) = R_0(k)e^{-8ik^3t}, \qquad T_t(k) = T_0(k), \qquad c_\ell(t) = c_\ell(0)e^{4\kappa_\ell^3 t}, \tag{2.20}$$

where $\kappa_\ell^2 = -\varepsilon_\ell$. (2.20) can be more easily obtained from the Lax pair which is behind the integrability. Note that we have found an infinite number of conserved quantities. Scattering data are the analog of action-angle variables. Similarly to the drum problem we may expand $\mathrm{Tr}\exp(-\beta H) \overset{\beta\searrow 0}{\simeq} \sum_n \beta^n I_n(v)/\beta^{1/2}$ and obtain a hierarchy of invariants.

2.6 Lax Pairs

We are familiar with invariant spectra if we think of unitary transformations of operators in Hilbert space. Assume that there exists a pair (L,B) such that $\dot{L} = [B,L]$ is equivalent to a nonlinear evolution equation. The pure point spectrum of $L(t)\psi(t) = \lambda(t)\psi(t)$, with $L^\dagger = L$, will then remain invariant. We differentiate, use the Heisenberg type equation and obtain

$$(L-\lambda)(\dot{\psi} - B\psi) = \dot{\lambda}\psi. \tag{2.21}$$

Take the scalar product of (2.21) with ψ and conclude that $\dot{\lambda} = 0$. As for the continuous spectrum we assume that there exists a unitary operator U such that

$$\frac{\partial U}{\partial t} = BU, \qquad B^\dagger = -B, \qquad \dot{L} = [B,L]. \tag{2.22}$$

Differentiating $\tilde{L}(t) = U^\dagger(t)L(t)U(t)$ and doing simple algebra proves that $\tilde{L}$ is time independent and equals $L(0)$. $L(t)$ is therefore obtained from $L(0)$ by a unitary transformation.

As a first example we evaluate the commutator between $L = -\frac{d^2}{dx^2} + v$ and $B = -4\frac{\partial^3}{\partial x^3} + 3v\frac{\partial}{\partial x} + 3\frac{\partial}{\partial x}v$ and obtain the KdV equation. From the operator B we get the k^3 dispersion law which determines the time evolution of $R_t(k)$ and $c_\ell(t)$.

This scheme works for hierarchies connected to the Toda lattice, the Sine-Gordon equation and the nonlinear Schrödinger equation

$$i\dot{\psi} = -\psi_{xx} + 2|\psi|^2\psi. \tag{2.23}$$

The appropriate Lax operator for (2.23) becomes a Dirac operator

$$L = \begin{pmatrix} \frac{1}{i}\frac{d}{dx} & \psi \\ \psi^* & -\frac{1}{i}\frac{d}{dx} \end{pmatrix}, \qquad L\chi = \lambda\chi, \tag{2.24}$$

where λ denotes the spectral parameter. Jost solutions with asymptotic behaviour

$$F(k,x) \overset{x\to\infty}{\longrightarrow} \begin{pmatrix} e^{ikx} & 0 \\ 0 & e^{-ikx} \end{pmatrix}, \qquad G(k,x) \overset{x\to-\infty}{\longrightarrow} \begin{pmatrix} e^{ikx} & 0 \\ 0 & e^{-ikx} \end{pmatrix} \tag{2.25}$$

are connected through the transition matrix $T(\lambda)$

$$F(\lambda,x) = T(\lambda)G(\lambda,x), \qquad T(\lambda) = \begin{pmatrix} a_\lambda & b_\lambda \\ b^*_\lambda & a^*_\lambda \end{pmatrix}. \tag{2.26}$$

2.7 Remarks

a) In the latter case a possible Poisson bracket is defined by

$$\{f(\psi,\psi^*), g(\psi,\psi^*)\} = i\int dx \left(\frac{\partial f}{\partial\psi}\frac{\partial g}{\partial\psi^*} - \frac{\partial f}{\partial\psi^*}\frac{\partial g}{\partial\psi}\right). \tag{2.27}$$

A direct calculation of the Poisson brackets between a_λ and b_λ is tedious [1]:

$$\begin{aligned} &\{a_\lambda, a_\mu\} = \{a_\lambda, a^*_\mu\} = \{b_\lambda, b_\mu\} = 0 \\ &\{a_\lambda, b_\mu\} = \frac{1}{\lambda-\mu+i\varepsilon}a_\lambda b_\mu, \qquad \{a_\lambda, b^*_\mu\} = -\frac{1}{\lambda-\mu+i\varepsilon}a_\lambda b^*_\mu \\ &\{b_\lambda, b^*_\mu\} = 2\pi i|a_\lambda|^2\delta(\lambda-\mu). \end{aligned} \tag{2.28}$$

This indicates that the new variables play almost the role of action-angle variables. The calculation of (2.28) simplifies if we introduce the monodromy matrix $M(x,y|\lambda)$ as the solution of

$$\frac{\partial}{\partial x}M(x,y|\lambda) = \ell(x,\lambda)M(x,y|\lambda), \qquad \ell(x,\lambda) = i\begin{pmatrix} -\lambda & \psi \\ -\psi^* & \lambda \end{pmatrix} \tag{2.29}$$

with $M(x,x|\lambda) = 1$. The calculation of the Poisson brackets of elements of $\ell(x,\lambda)$ with elements of $\lambda(y,\mu)$ is simple. They imply the relation

$$\{M(x,y|\lambda)\overset{\otimes}{,}M(x,y|\mu)\} = [r(\lambda-\mu), M(x,y|\lambda)\otimes M(x,y|\mu)],$$

$$r(\lambda) = -\frac{1}{\lambda}\left(\begin{array}{cc|cc} 1 & 0 & 0 & 0 \\ 0 & 0 & 1 & 0 \\ \hline 0 & 1 & 0 & 0 \\ 0 & 0 & 0 & 1 \end{array}\right). \tag{2.30}$$

On the l.h.s. the bracket between elements of the tensor product is meant. Carefully taking the limits which define $T(\lambda)$ finally yields (2.28). (2.30) follows from the relation which yields the Yang-Baxter algebra. $r(\lambda)$ solves the classical Yang-Baxter relation (see next chapter).

b) More recently integrable models in 2 + 1-dimensions have been found (Kadomtsev-Petviashvili hierarchies). For this generalization the Bose-Fermion correspondence plays a crucial role. It helps to relate subspaces as well as operators of the fermionic Fock space to subspaces and operators in the bosonic one.

If one studies the orbit of the vacuum vector under the group GL_∞ on the fermionic side it turns out that the appropriate subspaces on the bosonic side are characterized by nonlinear integrable models. Solitons turn out to be created by a particular vertex operator $\Gamma(u,v)$ depending on parameters u, v. The n-soliton solutions are built up by applying the operator

$$(1+a_1\Gamma(u_1,v_1))\dots(1+a_n\Gamma(u_n,v_n))\Omega \tag{2.31}$$

to the vacuum.

3. Integrable Lattice Systems

3.1 Introduction

Many phase transitions occur in nature. The ferromagnetic one may serve as an example. Fe, Co, Ni and certain alloys show a spontaneous magnetization M_s below the Curie temperature T_c. Let $M(T,h)$ be the magnetization as a function of temperature T and magnetic field h. $M_s(T) = \lim_{h\searrow 0} M(T,h)$. M_s serves as an order parameter distinguishing between a disordered phase for $T \geq T_c$ and ordered phases for $T < T_c$. It vanishes for $T \geq T_c$.

In statistical mechanics we may describe properties within the canonical ensemble. We introduce a density matrix $\rho(q,p) = e^{-\beta H(q,p)}/Z$ for inverse temperature $\beta = 1/T$ and the partition function $Z = \operatorname{Tr} e^{-\beta H} = e^{-\beta F}$. F denotes the free energy of the system. Expectation values of observables are defined by $\langle A\rangle = \operatorname{Tr} A\rho/Z$. Differentiating Z with respect to β yields $\langle H\rangle = U = F + TS$, $S = -\partial F/\partial T$ and standard thermodynamic relations result. Classically Tr means integration over phase space. Quantum mechanically ρ becomes a positive trace class operator in Hilbert space.

The simplest model to describe ferromagnetism is the Ising model. Lenz asked his student Ising in 1925 to study a system supposed to describe phase transitions. Start from a hypercubical lattice and assign to each lattice point i a spin variable $s_i \in \{1,-1\}$ with possible values ± 1. The set $\{s_i\}_{i=1}^N$ form a spin field configuration. A model is defined through the definition of the energy of the configuration. The "exchange" interaction term suggests to take

$$H_N(\{s_i\}) = -J \sum_{\langle ij \rangle} s_i s_j - h \sum_i s_i, \tag{3.1}$$

where $\langle ij \rangle$ denotes nearest neighbours and h the magnetic field. The dipol-dipol interaction would give a term of the form $s_i s_j$ too, but it is weaker and plays a role only for more detailed questions. The partition function for N spins is given by summing up Boltzmann factors over all configurations

$$Z_N = \sum_{\{s_i\}} e^{-\beta H_N(\{s_i\})} = e^{-\beta F_N}, \qquad \frac{F_N}{N} = -\frac{1}{\beta} \frac{\ln Z_N}{N} \stackrel{N\to\infty}{\longrightarrow} f(\beta, h) \tag{3.2}$$

and the mean free energy $f(\beta, h)$ is obtained in the thermodynamic limit. Phase transitions show up through points of nonanalyticity of f.

As the simplest example we may solve the one-dimensional Ising model with the help of the transfer matrix T. We rewrite Z_N as

$$Z_N = \sum_{s_1,\dots,s_N} T_{s_1,s_2} \dots T_{s_N s_1}, \quad T = V^{1/2} W V^{1/2},$$
$$V_{s_1' s_1} = \delta_{s_1' s_1} e^{h s_1/2}, \quad W_{s_1 s_2} = e^{J s_1 s_2}. \tag{3.3}$$

If we denote the eigenvalues of T by Λ_i, $\Lambda_1 \geq \Lambda_2$, we obtain

$$f(\beta, h) = \lim_{N\to\infty} \frac{F_N}{N} = -\frac{1}{\beta} \ln \Lambda_1 \tag{3.4}$$

that the largest eigenvalue of the transfer matrix determines the thermodynamic properties. For $d = 1$ no nontrivial phase transition occurs ($T_c = 0$). Even a simple energy-entropy argument shows this. Configurations where all spins are aligned become unstable even for very low temperature. Since the free energy $F = U - TS$ should be minimal $\delta F = \delta U - T\delta S$ should be positive. But changing a part of the chain of N spins leads to δU independent of N while δS becomes proportional to $\ln N$. In higher dimensions we might switch inlands and thus δU depends on the length of the contour. Estimation of the probability of occurrence of such Peierls contours allows to conclude that a nontrivial phase transition occurs for $d \geq 2$. In case there is an internal symmetry (like in (1.2)), Bloch walls show up and the phase transition occurs only in dimensions $d \geq 3$. If we allow the classical spins to vary continuously on a circle we obtain the plane rotator for which vortices configurations occur. We remark that properties of these systems are determined by the occurrence of topological configurations like kinks, solitons, vortices (monopoles and instantons).

Polyacetylen is an example of a system for which solitons occur. It forms long $(CH)_x$-chains whose trans-form is stable. Denote by u_n deviations of the n-th (CH) molecule from its equilibrium position. Besides lattice vibrations there is hopping of electrons along the chain. A simple Yukawa type interaction yields the Su-Schrieffer-Heeger Hamiltonian

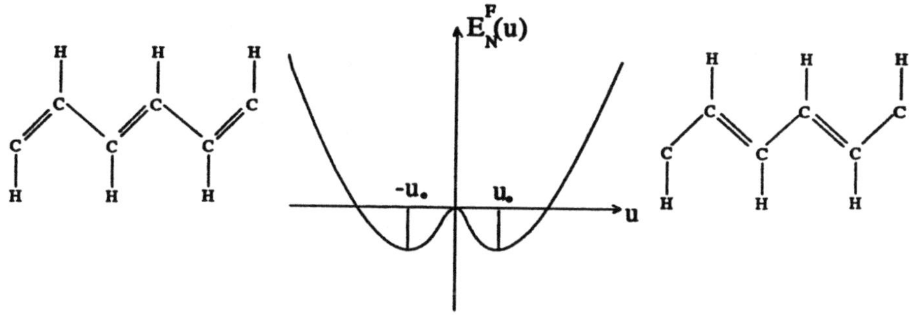

$$H_N = \sum_{n=1}^{N}(c_{n+1}^{\dagger}c_n + c_n^{\dagger}c_{n+1})(t + \alpha(u_{n+1} - u_n)) + \frac{\omega^2}{2}\sum_{n=1}^{N}(u_{n+1} - u_n)^2. \quad (3.5)$$

If one takes $u_n = (-)^n u$ with constant u and evaluates the fermionic ground state energy $E_N^F(u)$ one obtains a double well potential indicating a phase transition and spontaneous symmetry breaking. This Peierls instability leads to a gap and soliton sectors occur. Kink solutions interpolate between the degenerate minima of $E_N^F(u)$. Charged solitons may be responsible for the enhancement in conductivity under doping.

3.2 Ising and Potts Models

For the one-dimensional Ising model we introduced matrices $V = e^{h\sigma^3}$ describing the interaction at a point and $W = e^J + \sigma^1 e^{-J}$ describing the interaction between points. For $d = 2$ and a $M \times N$ lattice we write

$$\begin{aligned} Z_{M,N} &= \sum_{\{s_{m,n}\}} \exp[\frac{K_h}{2}\sum_{m,n} s_{m,n}s_{m,n+1} + \frac{K_v}{2}\sum_{m,n} s_{m,n}s_{m+1,n}] \\ &= \sum_{S_1',S_1\ldots S_M} V_{S_1' S_1} W_{S_1 S_2'} V_{S_2' S_2} \cdots \end{aligned} \quad (3.6)$$

where $S_m = \{s_{m_1}, \ldots s_{m_N}\}$. V is again diagonal, whereas W is not. They can be represented by the $2^N \times 2^N$ matrices $\sigma_n^j = \mathbf{1} \otimes \ldots \otimes \sigma^j \otimes \mathbf{1} \ldots \otimes \mathbf{1}$ where σ^j is put to the n-th factor. This yields

$$V = \exp[\frac{K_h}{2}\sum_n \sigma_n^3 \sigma_{n+1}^3], \qquad W = \prod_n (e^{K_v/2} + e^{-K_v/2}\sigma_n^1). \quad (3.7)$$

Diagonalization of $T = V^{1/2} W V^{1/2}$ is possible, but cumbersome. We may add a constant term to each $s_i \cdot s_j$ term of (3.6) and use the identity $(1+s_is_j)/2 = \delta_{s_is_j}$ to rewrite

$$Z_{M,N} = \text{const.} \sum_{\{s_i\}} \exp[K_v \sum_{\langle i,j\rangle^v} \delta_{s_is_j} + K_h \sum_{\langle i,j\rangle^h} \delta_{s_is_j}] \tag{3.8}$$

where $\langle i,j\rangle^{h,v}$ means nearest neighbour pairs in the horizontal and vertical direction. If we allow $s_j \in \{1,\dots,Q\}$ to take values one up to Q we obtain the Potts models; $Q=2$ gives Ising. V and W may be expressed in terms of $Q \times Q$ matrices $(E_{ij})_{\alpha\beta} = \delta_{\alpha i}\delta_{\beta j}$ which fulfill the algebra $E_{ij}E_{k\ell} = \delta_{jk}E_{i\ell}$. We define matrices e_i acting in $\bigotimes_{j=0}^{N-1} \mathbb{C}^Q$ with $1 \le i \le 2N-1$ [2]

$$\begin{aligned} e_{2j} &= Q^{1/2}\mathbf{1} \otimes \dots \otimes \sum_{k=1}^{Q} \underbrace{E_{kk}}_{j\text{-th place}} \otimes E_{kk} \otimes \mathbf{1} \dots \\ e_{2j-1} &= Q^{-1/2}\mathbf{1} \otimes \dots \otimes \sum_{k,\ell=1}^{Q} \underbrace{E_{k\ell}}_{j\text{-th place}} \otimes \mathbf{1} \dots , \end{aligned} \tag{3.9}$$

and obtain

$$V = \prod_{j=1}^{N-1} (\mathbf{1} + \xi_h e_{2j}), \qquad W = \prod_{j=1}^{N} Q^{1/2}(\xi_v \mathbf{1} + e_{2j-1}) \tag{3.10}$$

with $\xi_{h,v} = Q^{-1/2}(e^{K_{h,v}} - 1)$, as a simple calculation shows.

The matrices e_j fulfill the Temperley-Lieb algebra

$$e_j^2 = \sqrt{Q}\, e_j, \qquad e_ie_j = e_je_i \text{ for } |i-j| \ge 2, \qquad e_ie_{i\pm1}e_i = e_i, \tag{3.11}$$

and the partition function is completely determined by these algebraic relations. We shall come back to (3.11) later on.

3.3 The Vertex Model

There exists a class of "ice"-like models. Ice forms a lattice of H_2O molecules. A simplified $d=2$ model is obtained by placing O-atoms at the lattice points and H-atoms in between. Assume two possible positions for the H-atoms and indicate them by an arrow: —→— or —←—. Surround each O-atom by exactly two H-atoms. This allows six possible vertices to which we assign weights $\omega_j = \exp(-\beta\varepsilon_j)$:

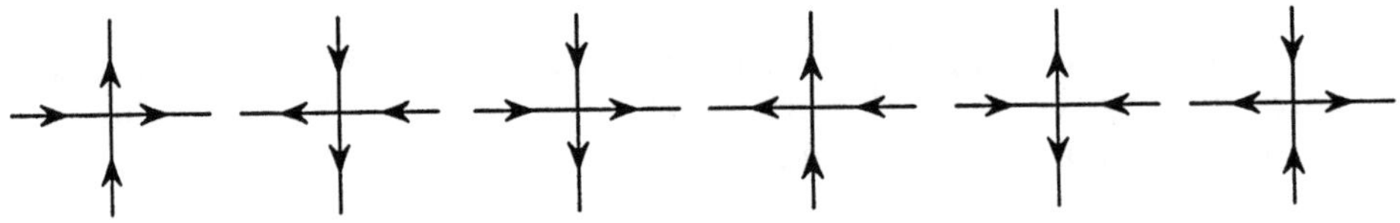

$\omega_1 = \omega_2 = a$ $\qquad$ $\omega_3 = \omega_4 = b$ $\qquad$ $\omega_5 = \omega_6 = c$

We assumed a symmetry under reflections and obtained three independent constants. The partition function is given by $Z = \sum_{\text{Conf.}} \Pi_{n,m}\omega_j(n,m)$.

The model becomes critical for $|a^2 + b^2 - c^2| \leq 2|ab|$. Correlation functions decay then algebraically. If we add the vertices

we obtain the eight-vertex model.

For the 6-vertex model we define the matrix

$$L^{\alpha_1\alpha_2}_{\beta_1\beta_2'} = $$

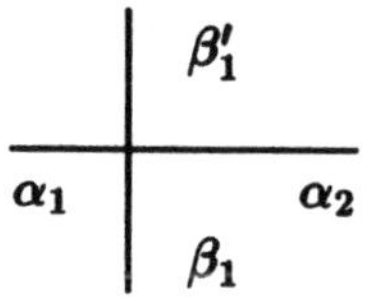

$$= \begin{pmatrix} a & 0 & 0 & 0 \\ 0 & b & c & 0 \\ 0 & c & b & 0 \\ 0 & 0 & 0 & a \end{pmatrix}$$

acting in $h \otimes \mathbb{C}^2$, where the auxiliary space $h = \mathbb{C}^2$ too. The monodromy matrix describes

$$M_N = $$

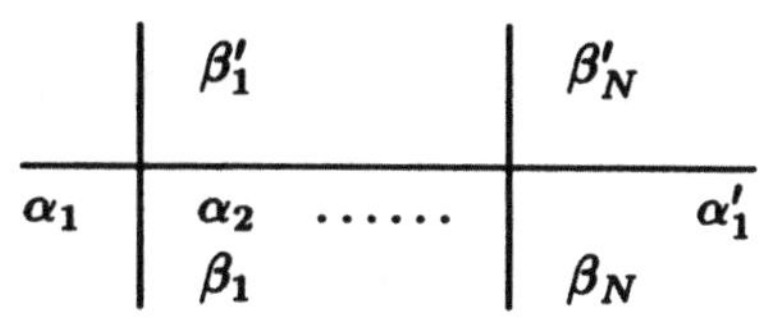

$$= L_N \ldots L_1 \equiv \begin{pmatrix} A & B \\ C & D \end{pmatrix} \qquad (3.12)$$

and acts in $h \otimes (\mathbb{C}^2)^N$. Finally the transfer matrix is defined by $\mathrm{tr}_h\, M_N = T_N$ and $Z = \mathrm{Tr}\, T_N^M$.

As a suitable parameterization we choose $a = \rho \sin(\gamma - \lambda)$, $b = \rho \sin \lambda$, $c = \rho \sin \gamma$, put $\rho = 1$ and denote λ as the spectral parameter.

3.4 Connection to Quantum Spin Models

Proposition: We introduce the Hamiltonian of the XXZ model

$$H_{XXZ} = \frac{1}{2} \sum_{j=1}^{N} (\sigma_j^1 \sigma_{j+1}^1 + \sigma_j^2 \sigma_{j+1}^2 + \Delta(\sigma_j^3 \sigma_{j+1}^3 + 1)) \qquad (3.13)$$

and claim that

$$H_{XXZ} = \sin\gamma \frac{d}{d\lambda} \ln T_N(\lambda)\Big|_{\lambda=0} \qquad \text{with } \Delta = -\cos\gamma. \qquad (3.14)$$

Proof: Note that $L_n(0) = \sin\gamma P_{0n}$, where P_{0n} denotes the permutation operator between vector space h and the n-th $\mathbb{C}^2$. $\sin\gamma^{-N} T_N(0) = \mathrm{tr}_h P_{0N} \ldots P_{01}$ $= \mathrm{tr}_h P_{12} P_{23} \ldots P_{N-1,N} P_{N0} = U$, where U denotes the shift operator. Let $(\partial L_n / \partial\lambda)|_{\lambda=0} \equiv \bar{L}_{0n}$. We calculate the logarithmic derivative of the transfer matrix at $\lambda = 0$ and obtain

$$\sin\gamma\frac{d}{d\lambda}\ln T(\lambda)\bigg|_{\lambda=0} = U^{-1}\sum_{n=1}^{N}\mathrm{tr}_h(P_{01}\dots P_{0,n-1}\bar{L}_{0n}P_{0,n+1}\dots P_{0N})$$

$$= \sum_{n=1}^{N} P_{n-1,n}\bar{L}_{n-1,n} = \sum_{n=1}^{N}\left(\begin{array}{cc|cc} -\cos\gamma & 0 & 0 & 0 \\ 0 & 0 & 1 & 0 \\ \hline 0 & 1 & 0 & 0 \\ 0 & 0 & 0 & -\cos\gamma \end{array}\right)_{n-1,n} . \tag{3.15}$$

The indices indicate the vector space in which the matrices act.

Remark: L_n corresponds to a Lax operator; let $L_n\phi_n = \phi_{n+1}$. There exists an operator M_n with $\dot{\phi}_n = M_n\phi_n$ such that the integrability condition between the two equations $\dot{L}_n = M_{n+1}L_n - L_nM_n$ is equivalent to the equation of motion for spins $\dot{\sigma}_n = i[H_{XXZ}, \sigma_n]$ in the XXZ model.

3.5 Integrability of the Lattice Model

We take γ fixed and consider $L_n(\lambda)$ as a function of λ. We take the tensor product of two subsidiary spaces and consider $L_n(\lambda)\otimes L_n(\mu)$ acting in $h\otimes h\otimes(\mathbb{C}^2)^N$. It is remarkable that there exists an operator $R \in \mathrm{End}(h\otimes h)$ such that the Yang-Baxter relation [2]

$$R(\lambda-\mu)L_n(\lambda)\otimes L_n(\mu) = L_n(\mu)\otimes L_n(\lambda)R(\lambda-\mu) \tag{3.16}$$

holds. A possible solution for the quantum R matrix is given by $R(\lambda) = PL(\lambda)$ where P denotes again the permutation operator. If we put indices one and two for the first two vector spaces h and index three for the remaining space we may rewrite (3.16) as

$$R_{12}(\lambda-\mu)P_{13}R_{13}(\lambda)P_{23}R_{23}(\mu) = P_{13}R_{13}(\mu)P_{23}R_{23}(\lambda)R_{12}(\lambda-\mu)$$

or

$$R_{23}(\lambda-\mu)R_{12}(\lambda)R_{23}(\mu) = R_{12}(\mu)R_{23}(\lambda)R_{12}(\lambda-\mu). \tag{3.17}$$

There are various other formulations. The S-matrix factorization conditions are identical to (3.17). The quantum space need not be identical to the subsidiary space. (3.16) may hold nevertheless.

Since L_n matrices to different indices commute it follows from (3.16) that

$$R(\lambda-\mu)M(\lambda)\otimes M(\mu) = M(\mu)\otimes M(\lambda)R(\lambda-\mu) \tag{3.18}$$

and by taking the trace in $\mathbb{C}^4$ we conclude that

$$[T(\lambda), T(\mu)] = 0. \tag{3.19}$$

We have therefore obtained an infinite number of conserved quantities, one signal of integrability. From the Yang-Baxter relation (3.18) we deduce "commutation" relations among the elements A, B, C, D of the monodromy matrix. They are used for the algebraic Bethe ansatz.

(3.17) can be considered as a sufficient condition such that associativity of the tensor product holds. We consider operators L^1, L^2 and L^3 acting in $h \otimes h \otimes h \otimes (\mathbb{C}^2)^N$. We transform from $L^1L^2L^3$ to $L^2L^1L^3$ with the help of R_{12}, apply next R_{23} and finally R_{12} again. We obtain $L^3L^2L^1$. The same result can be obtained by first transforming with R_{23}, then using R_{12} and finally R_{23} again. Consistency is obtained from (3.17).

3.6 Bethe States

Bethe obtained already in 1931 exact eigenstates for the one-dimensional Heisenberg chain (1.2) assuming periodic boundary conditions. The principal idea is simple. We start from a reference vector, e.g. $|\Omega\rangle = |\uparrow \dots \uparrow\rangle$ where all spins are up. The total spin $S = \sum_{n=1}^N \sigma_n/2$ is obviously conserved. $|\Omega\rangle$ is a first eigenstate of H. A new one can be obtained by taking superpositions of the form $\sum_{n=1}^N e^{ikn}\sigma_n^-|\Omega\rangle \equiv |k\rangle$. These eigenstates form spin waves (or magnons) to wave vectors k and are quasiparticle excitations. We next may try to flip two spins. Since $(\sigma_n^-)^2 = 0$ two magnons repel each other. This leads to a phase shift $\Theta_{k_1k_2}$ and an interaction between spin waves. The Bethe ansatz

$$|k_1, k_2\rangle = \sum_{n_1<n_2} \psi_{n_1,n_2}^{k_1,k_2}\sigma_{n_1}^-\sigma_{n_2}^-|\Omega\rangle \tag{3.20}$$

$$\psi_{n_1,n_2}^{k_1,k_2} = \exp[\frac{i}{2}\Theta_{k_1k_2}]\exp[i(k_1n_1 + k_2n_2)] + \exp[-\frac{i}{2}\Theta_{k_1,k_2}]\exp[i(k_2n_1 + k_1n_2)]$$

leads to an eigenstate of H iff $\Theta_{k_1k_2}$ solves

$$2\cot\frac{\Theta_{k_1k_2}}{2} = \cot\frac{k_1}{2} - \cot\frac{k_2}{2}. \tag{3.21}$$

Note that the effective repulsion is built in since $\exp(i\Theta_{kk}) = -1$ and $\psi_{n_1,n_2}^{k,k} = 0$. For M interacting magnons the ansatz

$$|k_1, \dots, k_M\rangle = \sum_{n_1<n_2<\dots<n_M} \psi_{\underline{n}}^{\underline{k}}\sigma_{n_1}^- \dots \sigma_{n_M}^-|\Omega\rangle \tag{3.22}$$

$$\psi_{\underline{n}}^{\underline{k}} = \sum_{\text{Perm }\pi} \exp[i\sum_\alpha k_{\pi_\alpha}n_\alpha + \frac{i}{2}\sum_{\alpha<\beta}\Theta_{k_{\pi_\alpha}k_{\pi_\beta}}]$$

leads to eigensolutions iff Θ and k solve the Bethe equations

$$2\cot\frac{\Theta_{k_\alpha k_\beta}}{2} = \cot\frac{k_\alpha}{2} - \cot\frac{k_\beta}{2}$$

and

$$Nk_\alpha + \sum_{\beta\neq\alpha}\Theta_{k_\alpha k_\beta} = 2\pi\lambda_\alpha, \qquad \lambda_\alpha = 0, \dots, N-1. \tag{3.23}$$

Lieb and Liniger solved the one-dimensional Bose gas using the Bethe ansatz. Let $\psi(x)$ denote a Bose field with $[\psi(x), \psi^\dagger(y)] = \delta(x-y)$. The Hamiltonian operator $H = \int_0^L dx(\psi_x^\dagger \psi_x + \lambda \psi^\dagger \psi^\dagger \psi \psi)$ leads on the N-particle sector to the many body interaction

$$(-\sum_{j=1}^{N} \frac{\partial^2}{\partial x_j} + 2\lambda \sum_{i<j} \delta(x_i - x_j))\psi_N = E\psi_N. \tag{3.24}$$

There exist again Bethe states of the form

$$\psi_N = \sum_{\text{Perm } \pi} \exp[i \sum_{j=1}^{N} k_{\pi_j} x_j + \frac{i}{2} \sum_{i<j} \Theta_{k_{\pi_i} k_{\pi_j}}], \qquad x_1 < x_2 < \dots x_n, \tag{3.25}$$

$$\exp[i\Theta_{k_i k_j}] = \frac{k_i - k_j + i\lambda}{k_i - k_j - i\lambda}$$

where periodic boundary conditions require that

$$k_\ell L = 2\pi I_\ell - 2 \sum_{j=1}^{N} \arctan(\frac{k_\ell - k_j}{\lambda}). \tag{3.26}$$

In the thermodynamic limit $L \to \infty$, $N \to \infty$, with $\rho = N/L$ fixed, one introduces a distribution function for the roots of Equ. (3.26). This leads to an integral equation.

3.7 The Algebraic Bethe Ansatz

The Yang-Baxter relations allow to establish an algebraic procedure such that the Bethe states are built up. Among the 16 relations which determine the algebra of operators A, B, C, D we quote

$$\begin{aligned} A_\lambda B_\mu &= \frac{a(\mu - \lambda)}{b(\mu - \lambda)} B_\mu A_\lambda - \frac{c(\mu - \lambda)}{b(\mu - \lambda)} B_\lambda A_\mu \\ D_\lambda B_\mu &= \frac{a(\lambda - \mu)}{b(\lambda - \mu)} B_\mu D_\lambda - \frac{c(\lambda - \mu)}{b(\lambda - \mu)} B_\lambda D_\mu . \end{aligned} \tag{3.27}$$

We intend to solve the family of eigenvalue problems $T(\lambda)\psi = (A_\lambda + D_\lambda)\psi = E\psi$, since the free energy is determined by the largest eigenvalue. We observe that the reference vector $|\Omega\rangle_N = \bigotimes_{j=1}^{N} |\uparrow\rangle_j$ acts like a "pseudo"vacuum since

$$L_n|\uparrow\rangle_n = \left(\begin{array}{c|c} a_\lambda|\uparrow\rangle_n & c_\lambda|\downarrow\rangle_n \\ \hline 0 & b_\lambda|\uparrow\rangle_n \end{array} \right), \tag{3.28}$$

$$A_\lambda|\Omega\rangle_N = a_\lambda^N|\Omega\rangle_N, \qquad D_\lambda|\Omega\rangle_N = b_\lambda^N|\Omega\rangle_N, \qquad C_\lambda|\Omega\rangle_N = 0.$$

$|\Omega\rangle_N$ is eigenvector of $T(\lambda)$ to eigenvalue $a_\lambda^N + b_\lambda^N$. We may try to take B_λ as a creation operator of quasiparticles. If we consider only the first terms in Equs. (3.27) the state $B_{\lambda_1} \dots B_{\lambda_n}|\Omega\rangle_N$ would be an eigenstate of $T(\lambda)$ to eigenvalue

$$a_\lambda^N \prod_{j=1}^{N} \frac{a(\lambda_j-\lambda)}{b(\lambda_j-\lambda)} + b_\lambda^N \prod_{j=1}^{N} \frac{a(\lambda-\lambda_j)}{b(\lambda-\lambda_j)}.$$

But this holds true if we require vanishing of all other terms obtained from the last terms of (3.27). These Bethe equations become

$$\left(\frac{a(\lambda_\ell)}{b(\lambda_\ell)}\right)^N = \prod_{j\neq\ell}^{n} \frac{a(\lambda_\ell-\lambda_j)b(\lambda_j-\lambda_\ell)}{b(\lambda_\ell-\lambda_j)a(\lambda_j-\lambda_\ell)}, \tag{3.29}$$

if $c(\lambda)b(-\lambda) = -c(-\lambda)b(\lambda)$, as it is true in most examples. For the 6-vertex model (3.29) goes into

$$e^{ip_\ell N} \equiv \left(\frac{\sin\lambda_\ell}{\sin(\gamma-\lambda_\ell)}\right)^N = \prod_{j\neq\ell}^{N} \frac{\sin(\lambda_j-\lambda_\ell-\gamma)}{\sin(\lambda_j-\lambda_\ell+\gamma)} \equiv \prod_{j\neq\ell}^{n} e^{i\phi(\lambda_j-\lambda_\ell)} \tag{3.30}$$

or identically

$$Np_\ell + \sum_{j\neq\ell}^{N} \phi(\lambda_j-\lambda_\ell) = 2\pi I_\ell, \qquad I_\ell \in \mathbb{Z} \tag{3.31}$$

an equation similar to (3.26). We note that $b(\lambda_\ell - \lambda)$ vanishes at $\lambda = \lambda_\ell$, but the transfer matrix should be regular. Vanishing of the residuum at these poles explains (3.29) too.

3.8 Knots, Links and Braids

A knot is a closed line embedded into $\mathbb{R}^3$ without crossings (Fig. 4). A link is a set of knoted knots. A complete classification of links is complicated. In order to define braids we choose n points in $\mathbb{R}^2$ and consider mappings γ_i from $[0,1] \to \mathbb{R}^2 \times [0,1] \subset \mathbb{R}^3$, $i = 1,\dots,n$ with the properties that $\dot\gamma_i^3(t) > 0$ for $t \in [0,1]$, $\gamma_i(t) \neq \gamma_j(s)$ for $i \neq j$ and all $s,t \in [0,1]$ and $\gamma_i(0) = x_i$, $\gamma_i(1) = x_{\pi(i)}$. We identify all mappings with the mentioned properties which can be obtained by continuous deformations. The resulting object forms a braid of n strings $\in B_n$.

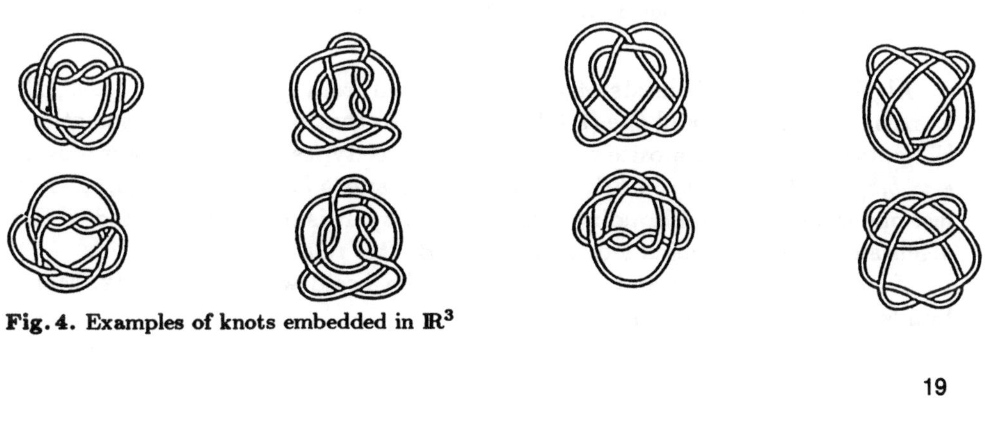

Fig. 4. Examples of knots embedded in $\mathbb{R}^3$

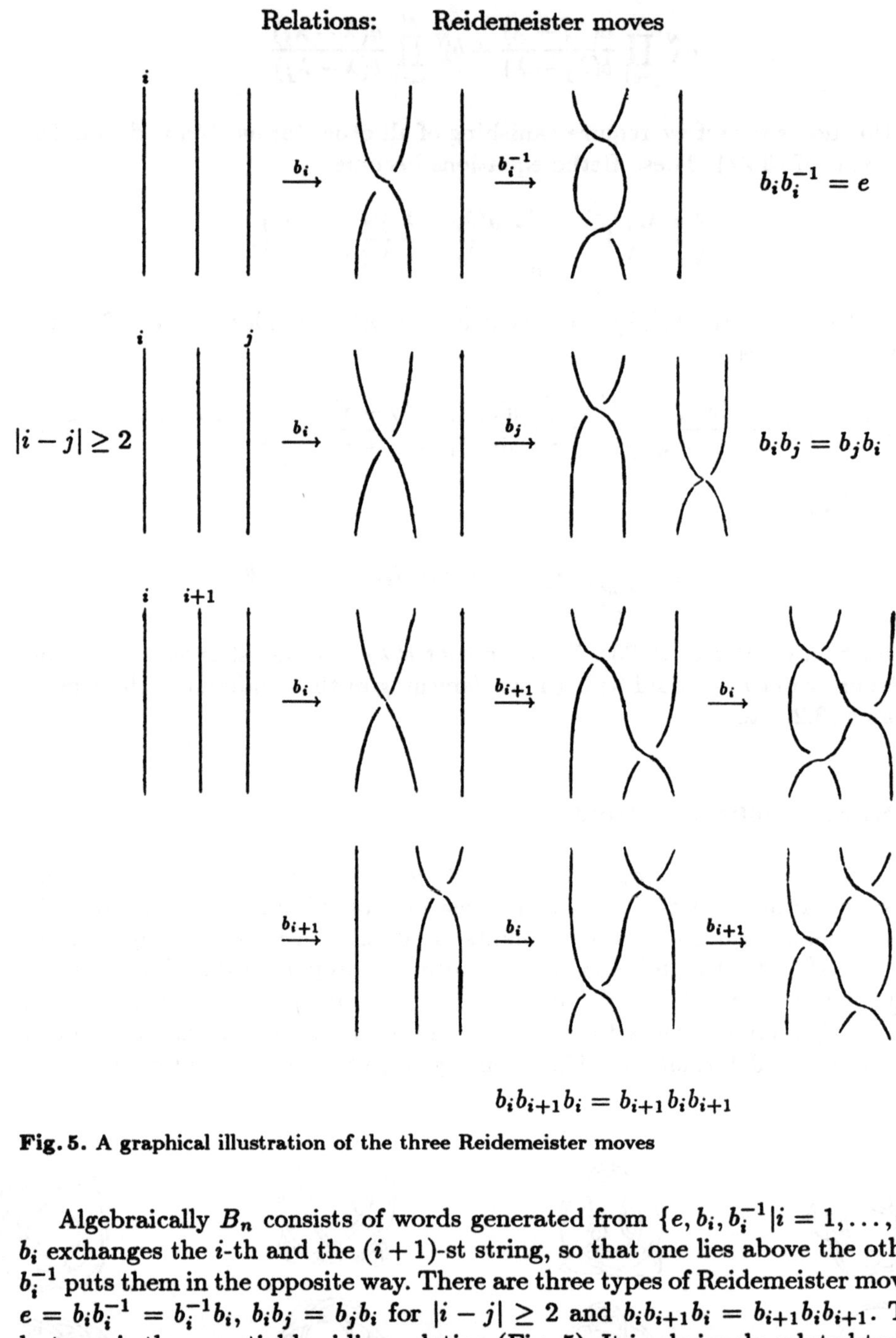

Fig. 5. **A graphical illustration of the three Reidemeister moves**

Algebraically B_n consists of words generated from $\{e, b_i, b_i^{-1} | i = 1, \ldots, n\}$. b_i exchanges the i-th and the $(i+1)$-st string, so that one lies above the other. b_i^{-1} puts them in the opposite way. There are three types of Reidemeister moves: $e = b_i b_i^{-1} = b_i^{-1} b_i$, $b_i b_j = b_j b_i$ for $|i - j| \geq 2$ and $b_i b_{i+1} b_i = b_{i+1} b_i b_{i+1}$. The last one is the essential braiding relation (Fig. 5). It is obviously related to the algebraic procedure which led to the Yang-Baxter relations [3].

If we identify upper and lower endings of a braid we obtain a link. Many braids lead to the same link.

Markov Theorem: Braids which corresponds to the same link can be obtained from each other by successive applications of Markov moves of type I and II (Fig. 6). If $A, B \in B_n$: $\mathcal{L}(AB) = \mathcal{L}(BA)$ and if $A \in B_n$, $b_n \in B_{n+1}$: $\mathcal{L}(Ab_n) = \mathcal{L}(A) = \mathcal{L}(Ab_n^{-1})$, $\mathcal{L}(\cdot)$ denotes an invariant function on links. The Burau representation of B_n is given by $n \times n$ matrices

$$b_j = \begin{pmatrix} 1 & & & & & & \\ & \ddots & & & & & \\ & & 1 & & & & \\ & & & 1-t & t & & \\ & & & 1 & 0 & & \\ & & & & & 1 & \\ & & & & & & \ddots \\ & & & & & & & 1 \end{pmatrix}$$

whose (j, j) element is $1 - t$. If we require that the eigenvalues of b_n are -1 and $t \in \mathbb{C}$, we restrict the algebra by imposing the condition $(g_i + 1)(g_i - t) = 0$. Together with $g_i g_j = g_j g_i$, $|i - j| \geq 2$ and $g_i g_{i+1} g_i = g_{i+1} g_i g_{i+1}$ the Hecke algebra is defined.

If we transform to e_i putting $g_i = (1 + t)e_i - 1$, we obtain

$$(1+t)^2 e_i e_{i+1} e_i - t e_i = (1+t)^2 e_{i+1} e_i e_{i+1} - t e_{i+1}.$$

If we put both sides to zero the Temperley-Lieb algebra results.

We note finally that we may obtain link-invariants from solutions of the Yang-Baxter equation.

Theory: Equivalent braids are obtained through application of Markov moves

Type I: $AB \longrightarrow BA$, $A, B \in B_n$

Type II: $A \longrightarrow Ab_n$, $A \in B_n$, $b_n \in B_{n+1}$

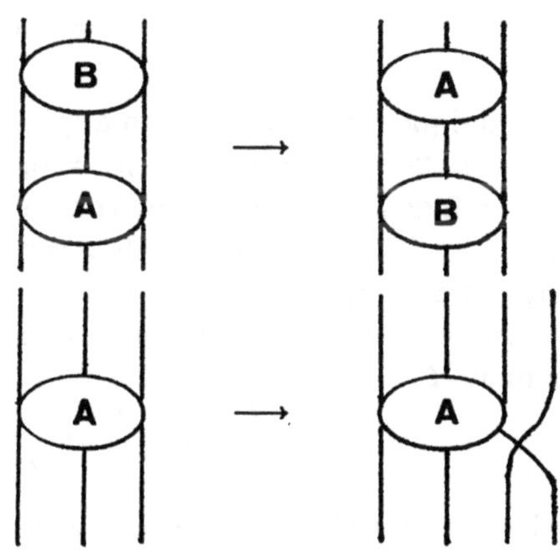

Inv. Linkpol.:

$$\mathcal{L}(AB) = \mathcal{L}(BA) \ , \ \mathcal{L}(Ab_n) = \mathcal{L}(A) = \mathcal{L}(Ab_n^{-1})$$

Fig. 6. A graphical illustration of Markov moves of type I and II

4. Conformal Field Theory

4.1 Introduction

Through the behaviour near the critical point we define critical exponents: $M_s \simeq |\tau|^\beta$, $\tau \equiv T - T_c$, specific heat $c_h \simeq |\tau|^{-\alpha}$ and susceptibility $\chi_T \simeq \tau^{-\gamma}$. The two point function of the Ising model decays away from T_c like an exponential (Ornstein-Zernike behaviour)

$$|\langle s_i s_j\rangle - \langle s_i\rangle\langle s_j\rangle|_{|i-j|\to\infty} \simeq c\frac{\exp[-\frac{|i-j|}{\xi}]}{|i-j|^{d-2}}, \qquad T \neq T_c \tag{4.1}$$

while at $T = T_c$ we obtain a second order phase transition and the correlation length diverges (long range order): $\xi(T) \simeq |\tau|^{-\nu}$, in addition

$$|\langle s_i s_j\rangle^c|_{|i-j|\to\infty} \simeq \frac{\text{const}}{|i-j|^{d-2+\eta}}. \tag{4.2}$$

Renormalization group ideas have been used to get reliable answers for non-trivial exponents. Mean field methods underestimate fluctuations and $\alpha = 0$, $\beta = 1/2$, $\gamma = 1$, $\nu = 1/2$, $\eta = 0$ result. For the two-dimensional Ising model we get $\alpha = 0$, $\beta = 1/8$, $\gamma = 7/4$, $\nu = 1$, $\eta = 1/4$ [4].

Already in '65 Widom suggested a scaling behaviour close to the critical point for the free energy $\lambda f(\tau, h) \simeq f(\lambda^a\tau, \lambda^b h)$. a and b determine all critical exponents. Block spin methods led to a justification of this ansatz, which we illustrate in one dimension. We may integrate out each second spin and observe that

$$\begin{aligned}\sum_{s_2} \exp[J(s_1s_2 - 1) + \frac{h}{2}(s_1 + 2s_2 + s_3) + J(s_2s_3 - 1) + c]\\ = \exp[\bar J(s_1s_3 - 1) + \frac{\bar h}{2}(s_1 + s_3) + \bar c].\end{aligned} \tag{4.3}$$

The unique fix point of the mapping $(J, h, c) \to (\bar J, \bar h, \bar c)$ determines the critical point $\mu \equiv e^{-J} = e^{-\bar J} \equiv \bar\mu = 0$, $h = \bar h = 0$. The flow around this point $e^{-\bar J} = \bar\mu \simeq \sqrt{2}\mu$, $\bar h \simeq 2h$ determines the behaviour of thermodynamic functions $\ln \Lambda_1 \simeq \mu^2\sqrt{1 + h^2/\mu^4}$ which agrees with the renormalization group behaviour $f(\mu, h) \simeq \mu^2 f(1, h/\mu^2)$.

In the continuum limit ($a \to 0$) the transfer matrix determines the Hamiltonian $\mathcal{H}$

$$T = \begin{pmatrix} e^h & e^{-2J} \\ e^{-2J} & e^{-h}\end{pmatrix} \simeq 1 + h\sigma^3 + \mu^2\sigma^1 = 1 + a\mathcal{H} \tag{4.4}$$

where $\mathcal{H} = \lambda\sigma^3 + \sigma^1$, $\mu^2 = a$, $h = \lambda\mu^2$. We note that the limit $a \to 0$ is connected to the limit $T \to T_c$. Keeping $\xi(a)\cdot a =$ const. requires that $\xi(a = 0) = \infty$. According to renormalization group ideas we have to follow renormalization group trajectories for $a \to 0$. This has been achieved explicitly only for the $d = 2$ Ising model. A massless Majorana field theory is obtained in the limit. This model is not only scale invariant but invariant under the conformal group. η

becomes determined by the anomalous scale dimension of a field $s(x)$. A fruitful hypothesis for $d=2$ turns out to be the assumption that the continuum model is conformal covariant. A classification of possible critical exponents results. Operators fall into representations of the Virasoro algebra.

4.2 Conformal Invariance

We start with $\mathbb{R}^d$ and metric $g = g_{\mu\nu}dx^\mu \otimes dx^\nu$, $g_{\mu\nu} = \eta_{\mu\nu}$, $p+q=d$ and signature (p,q). A mapping from $\mathbb{R}^d \cup \{\infty\}$ to $\mathbb{R}^d \cup \{\infty\}$ such that $g_{\mu\nu} \to f(x)g_{\mu\nu}$ is called a conformal transformation.

Examples of conformal transformations are translations $x^\mu \longrightarrow x^\mu + a^\mu$, $a^\mu \in \mathbb{R}^d$ and pseudorotations $x^\mu \to \Lambda^\mu_\nu x^\nu$, $\Lambda \in O(p,q)$. In both cases $f=1$. In addition dilatations $x^\mu \longrightarrow \lambda x^\mu$, $\lambda \in \mathbb{R}^+$ lead to $f(x) = \lambda^{-2}$. Finally there are special conformal transformations

$$x^\mu \xrightarrow{\text{refl.}} \frac{x^\mu}{x^2} \xrightarrow{\text{transl.}} \frac{x^\mu + b^\mu}{(x+b)^2} \xrightarrow{\text{refl.}} \frac{x^\mu + b^\mu x^2}{(1+2(bx)+b^2x^2)}, \qquad b^\mu \in \mathbb{R}^d \tag{4.5}$$

which yield $f(x) = (1+2(bx)+b^2x^2)^2$. The $(d+1)(d+2)/2$ generators of these transformations form the Lie-algebra of $so(p+1,q+1)$. These are the only transformations which can be globally defined on $\mathbb{R}^d \cup \{\infty\}$. Calculation of the Jacobi determinant yields $|\partial x'/\partial x| = (1+2(bx)+b^2x^2)^{-d}$ for the special conformal transformations. For N points we obtain $(N-3)$ invariant quantities which are the anharmonic quotients, for example $|x_1-x_3||x_2-x_4|/|x_1-x_2||x_3-x_4|$.

The hypothesis of covariance under global conformal transformations asserts that there exists quasiprimary fields $\{A_\ell\}$ which transform under conformal transformations $x \longrightarrow x'$ as

$$A_\ell(x) \to \left|\frac{\partial x'}{\partial x}\right|^{\Delta_\ell/d} A_\ell(x'). \tag{4.6}$$

All other fields are linear combinations of quasiprimary fields A_ℓ and their derivations. Correlation functions transform covariantly, the vacuum is invariant. Δ_ℓ is called the anomalous dimension. Correlation functions are therefore restricted. For example, two- and three-point functions are given by

$$\langle A_1(x_1)A_2(x_2)\rangle = \frac{c_{12}}{x_{12}^{\Delta_1+\Delta_2}}, \qquad x_{12} = |x_1 - x_2|, \qquad c_{12} = 0 \text{ if } \Delta_1 \neq \Delta_2,$$

$$\langle A_1(x_1)A_2(x_2)A_3(x_3)\rangle = \frac{c_{123}}{x_{12}^{\Delta_1+\Delta_2-\Delta_3}x_{23}^{\Delta_2+\Delta_3-\Delta_1}x_{13}^{\Delta_1+\Delta_3-\Delta_2}}. \tag{4.7}$$

The four-point function depends on the anharmonic quotient.

4.3 Local Conformal Transformations $d = 2$

As we know from hydrodynamics there exist conformal transformations in $\mathbb{R}^2$ for which certain singularities occur. Requiring that $g_{\mu\nu} \to h(x,y)g_{\mu\nu}$ implies that the mapping $(x,y) \to (x',y')$ fulfills the Cauchy-Riemann differential equations. In terms of $z = x + iy$, $\bar{z} = x - iy$, $z' = f(z)$ and $\bar{z}' = \bar{f}(\bar{z})$ have to be analytic resp. antianalytic: $\partial_{\bar{z}} z' = 0$, $\partial_z \bar{z}' = 0$. This group becomes infinite-dimensional. We remark that $(z, \bar{z})$ may be considered as a point in $\mathbb{C}^2$.

It is an easy exercise that the mapping $z \to z'$ is globally defined iff $f(z) = (az+b)/(cz+d)$. No essential singularities or branch point singularities may occur. $ad - bc$ has to be non-zero and is put to one. 6 real parameters occur. These transformations can be mapped by a homomorphism to $s\ell(2,\mathbb{C})$. The composition $(\tilde{f} \circ f)(z)$ corresponds to matrix multiplication. The six parameters correspond to translations:

$$z \to z + b \leftrightarrow \begin{pmatrix} 1 & b \\ 0 & 1 \end{pmatrix}, \qquad b \in \mathbb{C},$$

to dilatations

$$z \to \lambda z \leftrightarrow \begin{pmatrix} \sqrt{\lambda} & 0 \\ 0 & 1/\sqrt{\lambda} \end{pmatrix}, \qquad \lambda \in \mathbb{R}^+,$$

to rotations

$$z \to e^{i\theta} z \leftrightarrow \begin{pmatrix} e^{i\theta/2} & 0 \\ 0 & e^{-i\theta/2} \end{pmatrix}, \qquad \theta \in \mathbb{R},$$

and to special conformal transformations

$$z \to z/(cz+1) \leftrightarrow \begin{pmatrix} 1 & 0 \\ c & 1 \end{pmatrix}, \qquad c \in \mathbb{C}.$$

The infinitesimal transformations $z \to z' = z - \varepsilon_n z^{n+1}$ can be represented on functions by $F(z) \to F(z') \cong F(z) - \varepsilon_n z^{n+1} F'(z)$. The algebra of the generators $\ell_n = -z^{n+1}\partial_z$ becomes the Virasoro algebra $[\ell_n, \ell_m] = (n-m)\ell_{n+m}$. This is the algebra of the diffeomorphism group of S^1 too. $\bar{\ell}_n = -\bar{z}^{n+1}\partial_{\bar{z}}$ fulfills an isomorphic algebra. Everything becomes doubled.

The subalgebra $\{\ell_{-1}, \ell_0, \ell_1\}$ forms an $s\ell(2,\mathbb{C})$ and generates global conformal transformations. Assume that $|h, \bar{h}\rangle$ is an eigenvector of ℓ_0 and $\bar{\ell}_0$ to eigenvalues h and $\bar{h}$. h, $\bar{h}$ are called conformal weights. Since $i(\ell_0 - \bar{\ell}_0)$ generates rotations $|h - \bar{h}|$ may be called spin. Since $\ell_0 + \bar{\ell}_0$ generates dilatations $h + \bar{h}$ becomes the anomalous dimension.

4.4 Three Implications

Energy-Momentum Tensor: We start from the action functional $S = \int d^dx \mathcal{L}(\phi, \partial\phi)$ and vary ϕ by $\delta\phi = \varepsilon^\mu \partial_\mu \phi$. S is changed by $\delta S = \int d^dx \partial^\nu \varepsilon^\mu T_{\mu\nu}$, where $T_{\mu\nu}$ denotes the energy-momentum tensor $T_{\mu\nu} = -g_{\mu\nu}\mathcal{L} + \partial^\mu\phi(\partial\mathcal{L}/\partial\partial_\nu\phi)$. In two dimensions the traceless and symmetric tensor T becomes

$$T = \begin{pmatrix} H & -P \\ -P & H \end{pmatrix}.$$

Since according to Noether it is conserved $\partial^\mu T_{\mu\nu} = 0$, we obtain $(\partial_0 + \partial_x)(H + P) = 0$ and $(\partial_0 - \partial_x)(H - P) = 0$. In terms of $T_\pm = \frac{1}{2}(H \mp P)$ we get $\partial_+ T_- = \partial_- T_+ = 0$ where $\partial_\pm$ denote derivatives with respect to light cone coordinates. In the Euclidean formulation $t = -i\tau$, $x_+ = \bar{z} = x - i\tau$ etc. and the two components of the energy-momentum tensor $(T_{11} - iT_{12})/2 = T(z)$, $(T_{11} + iT_{12})/2 = \bar{T}(\bar{z})$ depend only on z resp. $\bar{z}$.

Ward Identity: The hypothesis that correlation functions transform covariantly under global conformal transformations $\delta_{\varepsilon(z)}\langle\phi_1(z_1, \bar{z}_1)\dots\phi_N(z_N, \bar{z}_N)\rangle = 0$ yields

$$\begin{aligned} &\sum_k \partial_{z_k}\langle\phi_1\dots\phi_N\rangle = 0 \\ &\sum_k (h_k + z_k\partial_{z_k})\langle\phi_1\dots\phi_N\rangle = 0 \\ &\sum_k (2z_k h_k + z_k^2\partial_{z_k})\langle\phi_1\dots\phi_N\rangle = 0 \end{aligned} \tag{4.8}$$

where $\phi_j \equiv \phi_j(z_j, \bar{z}_j)$. From conservation of the energy momentum tensor we deduce that

$$\frac{\partial}{\partial z_0^i}\langle T_{ij}(z_0)\phi_1\dots\phi_N\rangle = 0 \qquad \text{if } z_i \neq z_j,\ i = 0,\dots,N. \tag{4.9}$$

The next question concerns the behaviour of correlation functions under local transformations. The Ward identity asserts that the change is determined by T_{ij} generating (4.8):

$$\delta_{\varepsilon(x)}\langle\mathcal{F}(A)\rangle = \int d^2x\,\partial_i\varepsilon_j(x)\langle T_{ij}(x)\mathcal{F}(A)\rangle \tag{4.10}$$

where $\mathcal{F}(A)$ denotes a product of operators $A_\ell(z_\ell, \bar{z}_\ell)$. We have to be careful since such correlation functions are singular at coinciding points. Doing a partial integration modifies (4.10) to

$$\delta_\varepsilon\langle\mathcal{F}(A)\rangle = \sum_\alpha \oint_{C_\alpha} d\ell_i\varepsilon_j(x)\langle T_{ij}(x)\mathcal{F}(A)\rangle - \int d^2x\,\varepsilon_j(x)\partial_i\langle T_{ij}(x)\mathcal{F}(A)\rangle. \tag{4.11}$$

Here C_α denotes a circle surrounding the singular point. Next we introduce $T(z)$ and $\bar{T}(\bar{z})$ and obtain the final form of the Ward identity

$$\delta_\varepsilon\langle\mathcal{F}(A)\rangle = \sum_\alpha \oint_{C_\alpha} \frac{dz}{2\pi i}\varepsilon(z)\langle T(z)\mathcal{F}(A)\rangle + \sum_\beta \oint_{C_\beta} \frac{d\bar{z}}{2\pi i}\bar{\varepsilon}(\bar{z})\langle\bar{T}(\bar{z})\mathcal{F}(A)\rangle. \tag{4.12}$$

Schwinger Term: We may ask the question how $T(z)$ transforms under local conformal transformations $z \to z' = z + \varepsilon(z)$. Classically we would expect that

$$T(z)dz^2 \to T'(z)dz^2 \equiv T(z')dz^2 \cong (T(z) + \varepsilon(z)\frac{dT}{dz} + 2\frac{d\varepsilon}{dz}T)dz^2 + O(\varepsilon^2).$$

Quantum mechanically, due to normal ordering, an additional c-number term shows up:

$$T(z)dz^2 \to T'(z)dz^2 \cong T(z')dz'^2 + \frac{c}{12}\{z', z\}dz^2. \tag{4.13}$$

The c-number term is called a cocycle. It is restricted due to the Jacobi identity. Requiring that it should be a "local" expression determines its form to be the Schwarz derivative except for a constant c:

$$\{w, z\} = \frac{w_{zzz}}{w_z} - \frac{3}{2}\frac{w_{zz}^2}{w_z^2}. \tag{4.14}$$

Infinitesimally $\{z', z\} \cong \varepsilon'''(z)$ which vanishes for $\varepsilon(z) = -\varepsilon_0 - \varepsilon_1 z - \varepsilon_2 z^2$. The transformation law finally becomes

$$T(z) \to T(z) + \varepsilon(z)T'(z) + 2\varepsilon'(z)T(z) + \frac{c}{12}\varepsilon'''(z). \tag{4.15}$$

4.5 The Virasoro Algebra

Through the expansion $T(z) = \sum_{n=-\infty}^{\infty} L_n/z^{n+2}$ we may introduce operators L_n and similarly $\bar{T}(\bar{z})$ yields operators $\bar{L}_n$. We put T instead of $\mathcal{F}(A)$ in Equ. (4.12) and use the transformation law (4.15). This yields the celebrated Virasoro algebra with central extension term

$$[L_n, L_m] = (n - m)L_{n+m} + \frac{c}{12}\delta_{n,-m}n(n^2 - 1). \tag{4.16}$$

The Schwinger term vanishes for the $sL(2, \mathbb{C})$ algebra spanned by $\{L_{-1}, L_0, L_1\}$. Global transformations remain unbroken.

In order to proceed we have to quote a suitable conjugation. $z = 0$ and $z = \infty$ correspond to $t = -\infty$ and $t = \infty$ within the in-out formalism (radial quantization). Motivated by reflection positivity we require that

$$A(z, \bar{z})^+ = A(\frac{1}{\bar{z}}, \frac{1}{z})\frac{1}{\bar{z}^{2h}}\frac{1}{z^{2\bar{h}}}. \tag{4.17}$$

This conjugation yields for $T(z)$ that $L_{-m} = L_m^+$. We note that a nonvanishing value for c is very essential. If we require for the vacuum state that $L_{-1}|0\rangle = L_0|0\rangle = L_1|0\rangle = 0$ and $T(z)$ is regular so that $L_m|0\rangle = 0$ for $m \geq -1$, we obtain in addition that $\langle 0|L_m = 0$ for $m \leq 1$. Calculation of the correlation

$$\langle 0|T(z_1)T(z_2)|0\rangle = \frac{c/2}{(z_1 - z_2)^4} \tag{4.18}$$

shows the relevance of $c \neq 0$.

Example: The scaling limit of the $d = 2$ Ising model gives a Majorana field with $\mathcal{L} = \bar{\psi} i \gamma_i \partial_i \psi$ and equations of motion $\partial_{\bar{z}} \psi_1 = 0$, $\partial_z \psi_2 = 0$. The conformal weights of $\psi \equiv \psi_1$ are $(1/2, 0)$ and $T(z) = \frac{1}{2} i \psi \partial_z \psi_i$. The mode expansion

$$\psi(z) = \sum_{k \in \mathbb{Z}} a_k z^{-k} = \frac{1}{\sqrt{z}} \sum_{q=k-1/2} b_q z^{-q}, \qquad \begin{array}{ll} a_k|0\rangle = 0 & k > 0 \\ a_k^\dagger|0\rangle = 0 & k < 0 \end{array} \tag{4.19}$$

with $b_q^\dagger = b_{-q}$ fixes a representation of the Virasoro algebra and yields after explicit calculations $c = 1/2$.

It is remarkable that the Virasoro algebra admits unitary highest weight representations. All of them are classified. Let L_0 be diagonal $L_0|h\rangle = h|h\rangle$. The representation will be characterized by (c, h). We require not only that $L_1|h\rangle = 0$ but also $L_2|h\rangle = 0$ and therefore $L_j|h\rangle = 0$ for all $j > 0$.

A Verma modul is built up from the representation space which is spanned by $L_{-k_1} \ldots L_{-k_n}|h\rangle \equiv |k_1 \ldots k_n, h\rangle$ for $1 \leq k_1 \leq k_2 \leq \ldots \leq k_n$. We note that

$$L_0|k_1 \ldots k_n, h\rangle = L_{-k_1} L_0|k_2, \ldots\rangle + k_1|k_1 \ldots\rangle = (\sum_{i=1}^{n} k_i + h)|k_1 \ldots k_n, h\rangle. \tag{4.20}$$

This suggests to introduce the level of a vector $\nu(|\,\rangle) = \sum_{i=1}^n k_i$. All terms of (4.16) leave the level invariant. Let $P(N)$ be the number of basis vectors to level N. $P(N)$ equals the number of possible partitions of the integer N into positive integers. The generating function is given by

$$\prod_{N=1}^{\infty} \frac{1}{1 - x^N} = \sum_{M=0}^{\infty} x^M P(M). \tag{4.21}$$

We quote the first few examples: $N = 0$, $|h\rangle$; $N = 1$, $L_{-1}|h\rangle$; $N = 2$, $L_{-1}L_{-1}|h\rangle$, $L_{-2}|h\rangle$; $N = 3$, $L_{-1}L_{-1}L_{-1}|h\rangle$, $L_{-1}L_{-2}|h\rangle$, $L_{-3}|h\rangle$.

In order to define the notion of unitarity we need a scalar product:

$$\langle h|L_{\ell_m} \ldots L_{\ell_1} L_{-k_1} \ldots L_{-k_n}|h\rangle$$

becomes zero for vectors with different levels. The representation becomes unitary iff all matrices M_ν formed from scalar products on level ν (with $P^2(\nu)$ elements) are positive. The representations need not be irreducible. We calculate, for example, $\|L_{-1}|h\rangle\|^2 = 2h \geq 0$ and $\|L_{-n}|h\rangle\|^2 = 2nh + \frac{c}{12} n(n^2 - 1) \geq 0$ from which we conclude that $h \geq 0$ and $c \geq 0$. Next we calculate

$$M_2 = \begin{pmatrix} \langle h|L_1 L_1 L_{-1} L_{-1}|h\rangle & \langle h|L_2 L_{-1} L_{-1}|h\rangle \\ \langle h|L_1 L_1 L_{-2}|h\rangle & \langle h|L_2 L_{-2}|h\rangle \end{pmatrix} = \begin{pmatrix} 4h(2h+1) & 6h \\ 6h & 4h + \frac{c}{2} \end{pmatrix} \tag{4.22}$$

and obtain $\det M_2 = 32(h - h_{11}(c))(h - h_{12}(c))(h - h_{21}(c))$. The zeros of $\det M_2$ are given by $h_{11} = 0$ and

$$h_{\substack{12\\21}}(c) = \frac{5 - c \pm \sqrt{(1-c)(25-c)}}{16}.$$

If $\det M_2 < 0$ both eigenvalues cannot be positive. Next one studies $h_{p,q}(c)$ as a function of c: For $c \leq 1$ there exist two curves ending at $c = 1$ and $h = 1/4$. A part of the strip $0 \leq c \leq 1$ and $h \geq 0$ is excluded. For $1 < c < 25$ both zeros turn out to be complex. For $c > 25$ h_{12} and h_{21} are negative. One even is able to calculate

$$\det M_N(c,h) = \alpha_N \prod_{p,q \geq 1, pq \leq N} (h - h_{p,q}(c))^{P(N-pq)} \tag{4.23}$$

which are called Kac determinants. At each level parts of the above mentioned strip is excluded. Finally in the strip only certain points remain. A long argumentation reveals the

Theorem: There exist unitary heighest weight representations of the Virasoro algebra iff either $c \geq 1$, $h \geq 0$ or $c \in \{c_m\}$, $c_m = 1 - 6/(m(m+1))$, $m = 3, 4, \ldots$ and

$$h_{p,q}(m) = \frac{((m+1)p - mq)^2 - 1}{4m(m+1)} \quad \text{with } 1 \leq q \leq p \leq m-1. \tag{4.24}$$

As an example we quote the Ising case: $c = 1/2$ is obtained for $m = 3$. Three unitary representations to $h = 0, 1/16$ and $1/2$ are allowed. $(c,h) = (1/2, 0)$ corresponds to $T(z)$; $(1/2, 1/16)$ to $s(z,\bar{z})$ and $(1/2, 1/2)$ to the Majorana field $\psi(z)$. Note that

$$\langle s(z_1,\bar{z}_1) s(z_2,\bar{z}_2) \rangle = \frac{c_{12}}{|z_1 - z_2|^{h_1+h_2} |\bar{z}_1 - \bar{z}_2|^{\bar{h}_1+\bar{h}_2}} \tag{4.25}$$

and since $h_1 = h_2 = \bar{h}_1 = \bar{h}_2 = 1/16$ we obtain (!) the η-exponent to be $\eta = 1/4$.

4.6 Correlation Functions

We defined quasiprimary field operators by demanding the transformation law

$$A(z,\bar{z}) \to |f'(z)|^h |\bar{f}'(\bar{z})|^{\bar{h}} A(f(z), \bar{f}(\bar{z})) \tag{4.26}$$

or infinitesimally

$$\delta_\varepsilon A(z) = \varepsilon \partial_z A + h\varepsilon' A.$$

We note that the transformation law of the energy momentum tensor is different

$$\delta_\varepsilon T = \varepsilon \partial_z T + 2\varepsilon' T + \frac{c}{2} \varepsilon'''(z) \tag{4.27}$$

since the c-number term is added.

Following *BPZ* [6] we call ϕ a primary operator if

$$\delta_\varepsilon \phi = \varepsilon \partial_z \phi + h_\phi \varepsilon' \phi \tag{4.28}$$

for global *and* local conformal transformations. Otherwise the operator is called a secondary one. Note that primary implies quasiprimary. A secondary operator transforms as

$$\delta_\varepsilon A(z) = \sum_{k=0}^{N} B^{k-1}(z)\frac{d^k}{dz^k}\varepsilon(z), \qquad B^{-1} = \partial_z A, \qquad B^0 = hA. \tag{4.29}$$

Since the conformal weight of B^{k-1} is $h-k+1$, $k=0,\ldots,N$ and h_{k-1} has to be positive, $h_{k-1} > N$ and the sum in (4.29) has to be a finite one.

Given a primary field operator ϕ, there exists a family of secondary quasiprimary operators with conformal weights $(h+i, \bar{h}+j)$, $i,j \in \mathbb{N}$. They form a family $[\phi]$. The operator algebra is built up as $\mathcal{A} = \bigoplus_n [\phi_n]$. $[\phi]$ gives a representation of the Virasoro algebra. $\mathcal{A}$ can be studied with the help of the operator product expansion. As an example we note the expansion

$$T(\zeta)\phi(z) = \sum_{k=0}^{\infty} (\zeta - z)^{k-2}\phi^{-k}(z), \tag{4.30}$$

where ϕ^{-k} has conformal weight $(h+k)$. It follows that

$$L_{-k}\phi = \phi^{-k} \quad \text{with} \quad L_{-k}(z) = \oint \frac{d\zeta}{2\pi i}\frac{T(\zeta)}{(\zeta - z)^{k-1}}$$

and more generally elements of the family $[\phi]$ are given by

$$\phi^{(-k_1,\ldots,-k_N)}(z) = L_{-k_1}(z)\ldots L_{-k_N}(z)\phi(z), \qquad k_N \geq \ldots \geq k_1 \geq 1. \tag{4.31}$$

With the help of a primary operator we can obtain the heighest weight vector

$$|h\rangle = \phi(0)|0\rangle, \qquad \phi^{(-k_1,\ldots,-k_N)}(0)|0\rangle = L_{-k_1}\ldots L_{-k_N}|h\rangle. \tag{4.32}$$

Due to the proposed transformation laws correlation functions are severely restricted. Let $\phi_j \equiv \phi_j(z_j)$ be primary fields. The correlation function $\langle T(z)\phi_1 \ldots \phi_N\rangle$ can be expressed in terms of $\langle \phi_1 \ldots \phi_N\rangle$: From (4.28) we get for $\varepsilon_n(z) = -\varepsilon z^{n+1}$, $n \in \mathbb{Z}$

$$\delta_n\phi \equiv -\frac{1}{\varepsilon}\delta_{\varepsilon_n}\phi = [L_n, \phi] = z^{n+1}\partial_z\phi + h(n+1)z^n\phi, \tag{4.33}$$

and calculate

$$[\sum_{m=-1}^{\infty} \frac{L_m}{z^{m+2}}, \phi(z_1)] = \frac{1}{z-z_1}\partial_{z_1}\phi(z_1) - \frac{h}{(z-z_1)^2}\phi(z_1) \equiv \mathcal{L}_{zz_1}\phi(z_1). \tag{4.34}$$

We therefore deduce that

$$\langle T(z)\phi_1\ldots\phi_N\rangle = \sum_{j=1}^{N} \mathcal{L}_{zz_j}\langle\phi_1\ldots\phi_N\rangle. \tag{4.35}$$

For the special values of (c, j) (Equ. 4.24) we obtain differential equations for correlation functions. As an example we treat $m = 3$ for which three unitary representations exist. It follows that the determinants $\det M_N$ vanish for all

$N \geq 2$. This means that there exists a vector which is orthogonal to all vectors on this level and which has norm zero (the representation is reducible). Take for example level two. There exists a vector $|\chi\rangle$ such that $L_n|\chi\rangle = 0$ for all $n > 0$. $|\chi\rangle = (L_{-2} + aL_{-1}^2)|h\rangle$. Requiring that $L_1|\chi\rangle = L_2|\chi\rangle = 0$ gives two conditions and $a = -3/2(2h+1)$ and $h = [5 - c \pm \sqrt{(1-c)(25-c)}]/16$ which are the values we obtained before. It follows that $\langle\chi|\chi\rangle = 0$ with

$$|\chi\rangle = (L_{-2} - \frac{3L_{-1}L_{-1}}{2(2h+1)})|h\rangle. \tag{4.36}$$

It is implied that within the family $[\phi_h]$ there exists an operator $\chi(z)$ with $\chi(0)|0\rangle = |\chi\rangle$. $\chi(z)$ is called the null field and the family is called to be degenerate. It follows that the correlation functions of primary fields are given by solving differential equations. Start from

$$\langle 0|\phi_1 \dots \phi_N(L_{-2} - \frac{3L_{-1}L_{-1}}{2(2h+1)})\phi_h|0\rangle = 0, \tag{4.37}$$

and commute L_{-k} to the left and use finally $\langle 0|L_{-k} = 0$. The differential equation

$$(D_{-2}^N - \frac{3}{2(2h+1)}D_{-1}^{N\,2})\langle 0|\phi_1 \dots \phi_N\phi_h(0)|0\rangle = 0 \tag{4.38}$$

with

$$D_{-k}^N = \sum_{i=1}^N (-z_i^{k+1}\partial_{z_i} - (1-k)h_i z_i^{-k})$$

follows.

Acknowledgement: These notes summarize seminars in Vienna. I would like to thank the enthusiastic participants for their interest.

References

1 L. Faddeev: *Integrable Models in 1+1 Dimensional Quantum Field Theory*, Les Houches 1982 (Elsevier, Amsterdam 1984);
L. Faddeev and L.Takhtajan: *Hamiltonian Methods in the Theory of Solitons* (Springer 1987)

2 R. Baxter: *Exactly Solved Models in Statistical Mechanics* (Academic Press, London 1982)

3 M. Wadati and T. Deguchi: *Exactly Solvable Models and Knot Theory*, Phys. Rep. **180**, 247 (1989)

4 H. Grosse: *Models in Statistical Physics and Quantum Field Theory* (Springer 1988)

5 Y. Saint-Aubin: *Phéhomènes Critiques en deux dimensions et invariance Conforme*, lecture notes 1987

6 A.A. Belavin, A.M. Polyakov and A.B. Zamolodchikov: Nucl. Phys. **B241**, 333 (1984)

An Introduction to the Renormalization of Theories with Continuous Symmetries, to the Chiral Models and to Their Anomalies

C. Becchi

Dipartimento di Fisica, Università di Genova,
Istituto Nazionale di Fisica Nucleare, Sezione di Genova,
I – 16146 Genova, via Dodecaneso 33, Italy

Introduction

The role of symmetries, both discrete and continuous, in Quantum Mechanics is one of the main topics of a standard course for graduate students. In Quantum Field Theory symmetries are even more important, but any attempt of a systematic and unified treatment is made extremely difficult by the great variety of realizations that one encounters in the different dynamical frameworks. The mechanism of the spontaneous breakdown and the Goldstone theorem, the properties of the currents and their commutator algebras, the quantum anomalies, are three well known examples of the various aspects that symmetries assume in a quantum field theory.

In renormalized quantum field theory the situation described above becomes even more difficult because of the technical complexity of renormalization. The purpose of these lectures being an introduction to chiral models and anomalies, we have to begin with some preliminary consideration on the general properties of the renormalization procedure.

The recent developments in field theory, we refer in particular to the dominant role assumed by the dimensional schemes and to the great variety of unconventional models which have attracted the public attention, have, in a sense, overshadowed the foundations of renormalization, locality and power counting. The problems connected with the definition of the composite operators and of their time-ordering are hardly touched in the most diffused recent text-books. It is therefore difficult for a student to understand e.g. the nature of the current-algebraic relations and the origin of anomalies.

To oppose this trend we have decided to avoid the conventional introduction to renormalization based on the Gell-Mann Low formula and on the choice of a regularization procedure followed by that of the counter-terms. We have instead preferred to resume the old way starting from the field equations combining it with an introduction to the renormalization of local composite operators. The resulting point of view strengthens the role of locality and power counting at the price of some algebra. Indeed, for example in this framework the renormalization of a model with a rigid (space-time independent) continuous symmetry requires some algebraic manipulations which are not always needed

Fields and Particles Editors: H. Mitter · W. Schweiger

in a dimensional scheme. The advantage is however that the approach turns out to be almost completely self contained, which is far from being true for the dimensional scheme.

Concerning the analysis of the chiral models, it turns out that in two space-time dimensions, one has the great benefit of preserving, with the exception of the Goldstone theorem, the same properties of the 4-dimensional theory with however a great simplification on the algebraic side. Therefore we have chosen to discuss the technical aspects of renormalization in two dimensions comparing the results with the corresponding ones in four dimensions.

We shall then start from an analysis of the renormalization of field equations. Referring for simplicity to a (multicomponent) scalar field, we shall simultaneously construct the fully quantized Green functions of the basic fields and of the composite operators, putting into evidence their locality and power counting properties.

We shall then discuss the problem of renormalization of models with continuous symmetries, applying the method discussed in the first two lectures to a model with $O(n)$ symmetry.

In the last lectures we shall study the chiral models. Beginning from the simplest possible situation, a two-dimensional system of massless fermion fields, we shall discuss the renormalization of current-algebra and its anomalies. Coming back to four dimensions we shall consider the spontaneous symmetry breakdown and its consequences. We shall also describe the soft breaking due to the masses of the fermion fields (quarks).

For what concerns the choice of references whenever possible we have preferred lecture notes or review articles to the original papers whose references can, of course, always be found in the reviews.

1. The Renormalization of Field Equations

The existence of a theory of scattering is one of the fundamental outcomes of field theory. It is based on the classical axiomatic results on the asymptotic evolution of states and on the well known Lehmann-Symanzik- Zimmermann reduction formulae relating the S matrix elements to the time-ordered Green functions [1]. If ϕ_{in} is a set of asymptotic ingoing fields whose wave operator is K we can write the S operator in the Fock space of ϕ_{in} as:

$$S =: \exp\left(\int d^4x \phi_{\text{in}}(x) K_x z^{-1} \frac{\delta}{\delta j(x)}\right) : Z[j]|_{j=0} \,, \tag{1.1}$$

where z is the residue matrix of the Fourier transformed two-point function on the mass shell pole and the functional $Z[j]$ is the generator of the time-ordered Green functions:

$$Z[j] = (\Omega, T e^{i\int d^4x j(x)\phi(x)} \Omega) \,, \tag{1.2}$$

and Ω is the vacuum state.

Therefore the basic dynamical problem in the construction of a field theory is to compute the functional generator Z. In order to understand the main difficulties of this construction we shall take as starting point the field equations. To have the simplest possible notation we shall refer to a, in general multicomponent, scalar field theory in spite of its possible triviality [2].

The first step of our approach will be the translation of the field equations into a functional differential equation for Z. We shall begin with the free field case in which Z is completely determined from the propagator $\Delta_F(x)$ using Wick's theorem:

$$Z[j] = \exp\left(\frac{i}{2}\int d^4x d^4y j(x)\Delta_F(x-y)j(y)\right) . \tag{1.3}$$

The field equation is now:

$$(\Box + m^2)\phi(x) \equiv K_x\phi(x) = 0 \tag{1.4}$$

and hence

$$K_x\Delta_F(x) = \delta(x) . \tag{1.5}$$

The functional Z satisfies the differential equation:

$$-iK_x\frac{\delta}{\delta j(x)}Z[j] = j(x)Z[j] . \tag{1.6}$$

This equation can be taken as the starting point for the construction of the theory, indeed, taking into account the causal boundary conditions which are a direct consequence of the time-ordering in the definition of Z (1.2), it can be integrated giving back (1.3). In the non-free case a new interaction term will appear in (1.6). One would naively define this term as a local functional I' (e.g. a polynomial) of ϕ. In the equation for Z ϕ has to be replaced by $-i\frac{\delta}{\delta j(x)}$. One should then write:

$$\left[-iK_x\frac{\delta}{\delta j(x)} + I'\left[-i\frac{\delta}{\delta j(x)}\right]\right] Z[j] = j(x)Z[j] . \tag{1.7}$$

This equation is however not mathematically sound, indeed, already in the free field case the functional derivatives of Z are distributions, thus the interaction term would contain distributions with coinciding independent variables which are ill-defined mathematical objects.

We are going to see that this is one of the origins of the need of a renormalization theory. As mentioned above the quantum fields are so singular that even their square does not exist, as a matter of fact they are operator valued distributions. The traditional way to overcome this difficulty is to regularize the theory. This can be done, e.g. following Pauli-Villars, adding to the wave operator terms with higher derivatives and hence making the Green functions more regular in the origin. As an example consider

$$K_x^{(M)} = \frac{\Box^2 + (M^2+m^2)\Box + M^2m^2}{M^2 - m^2} , \tag{1.8}$$

where M has the role of a cut-off.

The regularized theory is physically unacceptable, since it violates unitarity, (it corresponds to an indefinite scalar product), but it renders our functional differential equation (1.7) much easier to handle. Indeed, assuming e.g. a cubic interaction term

$$I'\left[-i\frac{\delta}{\delta j(x)}\right]=i\frac{\lambda}{3!}\left[\frac{\delta}{\delta j(x)}\right]^3 , \tag{1.9}$$

we can write the solution of (1.7) in the form of a Feynman functional integral:

$$Z[j]=N\int\prod_y d\phi(y)\exp\left[-i\int d^4x\left[\phi\frac{K^{(M)}}{2}\phi+\frac{\lambda}{4!}\phi^4-j\phi\right](x)\right] , \tag{1.10}$$

as one can easily verify assuming the translation invariance of the measure $\Pi_y d\phi(y)$ and choosing N so that $Z[0]=1$. Then one has to return to the original wave operator taking the limit of the cut-off to infinity and keeping fixed as many quantities, e.g. the physical masses and coupling constants, as needed in order to specify the values of the parameters appearing in (1.10) for any value of the cut-off.

This procedure is the one which is actually followed in a computer analysis. There are cases, as the scalar field theories, in which one finds the surprising result of being able to reproduce only the free theory. There are however other cases where it is possible to obtain non-trivial physical results. This happens typically in the asymptotic free field theories where the short distance (high momentum) behaviour of the Green functions is as singular as that of a free theory.

In these lectures we shall study the short distance behaviour of some field theory models in the renormalized perturbation theory. The use of perturbation theory is justified from a formal point of view. Indeed we shall see in the following that perturbative renormalization gives a systematic method of analysis of field theories with a short distance behaviour as singular as the free ones. In the case of asymptotic free theories the perturbative results should also give a faithful description at short distances of the exact solution.

Perturbation theory is based on the recursive solution of the field equation (1.7). The recursive procedure turns out to be simpler if we consider instead of the generator of the Green functions Z, that of their connected parts Z_c. This is defined according:

$$Z=e^{iZ_c} . \tag{1.11}$$

The field equation (1.7) is easily translated in terms of Z_c:

$$K_x\frac{\delta}{\delta j(x)}Z_c[j]+I'\left[\frac{\delta Z_c}{\delta j(x)}-i\frac{\delta}{\delta j(x)}\right]=j(x) . \tag{1.12}$$

(1.12) considered as an ordinary differential equation can be transformed into an integral one:

$$\frac{\delta}{\delta j(x)}Z_c[j]=\int d^4y\Delta_F(x-y)\left[j(y)-I'\left[\frac{\delta Z_c}{\delta j(y)}-i\frac{\delta}{\delta j(y)}\right]\right] , \tag{1.13}$$

where we have taken into account the Feynman boundary conditions, resulting from the very definition of Z and hence of Z_c. The content of this equation is better put into evidence by its graphical representation. Introducing the graphic symbols:

$$\frac{\delta^n Z_c[j]}{\delta j(x_1),..,\delta j(x_n)} \equiv$$  , $\Delta_F(x-y) \equiv$ ——

(1.13) is written:

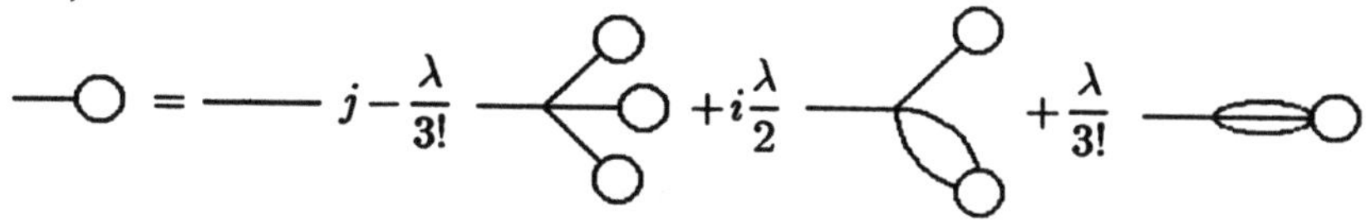

, (1.14)

where we have identified the interaction term with a cubic coupling.

It is obvious that the iteration of this equation generates the Feynman diagrammatic representation of the Green functions. It is also important to notice that all the diagrams generated by this procedure are connected. This justifies the name of Z_c.

Let us remark furthermore that the first two diagrams in the right-hand side of (1.14) do not show any loop while third and fourth exhibit one and two loops respectively. We notice also that neglecting the last two terms would be equivalent to replace (1.13) with:

$$K_x \frac{\delta}{\delta j(x)} Z_c[j] + I' \left[\frac{\delta Z_c}{\delta j(x)}\right] = j(x) \; , \tag{1.15}$$

which coincides with the classical field equation written for $\phi = \frac{\delta Z_c}{\delta j}$.

We shall not discuss in details the solution of this equation. We limit ourselves to remark that (1.15) can always be transformed into an integral equation which can be solved iteratively giving a formal power series in j. Starting from the solution of (1.15) we can solve (1.13) taking into account its last two terms recursively. In this way we get a loop ordered series of functionals each of which is a formal power series in j. The loop ordered series can be written as a formal power series in an auxiliary variable which is often identified with $\hbar$ owing to the fact that the zero loop term coincides with the classical field.

Now considering the first terms of the double series solution we have in the case with cubic interaction term I':

$$\begin{aligned} Z_c[j] = \frac{1}{2}\int d^4x d^4y j(x)\Delta_F(x-y)j(y) - \frac{\lambda}{4!}\int d^4x \left[\int d^4y \Delta_F(x-y)j(y)\right]^4 \\ + i\frac{\lambda^2}{4}\Delta_F(0)\int d^4x \left[\int d^4y \Delta_F(x-y)j(y)\right]^2 \\ - i\frac{\lambda^2}{16}\int d^4x d^4y \Delta_F^2(x-y)\left[\int d^4z \Delta_F(x-z)j(z)\right]^2 \left[\int d^4w \Delta_F(y-w)j(w)\right]^2 \\ + O(j^6, \hbar^2) \; . \end{aligned} \tag{1.16}$$

The mathematical sickness of our naive construction is put into evidence in the second and third terms in the right-hand side of this equation where one finds $\Delta_F(0)$ and Δ_F^2 respectively. Indeed $\Delta_F(0)$ is an indefinite, possibly infinite, constant. The propagator diverges at the origin as $(x^2)^{-1}$. For the same reason Δ_F^2 is not a distribution, indeed it is not absolutely integrable in any neighbourhood of the origin. Thus both quantities are indefinite, their mathematical meaning has to be specified in order to give a sense to (1.16).

Now $\Delta_F(0)$ can be interpreted as an arbitrary constant which can be fixed specifying the mass of the field up to one loop. Concerning $\Delta_F^2(x)$ let us notice that multiplying it with x^μ we obtain a distribution whose Fourier transform is:

$$2\int d^4k \frac{(k-p)^\mu}{\left[(k-p)^2+m^2-i\epsilon\right]^2(k^2+m^2-i\epsilon)} \,. \tag{1.17}$$

Therefore $\Delta_F^2(x)$ can be interpreted as the Fourier transform of a primitive of (1.17) which is, of course, defined up to an additive constant. We have therefore:

$$\Delta_F^2(x) = \frac{1}{(2\pi)^4}\int d^4p \int d^4k e^{ipx} \frac{(k-p)^2-k^2}{[(k-p)^2+m^2-i\epsilon](k^2+m^2-i\epsilon)^2} + C\delta(x) \,, \tag{1.18}$$

where C is arbitrary. It can be fixed specifying the S-wave scattering length up to one loop.

It is worth noticing at this point that our definition of (1.16) is in a sense minimal. Indeed, for example we could have multiplied $\Delta_F^2(x)$ by any positive integer power n of the components of x and, repeating the same construction, we would have found the Fourier transform of the nth primitive of a certain function. In this case the arbitrary additive constant would have been replaced by a polynomial in p which, after Fourier transformation, would have become a linear combination of partial derivatives of $\delta(x)$ up to the degree n. However this would have substantially changed the short distance behaviour of the contribution to the four point Green function coming from the fourth term in the right-hand side of (1.16) increasing by n the degree of its singularity at the origin. With our choice this degree differs from that of the second term (the tree term) only by a $\log(x^2)$.

Exactly the same effect would be produced by the introduction of corrections to the interaction term which are polynomials of degree higher than three in the field. Any increase of the **canonical dimension** (which, in d space-time dimensions, is $\frac{d-2}{2}$ for the scalar, $\frac{d-1}{2}$ for the fermionic fields and one for every partial derivative) leads to a drastic change of the power behaviour of the Green functions, which will also depend on the loop order. This is the case of a non-renormalizable theory.

The crucial problem of renormalization is to generalize the minimal procedure applied above to all loop orders and to every local operator which can be defined in the tree approximation. This requires, first of all, a characterization of the short distance behaviour of the Green functions that we want to maintain to all orders and that we call **canonical**. It can be defined as follows.

Let us call euclidean the region where all the points of a connected Green function are spacelike among each other. Scaling down in this region all the space-time variables we have canonical power counting for a connected Green function satisfying for any positive ϵ:

$$|G(\lambda x_1, ..., \lambda x_N)| < g(x_1, ..., x_N)\lambda^{-D_G+\epsilon} , \qquad (1.19)$$

where D_G is the sum of the canonical dimensions of the operators appearing in G.

Zimmermann has shown that this "canonical" power counting law holds true for a suitable class of composite operators. We shall call **power counting dimension** (p.c. dimension) of these operators their contribution to the exponent in (1.19), which cannot be lower than their canonical dimension. Following Zimmermann [3], for every M monomial in the fields and their derivatives it is possible to define an operator $N(M)$ which is local and finite. The set of these operators with all possible monomials with dimension $D(M)$ not exceeding D forms a basis for the local operators with p.c. dimension not exceeding D.

Let $P(x)$ and $Q(y)$ be two local operators of p.c. dimension p and q respectively, for $(x-y)^2 < 0$ the product of the two operators can be decomposed according:

$$P(x)Q(y) = \sum_{D(M)\leq p+q+r} f_M(x-y)N\left(M(\frac{x+y}{2})\right) + R(x,y) , \qquad (1.20)$$

where the remainder R vanishes as $(x-y)^r$, modulo logs, for $x \to y$. The coefficients f_M are C-numbers.

The definition of $N(M)$ is based on the prescription of a set of normalization conditions for the vertices, one-particle irreducible diagrams, containing it. The original recipe was to subtract at zero momentum all the potentially superficially divergent one-particle irreducible diagrams containing the operator $N(M)$. This means zero-momentum normalization conditions which are useless in the massless field case. However it is important to understand that the relevant properties of N operators are locality and the power counting rule following from (1.20).These properties do not depend on the particular choice of normalization conditions, the N operators can be introduced in a variety of equivalent ways, e.g. in a minimal dimensional subtraction scheme [4].

A possible approach to the problem of the construction of the interaction term could be based on (1.20). Indeed we can regularize the interaction operator by splitting its points and interpret every ill-defined quantity as an arbitrary parameter, as we have done for the first terms of (1.16). This point of view is usually justified by means of the following argument. When the cut-off, in our case the splitting, is removed, the Green functions can diverge. These divergences, which in our case are due to the singularity in the origin of the coefficients f_M in (1.20), require counter terms defined up to an additive constant.

It is evident from (1.20) that in a minimal set the counter-terms correspond to the operators $N(M)$ with divergent coefficient f_M, that is to the operators with canonical dimension not exceeding that of the splitted operator. These are the undetermined contributions in its definition.

Applying (1.20) to the splitted interaction term we find that it can be written as a linear combination of the N operators with p.c. dimension up to three, or equivalently, by $N(I')$, where I' is a polynomial in the fields and their derivatives of dimension not exceeding three. Of course, at the tree level ($\hbar = 0$) I' will coincide with the naive interaction. The quantum corrections contain finite contributions corresponding to the above mentioned counter-terms which have to be specified order by order.

By means of the point splitting regularization it is also possible to understand a very general result which is usually referred to as the weak form of the Quantum Action Principle (QAP)[5]. It asserts that all the equations involving local operators and which are true at the classical level (as e.g. Noether theorem and field equations) are possibly violated at the quantum level by local operator terms with the natural p.c. dimension. That is, whose p.c. dimension does not exceed the canonical one of the field functionals involved in the equation. We shall have many occasions to apply this QAP.

Coming back to our field equation we shall substitute the naive interaction term with the insertion of an operator $N(I')$. The polynomial I' is a Lorentz scalar, provided the normalization conditions defining the N-operators are not too crazy, and its coefficient are formal power series in the perturbative parameter ($\hbar$). Its classical limit coincides with the functional derivative of the interaction term in the action.

However this is not all concerning the composite operators. Indeed another difficulty comes from the fact that the T-ordering of the composite operators is not uniquely defined. To understand this it is sufficient to go back to the analysis of the first orders of the perturbative series and to notice that formally $\Delta_F^2(x-y)$ is proportional,in a free theory, to the vacuum expectation value of the T-product of the operator $:\phi^2:$ in the points x and y. In an approach consistent with the canonical power counting law this T-product is defined up to a term proportional to $\delta(x-y)$.

In order to push further our analysis of field equations in the functional framework, we have to accomplish one more formal duty. The introduction of every new operator, as $N(I')$ is, requires that of its source, say ϵ. After this the Green functional generator becomes:

$$Z[j] = (\Omega, T e^{i \int d^4x [j(x)\phi(x) + \epsilon(x) N(I'(x))]} \Omega) \ . \tag{1.21}$$

However if we introduce within the functional framework the operator $:\phi^2:$ through its source, σ, we transfer the indetermination of the two-point Green function into the functional generator Z. The freedom to add to the Green function a term proportional to $\delta(x-y)$ corresponds to that of redefining Z according:

$$Z[j,\sigma] \to Z[j,\sigma] e^{i\frac{c}{2} \int d^4x \sigma^2(x)} \ . \tag{1.22}$$

Indeed if we compute the Green function of two operators $:\phi^2:$ taking the second functional derivative with respect to σ, the factor appearing in (1.22) reproduces the possible addition of a contribution proportional to $\delta(x)$. Notice that the substitution (1.22) is analogous to the introduction of a counter term into the Lagrangian, this counter-term however in this particular case depends only on the sources.

In the following we shall introduce systematically composite operators. Thus we have to find an easy way to characterize the terms which are needed to identify uniquely the Green functions. The idea ensuing from the example above is that these terms should look like counter-terms that one can add to the action. In a framework with canonical power counting these counter-terms, involving both the quantized fields and the sources of composite operators, will be local and limited in number. Indeed, associating to every source of composite operator a canonical dimension equal to the difference between the space-time dimension and that of the operator, our pseudo-counter-terms will have total dimension limited by that of the space-time.

What has to be kept in mind from the above discussion is that whenever one introduces composite operators together with their sources, concerning the radiative corrections (finite counter-terms) to the local operators, these sources, which are often called classical fields, play a completely analogous role to that of the quantized fields . Indeed the correction terms are local functionals of both the quantized and the classical fields with a total p.c. dimension which does not exceed that of the original operator. The counter-terms involving sources correspond to the multilocal corrections to a time-ordering which have been described by Bogoliubov and Shirkov [6], that is to terms which are proportional to Dirac deltas, and their derivatives, in the space-time variables of the composite operators.

The problem of defining uniquely the T-ordering in the presence of composite operators will be of great importance in the discussion of anomalies.

Having so analyzed the role played by the local composite operators in the functional framework, we go back to the field equations considered as a tool to construct the functional generator. We introduce the interaction term through the operator I' with classical limit I'_0 and we write the field equation:

$$K_x \frac{\delta Z_c[j,\epsilon]}{\delta j(x)} + \frac{\delta Z_c[j,\epsilon]}{\delta \epsilon(x)} = j(x) \; . \tag{1.23}$$

However this equation will not be integrable for an arbitrary interaction term and even the existence of a non-trivial set of I' for which (1.23) has a solution does not appear totally obvious. To show this let us consider this equation at the classical level. As shown above, in the classical limit our equation is:

$$K_x \frac{\delta Z_c[j,\epsilon]}{\delta j(x)} + I'_0 \left[\frac{\delta Z_c[j,\epsilon]}{\delta j(x)}, \epsilon(x) \right] = j(x) \; . \tag{1.24}$$

Let us assume for the moment (1.24) to be integrable, we define the new variable

$$\varphi(x) = \frac{\delta Z_c[j,\epsilon]}{\delta j(x)} - \frac{\delta Z_c[0,0]}{\delta j(x)} \equiv \frac{\delta Z_c[j,\epsilon]}{\delta j(x)} - \varphi_0 \; , \tag{1.25}$$

and the new functional $\Gamma[\varphi, \epsilon]$ through:

$$Z_c[j, \epsilon] = \Gamma[\varphi, \epsilon] + \int d^4x j(x)[\varphi(x) + \varphi_0] . \tag{1.26}$$

Γ is the Legendre transform of Z_c. In the above equation φ_0 is the vacuum expectation value of the field. It is easy to verify, e.g. considering the solution of (1.24) as a formal power series in j, that Γ is the classical action and that the classical field equation (1.24) coincides with:

$$-\frac{\delta \Gamma[\varphi, \epsilon]}{\delta \varphi}\Big|_{\varphi(x)=\frac{\delta Z_c[j,\epsilon]}{\delta j(x)}} = j(x) . \tag{1.27}$$

It follows that Γ_0' must be a functional derivative, that of the interaction part of the action. In general this is a non-trivial constraint for Γ' (consider e.g. the case of a multicomponent scalar field).

Therefore we have to select the classical interaction among the field derivatives of integrated local functionals of dimension four. Still, for a generic definition of the quantum operator $N(\Gamma')$ (1.23) will not be integrable to all loop orders. Indeed the operator field equation which is true in the tree approximation receives radiative corrections which could break the integrability of (1.23). Suppose that this happens at the loop order n. At the same order we should have:

$$K_x \frac{\delta Z_c[j, \epsilon]}{\delta j(x)} + \frac{\delta Z_c[j, \epsilon]}{\delta \epsilon(x)} - j(x) = \Delta \left[\frac{\delta Z_c[j, \epsilon]}{\delta j(x)}, \epsilon\right] + O(\hbar^{n+1}) . \tag{1.28}$$

Here Δ stands for the breaking to the field equation which, according to the QAP, is finite and local, with p.c. dimension 3. $O(\hbar^{n+1})$ indicates possible further contributions beyond the n-loop order.

It remains however the possibility that the wanted field equation be recoverable by a change of the action. This is in fact the case if Δ is the field derivative of an integrated local functional of dimension four. Otherwise we shall speak of an ANOMALY to the field equation.

To exclude the existence of anomalies to the field equation let us perform again a Legendre transformation of Z_c, now to all loop orders. In this case we do not get anymore the action, but the functional generating the one-particle irreducible vertices [7].

Noticing that:

$$\frac{\delta \Gamma[\varphi, \epsilon]}{\delta \epsilon}\Big|_{\varphi(x)=\frac{\delta Z_c[j,\epsilon]}{\delta j(x)}} = \frac{\delta Z_c[j, \epsilon]}{\delta \epsilon(x)} . \tag{1.29}$$

We have:

$$K_x \varphi(x) - \frac{\delta \Gamma[\varphi, \epsilon]}{\delta \varphi(x)} - \frac{\delta \Gamma[\varphi, \epsilon]}{\delta \epsilon(x)} = \Delta[\varphi(x) + \varphi_0, \epsilon(x)] + O(\hbar^{n+1}) . \tag{1.30}$$

This equation implies the "consistency condition":

$$\frac{\delta \Delta[\varphi(x) + \varphi_0, \epsilon(x)]}{\delta \varphi(y)} + \frac{\delta \Delta[\varphi(x) + \varphi_0, \epsilon(x)]}{\delta \epsilon(y)} - \quad (x \leftrightarrow y) \quad = 0 . \tag{1.31}$$

Which guarantees the existence of an integrated local functional Δ' such that:

$$\Delta[\varphi(x)+\varphi_0,\epsilon(x)] = \frac{\delta\Delta'}{\delta\varphi(x)} + \frac{\delta\Delta'}{\delta\epsilon(x)} , \tag{1.32}$$

and hence the breaking Δ to loop order n can be reabsorbed as a correction to the action. Iterating this argument order by order we have shown that the field equation is renormalizable in the wanted form (1.23).

This completes our construction of the Green functional based on the recursive perturbative procedure and on locality and power counting together with their direct consequence the QAP.

Let us summarize the general strategy we have followed in the discussion of the renormalizability of the field equation. The first step has been the analysis of the classical limit where we have studied the integrability of the functional form of the equation. Then we have considered the possibility of quantum breakings, deducing from the QAP their locality and power counting properties. Finally, by means of a Legendre transformation, we have obtained a consistency condition for the possible terms breaking the equation. A consistency condition has guaranteed the compensability of the breaking by means of a suitable choice of finite counter-terms to the interaction.

Exactly the same strategy will be chosen in the following to discuss the renormalization of theories with continuous symmetries and to analyse the nature of possible anomalies.

2. The Renormalization of Models with Continuous Symmetries

The basic properties of theories with continuous symmetries are very well known. In the standard situation there is a conserved charge associated with every parameter of the symmetry group; in field theory this charge is the space integral of the time component of a divergenceless current density which is deduced from the lagrangian density according to Noether's theorem. Such current if often said to be conserved. The connection between a conserved current and a conserved charge can be violated in field theory owing to the fact that the vacuum state could not belong to the domain of definition of the space integral of the charge density. This is the case of spontaneous symmetry breakdown.

In renormalized quantum field theory one deals with Green functions rather than directly with operators. In this framework the presence of a continuous symmetry can be expressed as a family of relations among Green functions containing field components which are transformed into each other by the infinitesimal symmetry transformations. Other identities exhibit the existence of conserved currents and their commutation properties with local operators. All these identities are often called Ward-Takahashi (WTI) identities since they generalize those describing the consequences of charge conservation in QED.

The problem of the renormalization of a continuous symmetry reduces to that of proving the validity of the WTI to all loop orders.

The functional framework is particularly suitable to describe WTI and to discuss their full quantum extensions. In fact one takes into account first the tree approximation Green functional (or the corresponding functions) for which the WTI are deduced as a straightforward consequence of the action principle. Then one considers the effects of loop corrections.

In renormalized field theory it is also very important to distinguish between linear and non-linear symmetries. Indeed the full quantum description of a symmetry whose infinitesimal action on the basic fields is linear in them is much easier than that where an infinitesimal field variation is a composite operator. Indeed in this case one faces at the same time the problem of defining the symmetry transformations at the quantum level and that of verifying the invariance of the Green functions. The simplest strategy in the non-linear case is to limit strictly the analysis of the infinitesimal symmetry transformations to their Lie algebra. The formal trick consists in the substitution of the infinitesimal group parameters with the anticommuting generators of an exterior algebra. This leads to the so-called BRS transformations which are an essential tool in the study of gauge models.

In these lectures we shall focus our attention on the simplest linear case. We shall start considering, as an introductory model, a scalar theory with rigid (space-time independent) $O(n)$ symmetry. Then we shall study the chiral symmetries and the corresponding current-algebra referring to the simplest possible model, a system of free fermions in two space-time dimensions. The extension of chiral symmetry to 4-dimensional models will be discussed in the last part of this course.

Let us then discuss first of all a scalar model in 4 dimensions with $O(n)$ symmetry. We consider directly the functional framework and remember that the field equations after Legendre transformations are written:

$$-\frac{\delta\Gamma[\varphi]}{\delta\varphi_i}\Big|_{\varphi_i(x)+\varphi_{0i}=\frac{\delta Z_c[j]}{\delta j_i(x)}} = j_i(x) \ . \tag{2.1}$$

Here the field ϕ, its source j and the conjugate variable φ are n-component real vectors. We shall assume for the moment that the vacuum expectation value of the field φ_0 vanishes if we choose the fields so that they transform homogeneously under the infinitesimal action of the symmetry group. This is necessarily true if the vacuum of the theory is left invariant by the symmetry transformations. In the tree approximation the $O(n)$ invariance of the action which is written straightforwardly:

$$\lambda_a^{ij}\int d^4x\varphi_i(x)\frac{\delta\Gamma}{\delta\varphi_j(x)} \equiv \mathcal{W}_a[\varphi]\Gamma = 0 \ , \tag{2.2}$$

since Γ coincides with the action . (2.2) is immediately translated into a WTI,

$$-\lambda_a^{ij}\int d^4x\varphi_i(x)j_j(x) = \lambda_a^{ij}\int d^4x j_i(x)\frac{\delta Z_c}{\delta j_j(x)} \equiv \mathcal{W}_a Z_c = 0 \ , \tag{2.3}$$

which shows that the tree approximation connected functional has the same orthogonal invariance of the action. The matrices λ_a^{ij} above for $a = 1, .., \frac{n(n-1)}{2}$ are the infinitesimal generators of $O(n)$ in the vector representation. Of course, to get the WTI for the connected Green functions it is sufficient to take the functional derivatives in the origin of (2.3).

The analysis of the full quantum extension of the above identity is usually performed starting from a suitably regularized theory. From this point of view the validity of (2.3) to all loop orders is almost trivial. However we consider this problem on the basis of the QAP without reference to any specific regularization for two good reasons. The first reason is that the use of clever regularization procedures is not a general method and, in particular, it does not work in the chiral symmetry case. The second is that this trivial problem gives a perfect occasion to illustrate the use of the QAP.

Repeating the above derivation of the WTI at the quantum level we shall have in the left-hand side of (2.3) a composite operator, therefore its vanishing will depend on the particular renormalization scheme used to define the operator as discussed in the previous lectures. To put into evidence the ambiguity in its definition we split the points in the two factors of the operator version of (2.2). From the arguments presented before we get a quantized version of the WTI (2.3) which could be broken above the n loop level ($n \geq 1$) by the insertion of an integrated local operator Δ of dimension 4. This, in the functional framework, will affect the WTI as follows:

$$\mathcal{W}_a[j]Z_c = \int d^4x \Delta_a \left[\frac{\delta Z_c}{\delta j(x)}\right] + O(\hbar^{n+1}) \ . \tag{2.4}$$

Notice that considering the short distance expansion (1.20) one should be pushed to consider the breaking Δ_a as possibly infinite. This is however excluded since the left-hand side of (2.4) is finite in the limit of coinciding points.

Taking the Legendre transform of (2.4) we get:

$$\mathcal{W}_a[\varphi]\Gamma = \int d^4x \Delta_a [\varphi(x)] + O(\hbar^{n+1}) \ . \tag{2.5}$$

Now, in order to get a consistency condition for the breaking, we can use the commutation relations of the Lie algebra of $O(n)$:

$$[\mathcal{W}_a[\varphi], \mathcal{W}_b[\varphi]] = f_{abc}\mathcal{W}_c[\varphi] \ , \tag{2.6}$$

which, combined with (2.5), yield at the N-loop order:

$$\mathcal{W}_a[\varphi] \int d^4x \Delta_b [\varphi(x)] - \mathcal{W}_b[\varphi] \int d^4x \Delta_a [\varphi(x)] = f_{abc} \int d^4x \Delta_c [\varphi(x)] \ . \tag{2.7}$$

At this point a second important remark is necessary. If

$$\int d^4x \Delta_a[\varphi(x)] = \mathcal{W}_a[\varphi] \int d^4x L'[\varphi(x)] \ , \tag{2.8}$$

that is, if the breaking were equal to the infinitesimal variation of an integrated local functional L' of dimension 4, it would be trivial to recover the unbroken WTI by subtracting $\int d^4xL'$ from the action.

Therefore the anomalies to the $O(n)$ symmetry belong to the quotient space of the solutions of (2.7) modulo terms of the form (2.8). This coincides with the so called first Chevalley cohomology which is easily shown to be trivial in the case of semisimple compact groups. Indeed, let the Killing form of the Lie algebra be proportional to a Kronecker delta, combining (2.6) and (2.7) we have:

$$\mathcal{W}^2[\varphi]\int d^4x\Delta_a\left[\varphi(x)\right] = \mathcal{W}_a[\varphi]\mathcal{W}_b[\varphi]\int d^4x\Delta_b\left[\varphi(x)\right] . \tag{2.9}$$

Here $\mathcal{W}^2[\varphi] = \sum_a \mathcal{W}_a^2[\varphi]$. Now, the symmetry group being compact and the representation carried by the field completely reducible, we get:

$$\int d^4x\Delta_a\left[\varphi(x)\right] = \mathcal{W}_a[\varphi]\mathcal{W}^{-2}[\varphi]\mathcal{W}_b[\varphi]\int d^4x\Delta_b\left[\varphi(x)\right] , \tag{2.10}$$

so proving (2.8). Hence the WTI (2.3) are renormalizable to all perturbative orders.

Exactly the same argument can be generalized to prove the renormalizability of any theory with a continuous symmetry which is possibly broken at the classical levels by lagrangian terms of dimension strictly lower than four [8]. Of course, a case of particular interest is that of chiral models which were originally defined as fermionic field theories invariant under independent, isotopic, transformations of their two chiral components (the eigenvectors of γ_5).

3. The Chiral Models and Their Current Algebra

The simplest chiral model is a theory of free fermions in two space-time dimensions. Although of limited physical interest, the theory presents a current-algebra anomaly which is much easier to analyse than the corresponding four-dimensional case.

The Dirac matrices in two space-time dimensions can be identified with two Pauli matrices, e.g.:

$$\gamma^0 = \sigma_1 \quad , \quad \gamma^1 = i\sigma_2 . \tag{3.1}$$

Therefore

$$\gamma^5 = \sigma_3 . \tag{3.2}$$

With this choice the two chiral fields are identified with the upper, ψ_L, and lower, ψ_R, components of the Dirac bispinor. We consider a multicomponent spinor ψ^a for $a = 1, .., n$, that is our spinor is a complex vector in some isotopic space, but we shall omit the indices whenever not strictly necessary. A chiral $SU(n)$ transformation is defined as:

$$\psi_L \to u_L\psi_L \quad , \quad \psi_R \to u_R\psi_R , \tag{3.3}$$

where u_L and u_R are independent unimodular matrices in n dimensions. We also introduce the infinitesimal generators of the chiral transformations. These are two sets of n-dimensional, hermitian, traceless matrices τ^α for $\alpha = 1, .., n^2 - 1$ normalized according $\mathrm{Tr}(\tau^\alpha \tau^\beta) = \delta^{\alpha\beta}$. The action of the free spinor system is:

$$\Gamma = \int d^2 x i \bar{\psi} \gamma^\mu \partial_\mu \psi \; . \tag{3.4}$$

Using the light-cone variables:

$$z = \frac{x^0 + x^1}{2} \quad , \quad \bar{z} = \frac{x^0 - x^1}{2} \; , \tag{3.5}$$

and setting $\partial = \frac{\partial}{\partial z}, \bar{\partial} = \frac{\partial}{\partial \bar{z}}$, we have:

$$\Gamma = \int d^2 x i [\psi_L^\dagger \bar{\partial} \psi_L + \psi_R^\dagger \partial \psi_R] \; . \tag{3.6}$$

Thus the opposite chiral components of the field are dynamically independent. The invariance of this action under the chiral group is evident. According to Noether's theorem there are two systems of conserved currents, the left-handed current,

$$J_L^\alpha = \psi_L^\dagger \tau^\alpha \psi_L \; , \tag{3.7}$$

parallel to the $\bar{z}$ axis and the right-handed one,

$$J_R^\alpha = \psi_R^\dagger \tau^\alpha \psi_R \; , \tag{3.8}$$

along the z axis. Thus we have:

$$\bar{\partial} J_L^\alpha = 0 \; , \quad \partial J_R^\alpha = 0 \; . \tag{3.9}$$

We want to discuss the properties of these currents and in particular their commutation algebra. Therefore we introduce a system of sources, the classical vector fields a_L^α and a_R^α which are parallel to the z and $\bar{z}$ axis respectively. To simplify the notation we introduce the matrix valued fields:

$$a_{L/R} = a_{L/R}^\alpha \tau^\alpha \; , \tag{3.10}$$

and we write the new action:

$$\Gamma = \int d^2 x i [\psi_L^\dagger (\bar{\partial} - i a_L) \psi_L + \psi_R^\dagger (\partial - i a_R) \psi_R] \; . \tag{3.11}$$

Now the theory is invariant under a local, chiral gauge group whose left component corresponds to the infinitesimal action:

$$\begin{aligned} \delta a_L &= \bar{\partial} \lambda_L - i [a_L, \lambda_L] \\ \delta \psi_L &= i \lambda_L \psi_L \quad , \quad \delta \psi_L^\dagger = -i \psi_L^\dagger \lambda_L \; . \end{aligned} \tag{3.12}$$

From now on we shall limit our discussion to the left subgroup and fields omitting the completely analogous analysis for the right components. The above mentioned invariance of the action is expressed by the functional differential equation:

$$\mathrm{Tr}\left[\tau^\alpha\left[\left(\bar\partial\frac{\delta}{\delta a_L}-i\left[a_L,\frac{\delta}{\delta a_L}\right]+i\psi_L\frac{\delta}{\delta\psi_L}\right)(x)\Gamma+i\frac{\delta}{\delta\psi_L^\dagger(x)}\Gamma\psi_L^\dagger(x)\right]\right]$$
$$\equiv \mathcal{W}_L^\alpha(x)\Gamma=0\;, \qquad (3.13)$$

where ψ and $\psi^\dagger$ are Grassmann, anticommuting, variables. The quantum extension of (3.13) is interpreted as the Legendre transform of the WTI for the connected functional of the model. In general it will be affected with breakings. Following the line of the analysis of the previous section we shall verify if among these breakings there are possible anomalies. We notice first of all that the broken WTI can be written:

$$\mathcal{W}_L^\alpha(x)\Gamma=\Delta_L^\alpha(x)\;, \qquad (3.14)$$

where $\Delta_L^\alpha(x)$ is a local, Lorentz invariant, functional of dimension two. Here and from now on we shall forget the higher order corrections which were carefully taken into account in the previous section. These WTI, once integrated over x (z and $\bar z$), lead to identities which can be proved not to admit any anomaly exactly in the same way as for (2.3). This means that, once integrated over the space-time, the breaking vanishes, being a local functional of dimension two, it must be the derivative of another local functional of dimension one. Now we consider the commutation relations:

$$\left[\mathcal{W}^\alpha(x),\mathcal{W}^\beta(y)\right]=i\delta(x-y)f^{\alpha\beta\gamma}\mathcal{W}^\gamma(x)\;, \qquad (3.15)$$

and the ensuing (Wess-Zumino) consistency condition for the breaking

$$\mathcal{W}_L^\alpha(x)\Delta_L^\beta(y)-\mathcal{W}_L^\beta(y)\Delta_L^\alpha(x)=i\delta(x-y)f^{\alpha\beta\gamma}\Delta_L^\gamma(x)\;. \qquad (3.16)$$

The possible anomalies to the current-algebra WTI are identified with the elements of a quotient of two linear spaces. The first space is that of the local functionals of dimension two whose integral vanishes and which are solutions of (3.16). The second is the space of the local functionals of the form (3.14) for any Γ integrated local functional of dimension two. In practice Γ can be chosen among the functionals invariant under the rigid, that is space-time independent, action of the chiral group. Integrating (3.16) over the variable x leads to a further constraint for the breaking. Indeed on gets:

$$\int d^2x\mathcal{W}_L^\alpha(x)\Delta_L^\beta(y)=if^{\alpha\beta\gamma}\Delta_L^\gamma(y)\;. \qquad (3.17)$$

This means that the breaking transforms according to the adjoint representation of the rigid left-handed component of the chiral group. By a completely analogous treatment one shows that it must be invariant under the rigid action of the right-handed component.

This condition restricts the choice of the anomaly to the derivatives of a linear combination of a_L and $\psi^\dagger\tau\psi$. We must also take into account the constraint coming from the Lorentz invariance. In two space-time dimensions the Lorentz group is one dimensional, consequently the transformation properties of the fields and operators are characterized by a single quantum number, the helicity. a_L and ψ_L have helicity -1 and $\frac{1}{2}$ respectively, the right-handed ones have opposite helicities, ∂ changes the helicity of $+1$ and $\bar{\partial}$ of -1. Therefore we remain with the linear combination of ∂a_L and $\bar{\partial}(\psi^\dagger\tau\psi)$. However the second term can be reabsorbed since it can be written as the variation of a local counter-term and it corresponds to a rescaling of the left-handed current. The remaining candidate:

$$\Delta_L^\alpha = k\partial a_L^\alpha \ , \tag{3.18}$$

is a true anomaly since it cannot be written as in (3.14) for a local Γ. To have a better understanding of the extension of this result above two dimensions let us rewrite the left-handed anomaly in cartesian coordinates. We have:

$$\partial a_L = \frac{(\partial_0+\partial_1)(a_{L0}-a_{L1})}{2} = \frac{1}{2}(\partial_\mu a_L^\mu - \epsilon^{\mu\nu}\partial_\mu a_{L\nu}) \ . \tag{3.19}$$

The first term can be reabsorbed by the introduction of the counter-term $a_L^\mu a_{L\mu}$ thus the anomaly can be reduced to the second term, which is proportional to the skew-symmetric tensor $\epsilon^{\mu\nu}$.

It remains to compute the coefficient k. Its one-loop value is easily deduced from the current two point function. Indeed using the fermionic propagator

$$S(x) = \frac{i}{2\pi}\left[\frac{\theta(x^0)}{z-i0_+} + \frac{\theta(-x^0)}{z+i0_+}\right] \ , \tag{3.20}$$

one computes directly the two point function which, inserted into the WTI gives $k = -\frac{1}{2\pi}$. The fermions being free, there is no possibility of radiative corrections to the one-loop value of k. The absence of radiative corrections to the one loop coefficient of the current-algebra anomaly can however be proved in a much more general framework.

Before coming to the extension of our results to the four-dimensional theories let us give a short sketch of the main consequences of the WTI. As it is well known the commutation of a time derivative with the time-ordering generates commutators, in our case one gets the commutation relations on the light-cone

$$\left[J_L^\alpha(z), J_L^\beta(z')\right] = if^{\alpha\beta\gamma}\delta(z-z')J_L^\gamma(z) - \frac{ik}{2\pi}\delta^{\alpha\beta}\delta'(z-z') \ . \tag{3.21}$$

Indeed the left-handed currents, being conserved, are independent of $\bar{z}$. Now, defining

$$J_{nL}^\alpha = \int_{-\infty}^{+\infty}\left(\frac{1+iz}{1-iz}\right)^n J_L^\alpha(z) \ , \tag{3.22}$$

one gets the Kac-Moody algebra of the model:

$$\left[J_{nL}^\alpha, J_{mL}^\beta\right] = if^{\alpha\beta\gamma}J_{n+mL}^\gamma - km\delta^{\alpha\beta}\delta_{n+m,0} \ . \tag{3.23}$$

A completely analogous construction can be performed for the energy-momentum tensor leading to the whole Virasoro, Kac-Moody algebraic structure of our model, which is conformal invariant [9].

We now come to the generalization of our results to the four-dimensional models. The WTI (3.13) preserve exactly the same structure in four dimensions. The only formal difference is that the use of the light-cone variables is not helpful and the broken WTI are usually written in cartesian coordinates:

$$\begin{aligned} \mathrm{Tr}\left[\tau^{\alpha}\left[\left(\partial_{\mu}\frac{\delta}{\delta a_{L\mu}}-i\left[a_{L\mu},\frac{\delta}{\delta a_{L\mu}}\right]+i\psi_{L}\frac{\delta}{\delta\psi_{L}}\right)(x)\Gamma+i\frac{\delta}{\delta\psi_{L}^{\dagger}(x)}\Gamma\psi_{L}^{\dagger}(x)\right]\right] \\ \equiv \mathcal{W}_{L}^{\alpha}(x)\Gamma=\Delta^{\alpha}(x)\ . \end{aligned} \tag{3.24}$$

The anomaly in four dimensions can be characterized by means of the consistency condition (3.16), however the calculation is by no means simple as in two dimensions [8]. The final result can be reduced to a four-divergence of a local functional proportional to the skew-symmetric tensor. More precisely one has:

$$\begin{aligned} \Delta^{\alpha}=\epsilon^{\mu\nu\rho\sigma}\partial_{\mu}\Big[&d^{\alpha\beta\gamma}(\partial_{\nu}a_{\rho}^{\beta})a_{\sigma}^{\gamma} \\ &+\frac{1}{12}\left[d^{\alpha\beta\tau}f^{\tau\gamma\delta}+d^{\alpha\gamma\tau}f^{\tau\delta\beta}+d^{\alpha\delta\tau}f^{\tau\beta\gamma}\right]a_{\nu}^{\beta}a_{\rho}^{\gamma}a_{\sigma}^{\delta}\Big]\ , \end{aligned} \tag{3.25}$$

where $d^{\alpha\beta\gamma}$ is an invariant symmetric tensor on the Lie algebra of the left component of the chiral group.

An anomaly completely analogous to (3.25) breaks the conservation of the neutral component of the isovector axial current in QED. This is in fact the anomaly which has been discovered first; it accounts for the π^0 decay. The formal analogy between the current-algebra anomaly and the divergence of the axial current should be clear in our point of view which assigns a very similar role to quantized fields and sources. Notice however that the former anomaly introduces corrections to the time-ordering of currents, while the second one breaks explicitly a conservation law. Quite analogous to this second effect is the breakdown of the conservation of the baryonic axial current in QCD. This explains the mass difference between the singlet and the octet pseudoscalar mesons and the recently detected spin anomaly of the proton.

Analyzing the current-algebra WTI we have discussed in some detail the subtle phenomenon of the quantum anomalies, but we have disregarded the mechanism of spontaneous symmetry breakdown which has very important effects on the actual chirality. Indeed the vacuum of the strong interaction is not believed to be invariant under the action of the $SU(2)$ chiral group, which, on the contrary, is only weakly affected by the tiny mass of the first two quarks [10].

The lack of chiral invariance of the QCD vacuum, if real, is an infinite volume phenomenon which is far from being understood completely. What we shall do here is to take note, rather, to assume this fact and to see how it can be inserted into our formalism which is suitable to describe the short distance properties of a theory.

Owing to the fact that we study the Green functions, we characterize the lack of invariance of the vacuum by assigning at least one operator which is not invariant under the chiral symmetry and which assumes a non-vanishing vacuum expectation value. The operator will be given through its chiral and Lorentz transformation properties and its p.c. dimension. In particular if the dimension is chosen low enough the operator is specified up to a scale factor. A simple but non-trivial example is a chiral $SU(2)$ quark model in which we shall introduce the chiral currents:

$$\frac{1}{2}\bar{\psi}\gamma^{\mu}(1 \pm \gamma^5)\tau_i\psi \ , \tag{3.26}$$

and the scalar and pseudoscalar densities:

$$S = \bar{\psi}\psi \quad , \quad P_i = \bar{\psi}\gamma^5\tau_i\psi \ . \tag{3.27}$$

In the case of QCD it is convenient to combine the chiral currents into the axial and vector components since parity is conserved. Thus we assign the sources ω^i_μ and α^i_μ of the vector and axial current and Σ and Π^i of the scalar and pseudoscalar densities. The complete action, including also the source terms, is invariant under the infinitesimal chiral transformations which, for the fermion fields, are defined in (3.3), and for the sources are:

$$\begin{aligned}
\delta\omega^i_\mu &= \partial_\mu\lambda^i_+ - \epsilon^{ijk}[\lambda^j_+\omega^k_\mu - \lambda^j_-\alpha^k_\mu] \\
\delta\alpha^i_\mu &= \partial_\mu\lambda^i_- - \epsilon^{ijk}[\lambda^j_-\omega^k_\mu - \lambda^j_+\alpha^k_\mu] \\
\delta\Pi^i &= \epsilon^{ijk}\lambda^j_+\Pi^k - \lambda^i_-\Sigma \\
\delta\Sigma &= \lambda^i_-\Pi^i
\end{aligned} \ . \tag{3.28}$$

The corresponding current-algebra WTI are deduced in much the same way as in two dimensions. They are anomaly free since $SU(2)$ does not admit any rank three invariant symmetric tensor on its Lie algebra. Considering only the axial WTI and disregarding the fermion fields, we have:

$$\begin{aligned}
&\left[\partial_\mu\frac{\delta}{\delta\alpha_\mu} + 2\epsilon^{ijk}\left(\omega^j_\mu\frac{\delta}{\delta\alpha^k_\mu} - \alpha^j_\mu\frac{\delta}{\delta\omega^k_\mu}\right)\right. \\
&\left. +2\left(\Pi^i\frac{\delta}{\delta\Sigma} - \Sigma\frac{\delta}{\delta\Pi^i}\right)\right] Z_c\left[\omega,\alpha,\Sigma,\Pi\right] = 0 \ .
\end{aligned} \tag{3.29}$$

Notice that here we have written the WTI in terms of the connected functional instead of the vertex functional Γ which appears in the discussion of the two-dimensional model. Taking into account (1.29) the two functionals can in fact be identified.

Owing to the isotopic symmetry and to the parity conservation only the scalar density S, among the local operators that we have introduced, can acquire a non-vanishing vacuum expectation value. We introduce it as an external parameter, since we do not know how to compute it,

$$\frac{\delta}{\delta\Sigma}Z_c = F \ . \tag{3.30}$$

Taking the functional derivative of (3.29) with respect to Π_j and setting all the sources to zero leads to

$$\frac{\delta}{\delta \Pi_j}\partial_\mu \frac{\delta}{\delta \alpha^i_\mu} Z_c[0,0,0,0] = -2\delta^4(x)\delta^{ij} F \ . \qquad (3.31)$$

Taking the Fourier transform of (3.31), we can compute the vacuum expectation value of the T-ordered product of the axial current J^μ_i and of the pseudoscalar density P_j:

$$\int d^4x e^{iqx} < \Omega, T\left[J^\mu_i(x)P_j(0)\right]\Omega >= 2i\frac{q^\mu}{q^2}\delta^{ij}F \ . \qquad (3.32)$$

The singularity at $q^2 = 0$ shows that the operator P creates from the vacuum a massless single particle state. It is the pion, which in the chiral limit plays the role of a Goldstone particle, the massless particle associated with a spontaneously broken continuous symmetry according to the Goldstone theorem.

To understand how the pions are coupled among themselves and to the current and scalar density operators we have to take the partial Legendre transform of the connected functional with respect to the interpolating field P of the pion, indeed this will define the generator of the one-pion irreducible amplitudes, i.e. the wanted vertices.

The new functional is:

$$\Gamma[\pi,\omega,\alpha,\Sigma] = Z_c[\Pi,\omega,\alpha,\Sigma] - \int d^4x \Pi_i(x)\pi_i(x) \ , \qquad (3.33)$$

with:

$$\pi_i(x) = \frac{\delta}{\delta \Pi_i(x)} Z_c \ , \qquad (3.34)$$

It is easy to translate the chiral WTI in terms of Γ. We shall limit ourselves to the axial identity, which we shall write integrated over x and disregarding the sources of the currents:

$$\int d^4x \left[\pi_i \Sigma + \frac{\delta\Gamma}{\delta\pi_i}\frac{\delta\Gamma}{\delta\Sigma}\right](x) = 0 \ . \qquad (3.35)$$

This equation accounts for all the consequences of the chiral symmetry in the pion-pion interactions. An approximate way of extracting these consequences in the form of low energy theorems is to assume a local approximation for Γ. This corresponds to assume the analyticity at low momenta of the vertices generated by the functional Γ, setting:

$$\Gamma = \sum_{n=0}^{\infty} \int d^4x \Gamma_n[\Sigma,\pi,\partial] \ , \qquad (3.36)$$

where Γ_n contains n derivatives (∂). Of course this approximation can be accurate only for the first terms of this series, since for example one expects non-analytic contributions to the vertices from the pion loop diagrams.

Assuming a local approximation for Γ we can interpret (3.35) as the prescription of a non-linear symmetry, reading $\frac{\delta\Gamma}{\delta\Sigma}$ as the infinitesimal variation of the classical field π under an axial transformation.

To the lowest order in ∂ the general solution of (3.35) turns out to be

$$\Gamma = \int d^4x \left[\Sigma\sqrt{F^2 - \pi^2} + \frac{z}{2}\left((\partial\pi)^2 + \left(\partial\sqrt{F^2-\pi^2}\right)^2 \right) \right] . \qquad (3.37)$$

This result is known as the effective lagrangian of the $SU(2)$ chiral non-linear σ-model. To the second order in the momenta it accounts correctly for the multipion couplings in the chiral limit. It has also inspired the construction of the two-dimensional non-linear σ-models which play an important role e. g. in string theory.

To conclude this lecture let us notice that the scalar density operator S can also be considered as a quark mass term. Introducing it into the lagrangian with a strength m, the quark mass in the tree approximation, is equivalent to a translation of the source Σ by a space-time constant, m. Replacing Σ with $\Sigma + m$ in (3.29) and (3.35) accounts for the effects of the soft breaking of the chiral symmetry due to the quark mass. For example, the functional derivative with respect to $\Pi(0)$ of (3.29) computed for vanishing sources gives:

$$2F\delta^{ij} = m \int d^4x \frac{\delta}{\delta\Pi(x)} \frac{\delta}{\delta\Pi(0)} Z_c[0,0,0,0] . \qquad (3.38)$$

Now, the second functional derivative in the right-hand side of (3.38) is proportional to the pion propagator at zero momentum and hence to the inverse squared pion mass. Thus we have shown that the pion mass is proportional to the square root of the quark mass. This well known result was first found by Gell-Mann, Oakes and Renner.

This concludes our brief study of the renormalization of chiral symmetries and of the corresponding chiral anomalies. We have not tried to give a complete description of the many different aspect of this important theory, but rather to underline the fundamental ideas on which this and the whole theory of renormalized symmetry breakdown is based.

References

1 see e.g.: C.Itzykson and J.-B. Zuber: *Quantum Field Theory* (Mc.Graw-Hill B.C., Singapore 1980)

2 see e.g.: J.Fröhlich: *New Developments in Quantum Field Theory*, in: Proceeding of the XXIV International Conference on High Energy Physics, eds. R.Kotthaus and J.H.Kühn (Springer Verlag Berlin, Heidelberg 1989)

3 W.Zimmermann: *Local Operator Products and Renormalization*, in: Lectures on Elementary Particles and Fields, eds. S.Deser, M.Grisaru, H.Pendleton (MIT Press., Cambridge, Mass 1970)

4 G.Bonneau: Nucl. Phys. **167**, 261 (1980)

5 see e.g.: J.H.Lowenstein: *BPHZ Renormalization*, in: Renormalization Theory, eds. G.Velo and A.Wightman (D.Reidel, Dordrecht 1976)
D. Maison and P. Breitenlhoner: *Some Results on Dimensional Regularization*, ibid.

6 N.N.Bogoliubov, D.V.Shirkov: *Introduction to the Theory of Quantized Fields* (Interscience, New York 1959)
7 see e.g.: C.Becchi: *Lectures on the renormalization of gauge theories*, in: Relativity, Groups and Topology II (les Houches 1983),eds. B.S. DeWitt and R.Stora (Elsevier Science Pub.B.V. 1984)
8 C.Becchi, A.Rouet and R.Stora: *Renormalizable theories with symmetry breaking*, in: Field Theory, Quantization and Statistical Physics, ed. E.Tirapegui (D.Reidel Pub. Co., Dordrecht 1981)
9 see e.g.: P.Goddard, D.Olive: Int. J. Mod. Phys. 1, 415 (1986);
see also: G.Mack: Cargese Lectures July 1987, in: Eds., Non-Perturbative Quantum Field Theory, eds. G. 't Hooft et al.(Plenum Press, New York 1988)
10 see e.g.: S.Weinberg:*Dynamical and Algebraic Symmetries*, in: Lectures on Elementary Particles and Fields., eds. S.Deser, M.Grisaru, H.Pendleton (MIT Press., Cambridge, Mass 1970).

Quantum Field Theory in Low Dimensional Space Time

K. Fredenhagen

Institut für Theoretische Physik
Freie Universität Berlin, Germany

1. Introduction

Quantum field theory in 2 and 3 space time dimensions has some structural peculiarities which are absent in 4 space time dimensions. Far from being unphysical curiosities of toy models, these structures have turned out to be extremely interesting for applications to condensed matter physics, and one might speculate that they will even have an impact on elementary particle physics by providing completely new ideas for model building. Perhaps the most exciting feature is the generalization of the concept of symmetry for which quantum groups seem to be a promising candidate (cf. the lectures of Professor Fadeev). The dual aspect of internal symmetry in quantum field theory is the structure of charge sectors. It is the aim of these lectures to review the present status of the theory of charge sectors in quantum field theory in low dimensional space time.

In 4 dimensions the analysis has been completed, up to the infrared problem, by the recent result of Doplicher and Roberts [1] who were able to derive the group of internal symmetries from the structure of charge sectors, using essentially the intrinsic notion of particle statistics as representation of the permutation group associated to each sector. How nontrivial this result is becomes evident if one applies the same methods to 2 and 3 dimensional quantum field theories [18,24,34,40,42,48,51,52,53]. There the permutation group is replaced by the braid group, link invariants appear as statistical factors in cross sections and the "symmetry", defined as the dual of the structure of charge sectors, has still to be determined.

The lectures are organized as follows. After an introduction into the framework of algebraic field theory (Sect. 2) we discuss the multiplicative structure in the set of charge sectors (Sect. 3) and derive the braid group representation corresponding to commutativity properties of the composition of sectors (Sect. 4). In Sections 5-8 the structure is analysed in more detail. We find link invariants associated to each sector (Sect. 5) and show that the Jones polynomial, its generalizations and the associated braid group representations occur as the simplest nontrivial possibilities (Sect. 6). In Section 7 the structure is formulated in terms of R-matrices and exchange fields which have operator product expansions and satisfy all the relations which have been observed in 2d conformal field theory. Even the Verlinde algebra which has been derived in conformal field theory from modular invariance turns out to be a consequence

Fields and Particles Editors: H. Mitter · W. Schweiger

of the general structure (Sect. 8). In Section 9 we then describe why the analysis is nontrivial not only in 2 dimensions but also in 3 dimensions. This leads to an interesting spin statistics connection. The last 2 sections finally treat possible generalizations of the theory. In Section 10 the phenomena which occur for field theories on non simply connected spaces are discussed on the example of chiral theories on the circle. In Section 11 soliton sectors in 2 dimensions which do not really fit into the existing framework of the sector theory are considered and a groupoid like composition law is found.

2. The Algebraic Approach to Quantum Field Theory

Quantum field theory is usually formulated in terms of fields $\psi_\alpha(x)$ on D-dimensional Minkowski space which are operator valued distributions, i.e. for each test function $f \epsilon \mathcal{S}(\mathbb{R}^D)$

$$\psi_\alpha(f) := \int d^D x f(x) \psi_\alpha(x) \tag{2.1}$$

is an operator on some Hilbert space. Fields are covariant under Poincaré transformations, i.e. there is a unitary strongly continuous representation U of the covering group of the identity component $P_+^\uparrow$ of the Poincaré group such that

$$U(a,A)\psi_\alpha(x)U(a,A)^{-1} = \sum_\beta S_{\alpha\beta}(A)\psi((a,A)x) \tag{2.2}$$

where S is a finite dimensional representation of the covering group of the identity component of the Lorentz group. Moreover, fields at spacelike separated points either commute or anticommute, depending on whether the representation S is of integer or halfinteger spin. It is one of the great successes of quantum field theory, to explain why the opposite connection between spin and statistics is impossible. On the other hand, more general commutation relations for fields at spacelike distances may be envisaged, and it is difficult to decide whether they are compatible with physical principles.

Local commutativity of observables, however, has a direct physical meaning, since it simply expresses the fact that measurements at spacelike separated points cannot disturb each other (Einstein causality). Therefore the discussion of possible commutation relations of fields should be based on this fact.

Now the Hilbert space of state vectors decomposes into subspaces which are invariant under the action of observables. This decomposition has observable consequences. It means that relative phases between vector states in different subspaces (so-called superselection sectors) cannot be measured. This phenomenon has first been observed on the example of the relative phase between states with integer and halfinteger spin [2].

The occurrence of superselection sectors has been explained by Haag and Kastler [3]. They start from the observation that the expectation values of local observables can be approximated in an arbitrary sector, e.g. the vacuum sector, by adding suitable antiparticles "behind the moon", hence the information on the other sectors must already be contained in the vacuum sector.

The algebras generated by the local observables in different sectors are therefore isomorphic; only certain global observables ("charges") which are obtained as limits of local observables distinguish the sectors. Mathematically, this can be described in the following way: There is an abstract algebra generated by the local observables. Each sector provides a representation of this algebra by Hilbert space operators. Sectors are different if and only if the corresponding representations are not unitarily equivalent. The possibility of inequivalent faithful representations is a typical feature of systems with an infinite number of degrees of freedom whereas for a finite number of degrees of freedom the uniqueness theorem of von Neumann leads to a unique Hilbert space representation of the canonical commutation relations. [1]

It is the program of the theory of superselection sectors to analyse the relevant classes of representations and to understand the intrinsic meaning of statistics and symmetry. Before going into the details I want to describe the mathematical framework.

Observables are selfadjoint operators on some Hilbert space $\mathcal{H}$, i.e. a vector space with a positive definite scalar product

$$\phi, \psi \epsilon \mathcal{H} \rightarrow (\phi, \psi) \tag{2.3}$$

which is linear in the right and antilinear in the left factor such that $\mathcal{H}$ is complete with respect to the norm $\|\phi\| = (\phi, \phi)^{1/2}$. Selfadjoint operators have a spectral decomposition which allows to define functions of operators. Physically, this means a rescaling of the measured values of an observable. Mathematically, it allows to avoid unbounded observables (e.g. energy) by replacing them by their bounded functions (e.g. exp{−energy}). Bounded operators are linear mappings A from $\mathcal{H}$ to $\mathcal{H}$ such that

$$\|A\| := \sup_{\substack{\phi \epsilon \mathcal{H} \\ \|\phi\|=1}} \|A\phi\| < \infty. \tag{2.4}$$

Products and sums of bounded operators are again bounded operators, and one might look at algebras of operators.

There are two important notions of operator algebras in Hilbert space. The first one is that of a C^*-algebra. A C^*-algebra is defined to be an associative algebra with an involution $A \rightarrow A^*$ (transition to the adjoint) and a norm satisfying the C^*-condition

$$\|A^*A\| = \|A\|^2 \tag{2.5}$$

[1] For quantum mechanics on topologically nontrivial configuration spaces it is a useful point of view to consider the different values of topological charges as inequivalent representations of a universal algebra of observables. It remains, however, the crucial difference to quantum field theory that these representations are not faithful [4].

such that the algebra is complete as a normed space. A norm closed selfadjoint algebra of Hilbert space operators is a C^*-algebra; actually, any C^*-algebra is isomorphic to a norm closed selfadjoint algebra of Hilbert space operators.

C^*-algebras have, in general, a rich class of representations. A representation of a C^*-algebra $\mathcal{A}$ is a linear mapping π from $\mathcal{A}$ into the algebra of bounded operators $B(\mathcal{H}_\pi)$ on some Hilbert space $\mathcal{H}_\pi$ such that

$$\begin{aligned} \pi(AB) &= \pi(A)\pi(B) \\ \pi(A)^* &= \pi(A^*) \qquad , A, B \epsilon \mathcal{A}. \end{aligned} \tag{2.6}$$

In general, C^*-algebras have a huge number of inequivalent faithful representations.

States in quantum mechanics can be characterized by the expectation values of all observables. This makes it possible to give a purely algebraic definition of a state: A state ω is a linear complex valued functional on $\mathcal{A}$ such that

$$\begin{aligned} &(i) \quad \omega(A^*A) \geq 0 \\ &(ii) \quad \|\omega\| := \sup_{\substack{A \epsilon \mathcal{A} \\ \|A\|=1}} |\omega(A)| = 1. \end{aligned} \tag{2.7}$$

Provided $\mathcal{A}$ has a unit (which we will assume in the following), the supremum in (2.7) is reached on the unit, and the normalization condition can be replaced by

$$\omega(1) = 1. \tag{2.8}$$

Examples of states are vector states or density matrix states in some representation. Let π be a representation of $\mathcal{A}$, let $\psi \epsilon \mathcal{H}_\pi$ with $\|\psi\| = 1$ and let ρ be a density matrix in $\mathcal{H}_\pi$, i.e. a positive trace class operator with $tr\rho = 1$. Then the induced states ω_ψ, ω_ρ are

$$\begin{aligned} &(i)\ \omega_\psi(A) = (\psi, \pi(A)\psi) \\ &(ii)\ \omega_\rho(A) = Tr\ \rho\pi(A) \end{aligned} \tag{2.9}$$

Actually, all states of a C^*-algebra are of the form (2.9) (i) for a suitable representation π. This follows from the famous Gelfand-Naimark-Segal (GNS) construction which may be described as follows: Let ω be a state on $\mathcal{A}$. Because of the positivity condition (2.7) (i) it can be used to define a positive semidefinite scalar product on $\mathcal{A}$,

$$(A, B)_\omega := \omega(A^*B). \tag{2.10}$$

Let N_ω be the null space of this scalar product,

$$N_\omega = \{A \epsilon \mathcal{A} \mid \omega(A^*A) = 0\}. \tag{2.11}$$

By Schwartz' inequality, N_ω is a left ideal. The induced scalar product on the factor space $\mathcal{A}/N_\omega$ is positive definite, and a representation of $\mathcal{A}$ on $\mathcal{A}/N_\omega$ is given by left multiplication,

$$\pi_\omega(A)(B + N_\omega) = AB + N_\omega. \tag{2.12}$$

The Hilbert space $\mathcal{H}_\omega$ is now the completion of $\mathcal{A}/N_\omega$, and using the fact that $\mathcal{A}$ is a C^*-algebra one can show that $\pi_\omega(A)$ is bounded for all $A\epsilon\mathcal{A}$. Hence $\pi_\omega(A)$ has a unique continuous extension to $\mathcal{H}_\omega$. Finally, the vector inducing the state ω is given by

$$\Omega_\omega = 1 + N_\omega\ . \tag{2.13}$$

It satisfies the conditions

$$\begin{aligned} &(i) \quad (\Omega_\omega, \pi_\omega(A)\Omega_\omega) = \omega(A), \quad A\epsilon\mathcal{A} \\ &(ii) \quad \Omega_\omega \text{ is a cyclic vector for } \pi_\omega(\mathcal{A}) \\ &\qquad (i.e. \pi_\omega(\mathcal{A})\Omega_\omega \text{ is dense in } \mathcal{H}_\omega). \end{aligned} \tag{2.14}$$

The "GNS-triple" $(\mathcal{H}_\omega, \pi_\omega, \Omega_\omega)$ is (up to unitary equivalence) uniquely characterized by the conditions (2.14) (i) and (ii).

It is instructive to look at the GNS-triple in the case of a state of the form (2.9) (ii). Let $\mathcal{A}$ be the algebra $Mat_n(C)$ of $n \times n$-matrices, and let ρ be a positive definite $n \times n$-matrix with trace 1. The "density matrix" ρ induces the state

$$\omega(A) = Tr\rho A \quad , A\epsilon\mathcal{A} \tag{2.15}$$

on $\mathcal{A}$. Then the GNS-Hilbert space $\mathcal{H}_\omega$ can be identified with $Mat_n(C)$, equipped with the scalar product

$$(A, B) = TrA^*B\,, \tag{2.16}$$

π_ω is the representation by left multiplication, and the cyclic vector inducing ω is $\Omega_\omega = \rho^{1/2}$.

There is an important subclass of C^*-algebras, the so-called von Neumann algebras. These are defined as selfadjoint algebras $\mathcal{N}$ of Hilbert space operators which contain the unit operator and are closed with respect to the weak operator topology, i.e. if (A_λ) is a net of operators in N such that for all vectors $\phi, \psi\epsilon\mathcal{H}$

$$(\phi, A_\lambda\psi) \to (\phi, A\psi) \tag{2.17}$$

for some $A\epsilon B(\mathcal{H})$, then $A\epsilon\mathcal{N}$.

Von Neumann algebras in some Hilbert space $\mathcal{H}$ can be algebraically characterized. Let $\mathcal{N} \subset B(\mathcal{H})$ be a selfadjoint set of operators on $\mathcal{H}$. The commutant $\mathcal{N}'$ of $\mathcal{N}$ is defined by

$$\mathcal{N}' = \{A\epsilon B(\mathcal{H}) | [A, B] = 0 \qquad \text{for all } B\epsilon\mathcal{N}\}. \tag{2.18}$$

Now the bicommutant $\mathcal{N}''$ of $\mathcal{N}$, i.e. the commutant of the commutant of $\mathcal{N}$, is the smallest von Neumann algebra containing $\mathcal{N}$. This is the content of von Neumann's Bicommutant Theorem. Due to this Theorem, von Neumann algebras occur typically when algebras are defined by their commutation properties.

A further peculiarity of von Neumann algebras is the existence of a distinguished set of states, the so-called normal states. These are the states induced by density matrices in the Hilbert space; they are intrinsically characterized by the following condition. Let ω be a state on $\mathcal{N}$; ω is called to be normal if for any increasing net $(A_\lambda)\epsilon\mathcal{N}$ (i.e. $A_\lambda \geq A_{\lambda'}$ for $\lambda \geq \lambda'$) with least upper bound $A\epsilon\mathcal{N}$ the least upper bound of $(\omega(A_\lambda))$ is $\omega(A)$. The linear span $\mathcal{N}_*$ of normal states on $\mathcal{N}$ is a norm closed subspace of the space $\mathcal{N}^*$ of linear functionals on $\mathcal{N}$. Its dual can be identified with $\mathcal{N}$.

We close this mathematical interlude by a discussion of the notions of equivalence. Let $\mathcal{A}$ be a C^*-algebra and π a representation of $\mathcal{A}$. The density matrices on the representation space $\mathcal{H}_\pi$ induce a set of states $S(\pi)$ which is associated to π. Two representations π and π' are called to be unitarily equivalent , $\pi \simeq \pi'$, if there exists a unitary operator $U : \mathcal{H}_\pi \to \mathcal{H}_{\pi'}$, such that

$$U\pi(A) = \pi'(A)U \ , A\epsilon\mathcal{A}. \tag{2.19}$$

They are called quasiequivalent if the associated state spaces coincide, $S(\pi) = S(\pi')$. A representation π of a von Neumann algebra $\mathcal{N}$ is called to be normal if its state space contains only normal states. Von Neumann algebras have, up to quasiequivalence, only one normal faithful representation. Singular (i.e. non normal) representations of von Neumann algebras have pathological features and are usually disregarded. On the contrary, generic C^*-algebras have many faithful inequivalent representations which are to be treated on an equal footing.

We return now to the description of the algebraic approach to quantum field theory. Let $\mathcal{K}$ denote the set of open double cones 0 ("diamonds") in Minkowski space, i.e.

$$0 = V_+ + x \cap V_- + y \tag{2.20}$$

where $V_\pm$ denote the interior of the forward, resp. backward light cone and x, y are points in Minkowski space with $y - x\epsilon V_+$. To each $0\epsilon\mathcal{K}$ there is associated a von Neumann algebra $\mathcal{A}(0)$. Physically, $\mathcal{A}(0)$ is interpreted as the algebra which is generated by all observables which can be measured within 0. The assignment $0 \to \mathcal{A}(0)$ satisfies the following conditions (Haag-Kastler-axioms)

$$\begin{aligned}(i) \quad & 0_1 \subset 0_2 \Rightarrow \mathcal{A}(0_1) \subset \mathcal{A}(0_2), \text{ and the unit operators of} \\ & \mathcal{A}(0_1) \text{ and } \mathcal{A}(0_2) \text{ coincide} \qquad \text{(isotony).}\end{aligned} \tag{2.21}$$

Using isotony, one can introduce the algebra of quasilocal observables as the C^*-inductive limit of the net $(\mathcal{A}(0))_{0\epsilon\mathcal{K}}$,

$$\mathcal{A} = \overline{\bigcup_{0\epsilon\mathcal{K}} \mathcal{A}(0)}^{(\text{norm})} . \tag{2.22}$$

$\mathcal{A}$ is a C^*-algebra which is uniquely characterized by the system of von Neumann algebras $(\mathcal{A}(0))_{0\epsilon\mathcal{K}}$.

$$\begin{aligned}(ii) \quad & 0_1 \subset 0_2' \text{ (the spacelike complement of } 0_2) \\ \Rightarrow \quad & \mathcal{A}(0_1) \subset \mathcal{A}(0_2)' \qquad \text{(locality).}\end{aligned} \tag{2.23}$$

Locality may be considered as the incorporation of Einstein causality into the theory.

(iii) There is a representation of the identity component $P_+^\uparrow$ of the Poincaré group by automorphisms of $\mathcal{A}$, $L \to \alpha_L$, such that

$$\alpha_L(\mathcal{A}(0)) = \mathcal{A}(L(0)) \qquad \text{(covariance)}. \tag{2.24}$$

The "Haag-Kastler net" $(\mathcal{A}(0))_{0\epsilon\mathcal{K}}$ specifies the theory, including its interpretation in terms of measurements. Representations of $\mathcal{A}$ describe situations in which the physical system may be. For the purposes of elementary particle physics the most important representations are the positive energy representations (PER) as emphasized long ago by Borchers [5]. They are representations π of $\mathcal{A}$ together with a unitary, strongly continuous representation U of the translation group such that

$$AdU(x) \circ \pi = \pi \circ \alpha_x \tag{2.25}$$

where the adjoint action AdV of a unitary V on $B(\mathcal{H})$ is as usually defined by

$$AdV(B) = VBV^{-1} \, , B\epsilon B(\mathcal{H}). \tag{2.26}$$

As U is a strongly continuous representation there are commuting selfadjoint operators $P_\mu, \mu = 0, ..D-1$ generating U,

$$U(x) = e^{iPx}. \tag{2.27}$$

The operators P_μ are interpreted as energy and momentum, and they are assumed to satisfy the spectrum condition

$$spP \subset \{p\epsilon R^D | p_o \geq |\underline{p}|\} = \bar{V}_+. \tag{2.28}$$

The interpretation of P_μ is enforced by a Theorem of Borchers which states that due to the spectrum condition the representation U can always be chosen such that $U(x)$ is in the weak closure $\pi(\mathcal{A})''$ of the algebra of observables in the representation π [6].

Often, U can be extended to an implementation of $P_+^\uparrow$, but there are also PER's where this is not possible, e.g. the electrically charged sectors in Quantum Electrodynamics [7].

Special PER's are the vacuum representations. They contain a unique (up to a phase) translation invariant vector which is cyclic for the observable algebra. By a theorem of Wightman, they are automatically irreducible [8].

3. Composition of Sectors

A complete analysis of the class of PER's could be carried through up to now only in special situations (free field, Virasoro algebra, Kac-Moody-algebras). An easier accessible class is the class of locally generated representations which has been investigated by Doplicher, Haag and Roberts (DHR) [9]. Let π_0 be a fixed vacuum representation. A representation π of $\mathcal{A}$ is called to be locally generated with respect to π_o if

$$\pi|\mathcal{A}(0') \simeq \pi_o|\mathcal{A}(0') \quad , \quad 0\epsilon\mathcal{K} \tag{3.1}$$

where $\mathcal{A}(0')$ is the C^*- algebra generated by all algebras $\mathcal{A}(0_1)$ with $0_1 \subset 0', 0_1\epsilon\mathcal{K}$.

The idea behind this condition is that the (global) observables which distinguish π and π_0 are not measurable in $0'$. In other words, one can generate a charge within 0 without any influence on the measurements in $0'$.

This may be illustrated by the following example [10]. Consider the theory of a free charged scalar field φ such that $\varphi \to e^{i\alpha}\varphi, \alpha\epsilon\mathbb{R}$, is a global symmetry of the theory. The observable algebra $\mathcal{A}$ may be defined as the set of fixed points under this symmetry. Let π_0 denote the restriction of $\mathcal{A}$ to the charge zero sector and π_1 that to the charge one sector. The smeared field $\varphi(f)$ where f is a real test function with compact support maps the charge zero subspace into the charge one subspace and commutes with all observables localized in the spacelike complement of the support of f. The operator $\varphi(f)$ still creates correlations between *supp* f and its spacelike complement. We therefore perform the polar decomposition

$$\varphi(f) = U_f|\varphi(f)| \tag{3.2}$$

where $|\varphi(f)| = (\varphi(f)^*\varphi(f))^{1/2}$ and U_f is unitary. Then we find, for *supp* $f \subset 0\epsilon\mathcal{K}$,

$$\pi_1|\mathcal{A}(0') = AdU_f \circ \pi_0|\mathcal{A}(0'), \tag{3.3}$$

hence π_1 satisfies the DHR-condition (3.1).

Now let π be a representation which is locally generated with respect to a vacuum representation π_o. Let $0_o\epsilon\mathcal{K}$. Then there is a unitary V from $\mathcal{H}_{\pi_0}$ into $\mathcal{H}_\pi$ such that

$$\pi|\mathcal{A}(0_0') = Ad\ V \circ \pi_0|\mathcal{A}(0_0'). \tag{3.4}$$

V may be used to identify the representation space $\mathcal{H}_\pi$ with $\mathcal{H}_{\pi_o}$. Let

$$\pi_V = Ad\ V^{-1} \circ \pi. \tag{3.5}$$

Then $\pi_V \simeq \pi$ by construction, and by (3.1)

$$\pi_V|\mathcal{A}(0_0') = \pi_0|\mathcal{A}(0_0'). \tag{3.6}$$

Hence the operators $\pi_V(A), A\epsilon\mathcal{A}$ act in the vacuum Hilbert space, and they coincide with the operators $\pi_o(A)$ provided $A\epsilon\mathcal{A}(0_0')$. Now let $0_1\epsilon\mathcal{K}$ be arbitrary. There is some $0_{o1}\epsilon\mathcal{K}$ with $0_o \cup 0_1 \subset 0_{o1}$. Then

$$\pi_V(\mathcal{A}(0_1)) \subset \pi_V(\mathcal{A}(0_{01})) \subset \pi_V(\mathcal{A}(0_{01}'))' = \pi_0(\mathcal{A}(0_{01}'))' \tag{3.7}$$

where we used first isotony, then locality and finally (3.6).

In order to proceed one needs some information on the commutants of $\pi_0(\mathcal{A}(0')), 0 \epsilon \mathcal{K}$. From locality one has

$$\pi_0(\mathcal{A}(0)) \subset \pi_0(\mathcal{A}(0'))'. \tag{3.8}$$

As suggested by Haag one may impose the maximality condition

$$\pi_0(\mathcal{A}(0)) = \pi_0(\mathcal{A}(0'))'. \qquad \text{(Haag duality).} \tag{3.9}$$

Haag duality is known to hold in free field theories [11], in superrenormalizable field theories in 2 dimensions [12,13] and in conformal light cone theories [14]. It does not hold in theories with spontaneous breakdown of symmetries [15].

A weaker version of Haag duality is essential duality, a notion due to Roberts [15]. One introduces the dual net of observable algebras (relative to π_0)

$$\mathcal{A}_d(0) = \pi_0(\mathcal{A}(0'))' \tag{3.10}$$

and asks whether this net satisfies Haag duality. In general, one has

$$\mathcal{A}_d(0')' = \bigcap_{\substack{0_1 \subset 0' \\ 0_1 \epsilon \mathcal{K}}} \mathcal{A}_d(0_1)' = \bigcap_{\substack{0_1 \subset 0' \\ 0_1 \epsilon \mathcal{K}}} \pi_0(\mathcal{A}(0_1'))''$$

$$\subset \bigcap_{\substack{0_1 \subset 0' \\ 0_1 \epsilon \mathcal{K}}} \pi_0(\mathcal{A}(0_1))' = \pi_0(\mathcal{A}(0'))' = \mathcal{A}_d(0), \tag{3.11}$$

hence the dual net satisfies the opposite inclusion in (3.8). Thus essential duality holds if and only if the dual net is local. It is now gratifying that essential duality can be derived provided the local observable algebras are generated by Wightman fields. This remarkable result has been found by Bisognano and Wichmann [16]. It relies essentially on the identification of the TCP-operator with the modular involution of the Tomita-Takesaki-theory [49] which induces an antiisomorphism between a von Neumann algebra and its commutant.

As shown by Roberts [15] representations satisfying the DHR condition (3.1) can uniquely be extended to the dual net and satisfy there again (3.1) provided the dimension of space time is at least 3. In the following we will assume that our original net satisfies Haag duality (3.9).

It follows then from (3.7) that

$$\pi_V(\mathcal{A}(0_1)) \subset \pi_V(\mathcal{A}(0_{01})) \subset \pi_0(\mathcal{A}) \tag{3.12}$$

hence, since $0_1 \epsilon \mathcal{K}$ was arbitrary, $\pi_V(\mathcal{A})$ is a subalgebra of $\pi_0(\mathcal{A})$. In order to determine the position of this subalgebra, we use the DHR condition for another double cone $0_1 \epsilon \mathcal{K}$ and find a unitary V_1 such that

$$\pi_{V_1} = Ad\ V_1^{-1} \circ \pi, \qquad \pi_{V_1} | \mathcal{A}(0_1') = \pi_0 | \mathcal{A}(0_1') \tag{3.13}$$

and

$$\pi_{V_1} = Ad\ U \circ \pi_V, \qquad U = V_1^{-1} V \epsilon B(\mathcal{H}_{\pi_0}). \tag{3.14}$$

Therefore

$$\pi_V|\mathcal{A}(0_1') = Ad\ U^{-1} \circ \pi_0|\mathcal{A}(0_1'). \tag{3.15}$$

Since $0_1 \epsilon \mathcal{K}$ is arbitrary we conclude that

$$\pi_0(A) \to \pi_V(A), A \epsilon \mathcal{A} \tag{3.16}$$

is an isometric mapping. Thus, the kernel of a DHR representation coincides with the kernel of π_0, and we may identify $\mathcal{A}$ with its image under π_0. The map (3.16) is then given by an endomorphism ρ of $\mathcal{A}$ such that $\pi_V = \pi_0 \circ \rho$. ρ is localized in 0_0 in the sense that

$$\rho|\mathcal{A}(0_0') = id_{\mathcal{A}(0_0')}. \tag{3.17}$$

There is another endomorphism ρ_1, localized in 0_1, such that $\pi_{V_1} = \pi_0 \circ \rho_1$. From (3.14)

$$\pi_0 \circ \rho_1 = Ad\ U \circ \pi_0 \circ \rho. \tag{3.18}$$

Now let $0 \epsilon \mathcal{K}, 0 \supset 0_0 \cup 0_1$. Then ρ and ρ_1 act trivially on $\mathcal{A}(0')$, and

$$U \epsilon \pi_0(\mathcal{A}(0'))' = \pi_0(\mathcal{A}(0)) \tag{3.19}$$

by Haag duality. Thus there is a unitary operator $U_0 \epsilon \mathcal{A}(0)$ such that

$$\rho_1 = Ad\ U_0 \circ \rho. \tag{3.20}$$

U_0 is called an intertwiner from ρ to ρ_1. We conclude that localized endomorphisms induce unitarily equivalent representations if and only if they are related by an inner automorphism. This fact provides the basis for the following composition law for equivalence classes of representations. Let $\pi_i \simeq \pi_0 \circ \rho_i$. Then

$$[\pi_1] \times [\pi_2] := [\pi_0 \circ \rho_1 \rho_2] . \tag{3.21}$$

This composition law is automatically associative. In the next section we will see that, as a consequence of locality, it is also commutative. This will lead us to an intrinsic notion of statistics.

4. Statistics

Let us investigate the localization of charge transporters U in

$$\hat{\rho} = AdU \circ \rho \tag{4.1}$$

where $\hat{\rho}$ and ρ are endomorphisms which are localized in $\hat{0}$ and $0 \epsilon \mathcal{K}$, respectively. Let a path γ in $\mathcal{K}$ denote a finite sequence $0_1, ..., 0_n \epsilon \mathcal{K}$ such that for $i = 1, .., n-1$ there exists some $0_{i,i+1} \epsilon \mathcal{K}$ with $0_{i,i+1} \subset 0_i \cap 0_{i+1}$, and let

$$\mathcal{A}(\gamma) = \vee_i \mathcal{A}(0_i) \tag{4.2}$$

where the symbol $\vee$ denotes the generated von Neumann algebra. Choose a path $\gamma = (0, 0_1, .., 0_n, \hat{0})$ in $\mathcal{K}$ and endomorphisms $\rho_i, i = 0, ..n+1$ with $\rho_o = \rho, \rho_{n+1} = \hat{\rho}$ and ρ_i localized in $0_{i,i+1}, i = 1, ..n$. Choose unitary intertwiners U_i such that

$$\begin{aligned} \rho_{i+1} &= AdU_i \circ \rho_i \ , i = 0, ..n, \\ U &= U_n..U_o. \end{aligned} \tag{4.3}$$

Then $U_i \epsilon \mathcal{A}(0_{i+1})$ and

$$U \epsilon \mathcal{A}(\gamma) \tag{4.4}$$

for all paths γ from 0 to $\hat{0}$.

The localization (4.4) can now be used to show that endomorphisms which are localized in spacelike separated regions commute with each other.

4.1. Theorem [9]: Let ρ_i be localized in $0_i \epsilon \mathcal{K}, i = 1, 2$ with $0_1 \subset 0_2'$. Then

$$\rho_1 \rho_2 = \rho_2 \rho_1 \tag{4.5}$$

Proof: Let $0 \epsilon \mathcal{K}$ be arbitrary. There are $\hat{0}_i \epsilon \mathcal{K}$, $\hat{0}_i \subset 0'$ and paths γ_i in $\mathcal{K}$ from 0_i to $\hat{0}_i, i = 1, 2$ such that $\gamma_1 \subset \gamma_2'$. Let $\hat{\rho}_i = AdU_i \circ \rho_i$ be localized in $\hat{0}_i$. We have $U_i \epsilon \mathcal{A}(\gamma_i)$ from (4.4), $i = 1, 2$,

$$\hat{\rho}_i | \mathcal{A}(0) = id_{\mathcal{A}(0)} \tag{4.6}$$

because $0 \subset \hat{0}_i', i = 1, 2$,

$$\hat{\rho}_i(U_j) = U_j, \tag{4.7}$$

since $\hat{0}_i \subset \gamma_i \subset \gamma_j', i \neq j, i, j = 1, 2$, and, since $\gamma_1 \subset \gamma_2'$,

$$U_1 U_2 = U_2 U_1 \quad . \tag{4.8}$$

Thus

$$\begin{aligned} \rho_1 \rho_2 | \mathcal{A}(0) &= AdU_1^{-1} \circ \hat{\rho}_1 \circ AdU_2^{-1} \circ \hat{\rho}_2 | \mathcal{A}(0) = AdU_1^{-1} U_2^{-1} | \mathcal{A}(0) \\ &= AdU_2^{-1} U_1^{-1} | \mathcal{A}(0) = \rho_2 \rho_1 | \mathcal{A}(0). \end{aligned} \tag{4.9}$$

As $0 \epsilon \mathcal{K}$ was arbitrary we obtain (4.5) on the dense subalgebra of local observables and therefore everywhere.

q.e.d.

An immediate consequence of Thm.4.1 is the commutativity of the composition law (3.21). Let π_1, π_2 be DHR- representations. Choose $0_1 \subset 0_2', 0_1, 0_2 \epsilon \mathcal{K}$ and endomorphisms ρ_i localized in 0_i such that $\pi_i \simeq \pi_o \circ \rho_i, i = 1, 2$. Then

$$[\pi_1] \times [\pi_2] = [\pi_0 \circ \rho_1 \rho_2] = [\pi_0 \circ \rho_2 \rho_1] = [\pi_2] \times [\pi_1]. \tag{4.10}$$

For endomorphisms ρ_1 and ρ_2 which are localized in the same double cone $0 \epsilon \mathcal{K}$ the product may not be commutative. However, from (4.10) it follows that there is a unitary $\varepsilon \epsilon \mathcal{A}(0)$ such that

$$\rho_2 \rho_1 = Ad\varepsilon \circ \rho_1 \rho_2. \tag{4.11}$$

The computation of ε is straightforward. Choose $0_1, 0_2 \epsilon \mathcal{K}$ with $0_1 \subset 0_2'$ and endomorphisms $\hat{\rho}_i = AdU_i \circ \rho_i$ localized in $0_i, i = 1,2$. Then from Thm.4.1 $\hat{\rho}_1\hat{\rho}_2 = \hat{\rho}_2\hat{\rho}_1$ and

$$\begin{aligned} \rho_2\rho_1 &= AdU_2^{-1} \circ \hat{\rho}_2 \circ AdU_1^{-1} \circ \hat{\rho}_1 \\ &= AdU_2^{-1}\hat{\rho}_2(U_1^{-1}) \circ \hat{\rho}_2\hat{\rho}_1 \\ &= Ad\rho_2(U_1^{-1})U_2^{-1} \circ \hat{\rho}_1\hat{\rho}_2 \\ &= Ad\rho_2(U_1^{-1})U_2^{-1} \circ AdU_1 \circ \rho_1 \circ AdU_2 \circ \rho_2 \\ &= Ad\rho_2(U_1^{-1})U_2^{-1}U_1\rho_1(U_2) \circ \rho_1\rho_2 \end{aligned} \tag{4.12}$$

hence

$$\varepsilon = \rho_2(U_1^{-1})U_2^{-1}U_1\rho_1(U_2). \tag{4.13}$$

ε is called the statistics operator. It has remarkable invariance properties.

First replace U_2 by $\hat{U}_2$ such that $Ad\hat{U}_2 \circ \rho_2$ is also localized in 0_2. Then $\hat{U}_2 = V_2U_2$ with $V_2 \epsilon \mathcal{A}(0_2)$, and we have

$$U_1\rho_1(V_2) = AdU_1 \circ \rho_1(V_2)U_1 = \hat{\rho}_1(V_2)U_1 = V_2U_1. \tag{4.14}$$

Hence we obtain

$$\begin{aligned} \hat{\varepsilon} := \rho_2(U_1^{-1})\hat{U}_2^{-1}U_1\rho_1(\hat{U}_2) &= \rho_2(U_1^{-1})U_2^{-1}V_2^{-1}U_1\rho_1(V_2)\rho_1(U_2) \\ &= \rho_2(U_1^{-1})U_2^{-1}V_2^{-1}V_2U_1\rho_1(U_2) = \varepsilon. \end{aligned} \tag{4.15}$$

Equally well we may replace U_1 by $\hat{U}_1$ such that $Ad\hat{U}_1 \circ \rho_1$ is localized in 0_1, and again we obtain the same operator ε.

In the second step we replace 0_i by $\hat{0}_i, i = 1,2$, such that $\hat{0}_1 \subset \hat{0}_2'$. Then the same operator U_i can be used in the definition of ε.

We now iterate this change of regions 0_i in the following way. Let

$$\mathcal{K}_{sp}^{(2)} = \{(0_1, 0_2)\epsilon \mathcal{K} \times \mathcal{K} | 0_1 \subset 0_2'\}. \tag{4.16}$$

ε may depend on the choice of $(0_1, 0_2)\epsilon\mathcal{K}_{sp}^{(2)}$. A path γ in $\mathcal{K}_{sp}^{(2)}$ is defined as a finite sequence $(0_1^{(i)}, 0_2^{(i)}), \epsilon\mathcal{K}_{sp}^{(2)}, i = 1,..n$ such that there are $0_j^{(i,i+1)}\epsilon\mathcal{K}$ such that $0_j^{(i,i+1)} \subset 0_j^{(i)} \cap 0_j^{(i+1)}, j = 1,2, i = 1,..n-1$. By the argument above, ε is constant on each connected component of $\mathcal{K}_{sp}^{(2)}$.

In $D \geq 3$ dimensional Minkowski space $K_{sp}^{(2)}$ has only one connected component. Hence the statistics operator $\varepsilon = \varepsilon(\rho_1, \rho_2)$ is unique. In D=2 dimensionsal Minkowski space $K_{sp}^{(2)}$ has two connected components, consequently there are two statistics operators which might differ from each other. We adopt the convention

$$\varepsilon = \varepsilon(\rho_1, \rho_2) \tag{4.17}$$

if 0_1 is in the right spacelike complement of 0_2. For 0_1 in the left spacelike complement of 0_2 one gets

$$\varepsilon = \varepsilon(\rho_2, \rho_1)^{-1} \quad . \tag{4.18}$$

We now look at the special case $\rho_1 = \rho_2 = \rho$ and use the notation

$$\varepsilon_\rho = \varepsilon(\rho, \rho). \tag{4.19}$$

ε_ρ has the following properties

$$\begin{aligned} &(i)\ \varepsilon_\rho \rho(\varepsilon_\rho)\varepsilon_\rho = \rho(\varepsilon_\rho)\varepsilon_\rho\rho(\varepsilon_\rho) \\ &(ii)\ \varepsilon_\rho \epsilon \rho^2(\mathcal{A})'. \end{aligned} \tag{4.20}$$

Property (ii) is nothing else than the intertwining property (4.11). To prove (i) we choose in the definition of ε_ρ the intertwiners $U_1 = 1, U_2 = U$ and 0_2 in the (left)spacelike complement of 0 (in D=2 dimensions). Then

$$\varepsilon_\rho = U^{-1}\rho(U), \quad \varepsilon_\rho\rho(\varepsilon_\rho) = U^{-1}\rho^2(U) \tag{4.21}$$

and

$$AdU \circ \rho(\varepsilon_\rho) = \varepsilon_\rho \tag{4.22}$$

since $\varepsilon_\rho \epsilon \mathcal{A}(0)$ and $AdU \circ \rho$ is localized in $0_2 \subset 0'$. Therefore

$$\varepsilon_\rho\rho(\varepsilon_\rho)\varepsilon_\rho = U^{-1}\rho^2(U)\varepsilon_\rho = U^{-1}\varepsilon_\rho\rho^2(U) \tag{4.23}$$

by (4.20) (ii) and

$$\rho(\varepsilon_\rho)\varepsilon_\rho\rho(\varepsilon_\rho) = \rho(\varepsilon_\rho)U^{-1}\rho^2(U) = U^{-1}AdU \circ \rho(\varepsilon_\rho)\rho^2(U). \tag{4.24}$$

Thus (4.20)(i) follows from (4.22).

Given a unitary $\varepsilon_\rho \epsilon \mathcal{A}$ and $\rho \epsilon End(\mathcal{A})$ such that the relations (4.20) hold one can define a unitary representation $\varepsilon^{(\rho)}$ of the braid group B_∞. According to Artin, the braid group with n strings B_n is generated by elements $\sigma_1, ..\sigma_{n-1}$ which satisfy the relations

$$\begin{aligned} &(i)\ \sigma_i\sigma_{i+1}\sigma_i = \sigma_{i+1}\sigma_i\sigma_{i+1} \quad , i = 1, ..n-2 \\ &(ii)\ \sigma_i\sigma_j = \sigma_j\sigma_i \quad , |i-j| \geq 2, i, j = 1, ..n-1. \end{aligned} \tag{4.25}$$

Therefore, a representation of B_∞ is obtained by setting

$$\varepsilon^{(\rho)}(\sigma_i) = \rho^{i-1}(\varepsilon_\rho) \quad . \tag{4.26}$$

Namely, for $j - i \geq 2$

$$[\varepsilon^{(\rho)}(\sigma_i), \varepsilon^{(\rho)}(\sigma_j)] = \rho^{i-1}([\varepsilon_\rho, \rho^{j-i}(\varepsilon_\rho)]) = 0$$

because of (4.20)(ii), and

$$\begin{aligned} \varepsilon^{(\rho)}(\sigma_i)\varepsilon^{(\rho)}(\sigma_{i+1})\varepsilon^{(\rho)}(\sigma_i) &= \rho^{i-1}(\varepsilon_\rho\rho(\varepsilon_\rho)\varepsilon_\rho)), \\ \varepsilon^{(\rho)}(\sigma_{i+1})\varepsilon^{(\rho)}(\sigma_i)\varepsilon^{(\rho)}(\sigma_{i+1}) &= \rho^{i-1}(\rho(\varepsilon_\rho)\varepsilon_\rho\rho(\varepsilon_\rho)), \end{aligned} \tag{4.27}$$

hence (4.25)(i) is respected by $\varepsilon^{(\rho)}$ due to (4.20)(i).

In $D \geq 3$ dimensions one finds in addition the relation

$$\varepsilon_\rho^{-1} = (U^{-1}\rho(U))^{-1} = \rho(U)^{-1}U = \varepsilon_\rho \tag{4.28}$$

by choosing $U_2 = 1$ and $U_1 = U$ in the definition of ε_ρ. This means that also the relation $\sigma_i^2 = 1$ is respected by $\varepsilon^{(\rho)}$, thus $\varepsilon^{(\rho)}$ defines a representation of the permutation group. Thus DHR sectors have permutation group statistics in $D \geq 3$ dimensions. There are, however, PER's which violate the DHR selection criterion. In Section 9 we will see that these sectors have permutation group statistics in $D \geq 4$ dimensions whereas in $D = 3$ dimensions they may have nontrivial braid group statistics.

5. Left Inverses, Markov Traces and the Possible Braid Group Representations

We now want to investigate the occurring braid group representations $\varepsilon^{(\rho)}$ for irreducible ρ (i.e. $\rho(\mathcal{A})' = C1$). The simplest case is present when ρ is an automorphism. Since ρ is isometric this is the case if $\rho(\mathcal{A}) = \mathcal{A}$. Then ρ^2 is again an automorphism, hence irreducible, and ε_ρ is a multiple of the identity due to (4.20)(ii). It has been proven in [9] that actually the following conditions are equivalent:

(i) ρ is an automorhism
(ii) ρ^2 is irreducible
(iii) ε_ρ is a multiple of the identity.

A sector $[\rho]$ for which one (hence all) of the conditions (5.1) holds is called abelian. The representation $\varepsilon^{(\rho)}$ is in the abelian case completely characterized by a phase $\Theta \epsilon [0, 2\pi)$

$$\varepsilon_\rho = e^{i\Theta} 1. \tag{5.2}$$

Θ may assume any value, therefore particles described by abelian sectors have been termed "anyons" by Wilczek [17].

More interesting from the point of view of the theory of superselection sectors is the case of non abelian sectors. Then $\rho(\mathcal{A}) \neq \mathcal{A}, \rho^2(\mathcal{A})' \neq C1$ and $\varepsilon_\rho \notin C1$ by (5.1). The main technical tool for an analysis of nonabelian sectors is the so called left inverse of ρ. A left inverse ϕ of ρ is defined to be a positive linear mapping from $\mathcal{A}$ to $\mathcal{A}$ such that $\phi\rho = id$ and $\rho\phi$ is a conditional expectation from $\mathcal{A}$ onto $\rho(\mathcal{A})$, i.e.

$$\phi(\rho(A)B\rho(C)) = A\phi(B)C \quad , \quad A, B, C \epsilon \mathcal{A}. \tag{5.3}$$

Left inverses always exist due to compactness arguments. If one applies a left inverse ϕ to the statistics operator ε_ρ one gets a multiple of the identity,

$$\phi(\varepsilon_\rho) = \lambda_\rho 1 \tag{5.4}$$

since $\phi(\varepsilon_\rho) \epsilon \phi(\rho^2(\mathcal{A})') \subset \rho(\mathcal{A})'$ and ρ is irreducible. In [18] it has been shown that (provided ρ is a PER) the "statistics parameter" λ_ρ is independent of ϕ; moreover, if $\lambda_\rho \neq 0$, ϕ is unique. The case $\lambda_\rho = 0$ ("infinite statistics") is pathological in the sense that $\phi|\mathcal{A}(0)$ is singular; it cannot occur for particles in massive theories [19]. In the following we will assume $\lambda_\rho \neq 0$ ("finite statistics"). By polar decomposition of λ_ρ

$$\lambda_\rho = \omega_\rho d_\rho^{-1} \,, |\omega_\rho| = 1, d_\rho \geq 1 \tag{5.5}$$

we find two numbers associated to the sector of ρ. d_ρ has been called the 'statistical dimension of ρ' in [9]. In theories with a compact group of internal symmetries d_ρ is the dimension of the group representation associated to the sector of ρ. ω_ρ is called the statistics phase. In theories with permutation group statistics $\omega_\rho = \pm 1$ depending on whether the particles are bosons or fermions. In a 2d conformal chiral field theory ω_ρ is related to the lowest weight h_ρ of the conformal light cone Hamiltonian L_o,

$$\omega_\rho = e^{2\pi i h_\rho} \tag{5.6}$$

in 3d quantum field theory, ω_ρ is related to the (fractional) spin of particles,

$$\omega_\rho = \pm e^{2\pi i s_\rho}. \tag{5.7}$$

Left inverses are related to the conjugate sector. An irreducible localized endomorphism $\bar{\rho}$ such that

$$\bar{\rho}\rho \supseteq id \tag{5.8}$$

is called conjugate to ρ. (5.8) means that there is an isometry $R \epsilon \mathcal{A}$ such that

$$\bar{\rho}\rho(A)R = RA \,, A \epsilon \mathcal{A}. \tag{5.9}$$

The left inverse of ρ is then given by the formula

$$\phi(A) = R^* \bar{\rho}(A) R. \tag{5.10}$$

The left inverse ϕ can be used to define a function φ on the braid group,

$$\varphi(b)1 = \lim_{n\to\infty} \phi_n(\varepsilon^{(\rho)}(b)) \,, b \epsilon B_\infty. \tag{5.11}$$

φ is a function of positive type and satisfies

$$\begin{aligned} &(i) \quad \varphi(b_1 b_2) = \varphi(b_2 b_1) \\ &(ii) \quad \varphi(b\sigma_n^{\pm 1}) = \varphi(b)\varphi(\sigma_n^{\pm 1}) \,, b \epsilon B_n \\ &(iii) \quad \varphi(b_1 b_2) = \varphi(b_1)\varphi(b_2) \,, b_1 \epsilon B_n, b_2 \text{ word in } \sigma_n, \ldots \sigma_{n+k}. \end{aligned} \tag{5.12}$$

A function of B_∞ satisfying (i) and (ii) is called a Markov trace. (iii) implies of course (ii); if φ satisfies in addition (iii) it is called a strong Markov trace [20].

Markov traces can be used to define link invariants. A braid can be made into a link by binding the opposite ends together. On the other hand, each link can be made into a braid by cutting some strings. The latter procedure, however, is highly nonunique. Due to Markov's Theorem [21], braids which lead to the same link can be transformed into each other by a finite number of so called Markov moves

$$\begin{aligned} b_1 b_2 &\rightarrow b_2 b_1 \\ b &\leftrightarrow b\sigma_n^{\pm 1}, b \epsilon B_n \end{aligned} \quad (5.13)$$

Therefore one obtains a link invariant from a Markov trace φ by rescaling,

$$\tilde{\varphi}(b) = d_\rho^{-(n-1)} \omega_\rho^{n_- - n_+} \varphi(b) \quad (5.14)$$

where $b \epsilon B_n$ and $n_\pm$ denote the number of $\sigma_i^{\pm 1}$ in b. The representations $\varepsilon^{(\rho)} | B_n$ are multiples of finite dimensional representations with dimension $\leq d_\rho^n$. The finite dimensional representations of B_n have not yet been classified. An interesting class has been found by Jones, Ocneanu and Wenzl [22]. We will show in the next section that this class occurs naturally in the present framework.

We close this section by exhibiting a further relation to the work of Jones [23], concerning the Temperley-Lieb-Algebra and the inclusion of von Neumann-algebras.

Let $\bar{\rho}$ be conjugate to ρ such that (5.9) is satisfied. Then

$$\rho\bar{\rho}(A)\bar{R} = \bar{R}A \, , A \epsilon \mathcal{A} \quad (5.15)$$

with

$$\bar{R} = \varepsilon(\bar{\rho}, \rho) R. \quad (5.16)$$

One has the relation

$$\bar{\rho}(\bar{R})^* R = \lambda_\rho 1 = \bar{R}^* \rho(R). \quad (5.17)$$

Since R and $\bar{R}$ are isometrics, $R\bar{R}^* = E$ and $\bar{R}\bar{R}^* = F$ are projections. We now define for $i \epsilon \mathbb{N}$

$$E_{2i-1} = (\bar{\rho}\rho)^{i-1}(E), E_{2i} = (\bar{\rho}\rho)^{i-1} \bar{\rho}(F) \quad (5.18)$$

and obtain a sequence of projections which satisfy the Temperley-Lieb relations

$$\begin{aligned} &(i) \; E_n E_m = E_m E_n \, , |n - m| \geq 2 \\ &(ii) \; E_n E_{n \pm 1} E_n = d_\rho^{-2} E_n. \end{aligned} \quad (5.19)$$

An immediate consequence of the work of Jones is that d_ρ is restricted to the set

$$\{2 \cos \frac{\pi}{m}, m \epsilon \mathbb{N}, m \geq 3\} \cup [2, \infty). \quad (5.20)$$

Jones discovered his results by investigating embeddings of von Neumann algebras. As Longo has shown [24] the sequence of embeddings ("Jones tunnel") corresponding to the sequence of projections (5.18) is

$$(\bar{\rho}\rho)^2(\mathcal{A}) \subset \bar{\rho}\rho\rho(\mathcal{A}) \subset \bar{\rho}\rho(\mathcal{A}) \subset \bar{\rho}(\mathcal{A}) \subset \mathcal{A}. \quad (5.21)$$

In particular, $\mathcal{A}$ is generated by $\bar{\rho}(\mathcal{A})$ and E. One finds [18]

$$A = \bar{\rho}(\bar{R}^* \rho(A)) E \bar{\rho}(\bar{R}) d_\rho^2. \quad (5.22)$$

d_ρ^2 is the index of the inclusions $\bar{\rho}(\mathcal{A}) \subset \mathcal{A}$ and $\rho(\mathcal{A}) \subset \mathcal{A}$.

6. The 2-Channel Situation

The simplest nonabelian sectors $[\rho]$ have the property that

$$\rho^2 \simeq \rho_1 \oplus \rho_2 \tag{6.1}$$

with ρ_1, ρ_2 irreducible and mutually inequivalent. In this case ε_ρ has two different eigenvalues λ_1, λ_2; so the situation is similar to the case of permutation group statistics where due to $\varepsilon_\rho^2 = 1$ one has $\lambda_{1,2} = \pm 1$. In the latter case, the arising permutation group representations have been completely classified by Doplicher, Haag and Roberts [9]. It is gratifying that essentially the same methods work in the case of braid group statistics and lead to a new class of braid group representations and link invariants [18].

The idea is the evaluation of the left inverse ϕ on the projections

$$E_i^{(n)} = E_i \wedge \rho(E_i) \wedge \ldots \wedge \rho^{n-2}(E_i) \tag{6.2}$$

where E_i is the spectral projection of the unitary operator ε_ρ with eigenvalue λ_i,

$$E_i = (\lambda_i - \lambda_j)^{-1}(\varepsilon_\rho - \lambda_j 1)\,, i \neq j, \tag{6.3}$$

and $\wedge$ denotes the projection onto the intersection of the corresponding subspaces. In the case of the permutation group, $E_i^{(n)}$ denote the projections onto the totally symmetric and antisymmetric subspaces.

The projections $\rho^n(E_i)$ satisfy the following modified Temperley-Lieb relations

$$\begin{aligned} E_i\rho(E_i)E_i - \tau E_i &= \rho(E_i)E_i\rho(E_i) - \tau\rho(E_i) \\ E_i\rho^n(E_i) &= \rho^n(E_i)E_i, n \geq 2 \end{aligned} \tag{6.4}$$

where

$$\tau = \frac{t}{(1+t)^2}\,, t = -\lambda_1\lambda_2^{-1}. \tag{6.5}$$

Wenzl [22] found the following recursion relation for $E_i^{(n)}$ which generalizes the well known relation for the permutation group. Let $t = e^{2i\alpha}, -\frac{\pi}{2} < \alpha < \frac{\pi}{2}$, and $m = \inf\{n\epsilon N, n|\alpha| \geq \pi\}(m = \infty$ for $\alpha = 0$ (permutation group case)). Then

$$(i) \quad E_i^{(n+1)} = \rho(E_i^{(n)}) - \frac{2\cos\alpha\sin n\alpha}{\sin(n+1)\alpha}\rho(E_i^{(n)})E_j\rho(E_i^{(n)}), i \neq j \tag{6.6}$$

for $n+1 < m$ and

$$(ii) \quad E_i^{(m)} = \rho(E_i^{(m-1)}). \tag{6.7}$$

As in [9] one now applies the left inverse ϕ to (6.6) and obtains

$$\phi(E_i^{(n+1)}) = E_i^{(n)}(1 - \frac{2\cos\alpha\sin n\alpha}{\sin(n+1)\alpha}\eta_j)\,, j \neq i \tag{6.8}$$

with

$$\eta_j 1 = \phi(E_j)\,, 0 \leq \eta_j \leq 1\,, \eta_1 + \eta_2 = 1\,, \lambda_\rho = \lambda_1\eta_1 + \lambda_2\eta_2. \tag{6.9}$$

Positivity of ϕ requires in the case of $\alpha \neq 0$ that the factor multiplying $E_i^{(n)}$ in (6.8) vanishes for some n. In the case of permutation group statistics DHR found as possible values of η_j $\frac{1}{2} \pm \frac{1}{d}$, $d \epsilon \mathbb{N} \cup \{\infty\}$, therefore the statistics parameter λ_ρ was restricted to $0, \pm\frac{1}{d}$. In the braid group statistics case one finds

$$\begin{aligned}
&\text{(i)} \quad \tfrac{\lambda_1}{\lambda_2} = -e^{\pm\frac{2\pi i}{m}}, m\epsilon\mathbb{N}, m \geq 4 \\
&\text{(ii)} \quad d_\rho = \tfrac{\sin d\pi/m}{\sin \pi/m}, d\epsilon\mathbb{N}, 2 \leq d \leq m-2 \\
&\text{(iii)} \quad \omega_\rho = -\lambda_1 e^{\pm\pi i(d+1)/m} \\
&\text{(iv)} \quad E_1^{(h)} = 0, h > d, E_2^{(\ell)} = 0, \ell > m-d
\end{aligned} \tag{6.10}$$

(v) $\varepsilon^{(\rho)}$ is an infinite multiple of the Jones- Ocneanu-Wenzl-representation of the braid group, tensored with a one dimensional representation.

(vi) The link invariant $\tilde{\varphi}$ in (5.14) is the HOMFLY link invariant [25] with parameter $t = -e^{2\pi i/m}$.

The permutation group results of DHR are recovered in the limit $m \to \infty$.

A similar analysis can be performed in the 3 channel situation provided one of the channels is abelian. One obtains the braid group representations via the Birman-Wenzl algebra, and the associated link invariant is the Kauffmann polynomial [24].

In the next section we will describe how the intrinsically defined braid group statistics in terms of observables is related to the R-matrices observed in field theory.

In general, it is difficult to find explicit formulae for endomorphisms and the associated statistics operators in a given model for nonabelian sectors. In a remarkable paper, Mack and Schomerus [26] recently succeeded in performing such an analysis for the conformal Ising field theory. It is not possible here, to give a full account of this work, so only the essential algebraic fact on which the analysis is based shall be presented.

We consider the algebra generated by selfadjoint elements $b_n, n\epsilon\mathbb{N}$ with the canonical anti- commutation relations

$$b_n b_m + b_m b_n = 2\delta_{nm}, n, m\epsilon\mathbb{N}. \tag{6.11}$$

There is a unique C^*-algebra $\mathcal{F}$ generated by $b_n, n\epsilon\mathbb{N}$. The even subalgebra $\mathcal{A}$ of $\mathcal{F}$ is defined to be the set of fixed points of $\mathcal{F}$ under the automorphism

$$b_n \to -b_n. \tag{6.12}$$

Now let ρ be the endomorphism of $\mathcal{F}$ with

$$\rho(b_n) = b_{n+1}, n\epsilon\mathbb{N}. \tag{6.13}$$

Then ρ induces an endomorphism ρ of $\mathcal{A}$. We find

$$\rho(\mathcal{A}) \neq \mathcal{A}. \tag{6.14}$$

hence ρ is not an automorphism,

$$\rho(\mathcal{A})' \cap \mathcal{A} = C1 \tag{6.15}$$

and

$$\rho^2(\mathcal{A})' \cap \mathcal{A} = C1 + Cb_1 b_2. \tag{6.16}$$

The left inverse ϕ of ρ is given by

$$\begin{aligned}\phi(b_n) &= b_{n-1}\ , n \geq 2 \\ \phi(b_1) &= 0.\end{aligned} \tag{6.17}$$

A statistics operator ε_ρ satisfying (4.20) is

$$\varepsilon_\rho = \frac{1-i}{2} b_1 b_2 + \frac{1}{2}(i-1) \tag{6.18}$$

hence the statistical dimension is

$$d_\rho = \sqrt{2}. \tag{6.19}$$

7. Exchange Algebras and R-Matrices

In integrable models and in conformal field theories so-called R-matrices occur which satisfy the braid relations (4.25) as a consequence of Yang-Baxter relations (cf. the lectures of Prof. Fadeev). In conformal field theory, they describe the algebraic structure of a field algebra ("exchange algebra") [27,28] whose correlation functions are the conformal blocks [29]. In the language of statistical mechanics, these R-matrices are associated to the RSOS model. R-matrices associated to the 6-vertex-model have also been discussed as structure constants for commutation relations of fields [30], there is however a problem with positivity related to a singularity of the vertex-SOS-transformation. We will see that the exchange algebra structure arises naturally from the algebraic formalism described in the preceeding sections. A vertex formulation of the commutation relations amounts to a solution of the symmetry problem for sectors with braid group statistics; this problem is already extremely nontrivial in the case of sectors with permutation group statistics and could been solved only by a generalization of the Tannaka-Krein-duality theory for nonabelian compact groups [31]; a corresponding theorem for quantum groups would be highly desirable. The results on models obtained so far are encouraging (see [26] and the lectures of Prof. Fadeev).

In order to find the relation between the representation of the braid group by unitary operators in the observable algebra to the R-matrices we choose one representative ρ_α from each class of irreducible endomorphisms. Let ρ be fixed. The representation $\rho_\alpha\rho$ can be decomposed into irreducible ones

$$\rho_\alpha\rho \simeq \bigoplus_\beta N_{\alpha\beta}(\rho)\rho_\beta \tag{7.1}$$

where $N_{\alpha\beta}(\rho)$ is a nonnegative integer; the matrix $(N(\rho))_{\alpha\beta} = N_{\alpha\beta}(\rho)$ has been called the fusion matrix. We may assume that ρ and ρ_α for all α are localized in a fixed $0\epsilon\mathcal{K}$. There are operators $T\epsilon\mathcal{A}(0)$ such that

$$\rho_\alpha\rho(A)T = T\rho_\beta(A), A\epsilon\mathcal{A}. \tag{7.2}$$

These intertwiners T span a linear subspace $\mathcal{H}_{\alpha\beta}$ of the algebra $\mathcal{A}(0)$. For $T_1, T_2\epsilon\mathcal{H}_{\alpha\beta}$ we have from (7.2)

$$T_1^*T_2\epsilon\rho_\beta(\mathcal{A})' , \tag{7.3}$$

hence from the irreducibility of ρ_β there is a complex number (T_1, T_2) with

$$T_1^*T_2 = (T_1, T_2)1. \tag{7.4}$$

$T_1, T_2 \to (T_1, T_2)$ has the property of a positive definite scalar product. So $\mathcal{H}_{\alpha\beta}$ gets the structure of a Hilbert space, and the dimension of $\mathcal{H}_{\alpha\beta}$ is nothing than the multiplicity $N_{\alpha\beta}(\rho)$ of ρ_β in $\rho_\alpha\rho$.

We may now iterate (7.2) and investigate the space of intertwiners $T\epsilon\mathcal{A}(0)$ with

$$\rho_\alpha\rho^n(A)T = T\rho_\beta(A). \tag{7.5}$$

The corresponding Hilbert space $\mathcal{H}_{\alpha\beta}^{(n)}$ has the following structure:

7.1 Theorem:

$$\begin{aligned} \mathcal{H}_{\alpha\beta}^{(n)} &= \sum_{\gamma_1,..\gamma_{n-1}} \mathcal{H}_{\alpha\gamma_1}..\mathcal{H}_{\gamma_{n-1}\beta} \\ &\cong \bigoplus_{\gamma_1,..\gamma_{n-1}} \mathcal{H}_{\alpha\gamma_1} \otimes .. \otimes \mathcal{H}_{\gamma_{n-1}\beta}. \end{aligned} \tag{7.6}$$

Here sum and product in the first line refer to the algebraic structure in $\mathcal{A}(0)$ whereas isomorphy, direct sum and product in the second line are to be understood in the sense of Hilbert spaces.

We now observe that the braid group representation $\varepsilon^{(\rho)}$ defined in Section 4 induces a representation of the braid group B_n with n strings on the finite dimensional Hilbert space $\mathcal{H}_{\alpha\beta}^{(n)}$. Namely, for $b\epsilon B_n$,

$$\varepsilon^{(\rho)}(b)\epsilon\rho^n(\mathcal{A})', \tag{7.7}$$

hence for $T\epsilon\mathcal{H}_{\alpha\beta}^{(n)}$ we have

$$\rho_\alpha(\varepsilon^{(\rho)}(b))T\epsilon\mathcal{H}_{\alpha\beta}^{(n)} \tag{7.8}$$

thus $\rho_\alpha(\varepsilon^{(\rho)}(B_n))$ acts by left multiplication on $\mathcal{H}_{\alpha\beta}^{(n)}$. According to the theorem, the dimension of $\mathcal{H}_{\alpha\beta}^{(n)}$ is

$$\dim \mathcal{H}_{\alpha\beta}^{(n)} = (N(\rho)_{\alpha\beta}^n). \tag{7.9}$$

A basis in $\mathcal{H}_{\alpha\beta}^{(n)}$ can be obtained by choosing an orthonormal basis in each Hilbert space $\mathcal{H}_{\gamma\delta}$; the products of these basis vectors then are a basis for $\mathcal{H}_{\alpha\beta}^{(n)}$ due to Thm.7.1.

For a convenient description we consider a graph with vertices α and $N_{\alpha\beta}$ edges from α to β. To each edge e from α to β we associate an intertwiner $T_e \epsilon \mathcal{H}_{\alpha\beta}$. Let $\mathrm{Path}_{\alpha\beta}^{(n)}$ denote the set of paths of length n from α to β, and let $T(\xi) = T_{e_1}..T_{e_n}$ if $\xi = e_1 o..o e_n \epsilon\ \mathrm{Path}_{\alpha\beta}^{(n)}$. Then

$$\{T(\xi), \xi \epsilon\ \mathrm{Path}_{\alpha\beta}^{(n)}\} \tag{7.10}$$

is a basis of $\mathcal{H}_{\alpha\beta}^{(n)}$ and by

$$\rho_\alpha(\varepsilon^{(\rho)}(b))T(\xi) = \sum_{\eta \epsilon \mathrm{Path}_{\alpha\beta}^{(n)}} R_{\xi\eta}(b)T(\eta) \tag{7.11}$$

we obtain the R-matrices with paths as indices as in the RSOS-model.

The exchange fields (sometimes also called generalized vertex operators) are obtained in the following way. Let

$$\mathcal{H} = \bigoplus_\alpha \mathcal{H}_\alpha \ , \ \mathcal{H}_\alpha = \mathcal{H}_o \forall \alpha, \tag{7.12}$$

and for $\phi \epsilon \mathcal{H}$, $A \epsilon \mathcal{A}$, let the α-component of $A\phi$ be

$$(A\phi)_\alpha = \rho_\alpha(A)\phi_\alpha. \tag{7.13}$$

Observables leave the subspaces $\mathcal{H}_\alpha$ invariant. Operators which interpolate between different subspaces are pairs (e, A) of an edge e and an observable A and act on $\mathcal{H}$ by

$$((e, A)\phi)_\delta = \delta_{\delta\beta} T_e^* \rho_\alpha(A)\phi_\alpha \tag{7.14}$$

if e is an edge from α to β, thus mapping $\mathcal{H}_\alpha$ into $\mathcal{H}_\beta$. An exchange field (e, A) is said to be localized in 0 if

$$AB = \rho(B)A \, , B \epsilon \mathcal{A}(0'). \tag{7.15}$$

Now exchange fields have the following commutation relations. Let (e_i, A_i) be localized in $0_i, i = 1, 2$ such that 0_2 is to the right of 0_1. Then

$$(e_2, A_2)(e_1, A_1) = \sum_{e_2' o e_1'} R(\sigma_1)_{e_1 o e_2, e_2' o e_1'} (e_1', A_1)(e_2', A_2). \tag{7.16}$$

These are exactly the relations found by Rehren and Schroer [28], up to the point that pointlike localized fields have been replaced by exchange fields localized in double cones.

There is also an "operator product expansion" in the general framework. In order to formulate it we add to our graph "coloured" edges corresponding to ρ_γ instead of the fixed ρ. Then

$$(e_2, A_2)(e_1, A_1) = \sum_{e,f} D_{e_1 o e_2; f, e}(e, A_f) \tag{7.17}$$

with $A_f = T_f^* \rho_1(A_2)A$ and

$$D_{e_1 \circ e_2; f, e} = T_{e_2}^* T_{e_1}^* \rho_\alpha(T_f) T_e \in C1. \tag{7.18}$$

Here e is an edge from the source of e_1 to the range of e_2. The colour of e is the range of f, and the colour and source of f are the colours of e_2 and e_1, respectively.

The coefficients D are the so called duality coefficients known from conformal field theory. The pentagon relations [32] e.g., are inferred from the relation

$$\varepsilon(\rho_\beta \rho_\alpha, \rho_\gamma) T_e = \rho_\gamma(T_e) \varepsilon(\rho_\delta, \rho_\gamma) \tag{7.19}$$

if e is an edge with colour α from β to δ.

8. Rehren's Derivation of the Verlinde Algebra

By the general analysis presented above, most of the relations which have been considered to be peculiar for $2d$ conformal field theory have been shown to be a property of superselection sectors in a generic field theory. Actually, this was more or less obvious for people familiar with algebraic quantum field theory. There is however one structural element of conformal field theory which seems to be of a different nature. This is the observation of Verlinde [33] that due to modular invariance of the partition function the structure of sectors in a conformal field theory has a certain selfduality property, similar to the selfduality of finite abelian groups. It is a remarkable result of Rehren [34] that even this structure can be derived in the general framework.

Let us assume that there are only finitely many sectors. Let ϕ_α be the unique leftinverse of ρ_α, let d_α be its statistical dimension and ω_α its statistical phase. The operator

$$\varepsilon(\rho_\alpha, \rho_\beta) \varepsilon(\rho_\beta, \rho_\alpha) = \varepsilon_M(\rho_\beta, \rho_\alpha) \tag{8.1}$$

measures the deviation of ρ_α, ρ_β from permutation group statistics (monodromy). We consider the matrix

$$Y_{\alpha\beta} = d_\alpha d_\beta \phi_\alpha \phi_\beta(\varepsilon_M(\rho_\beta, \rho_\alpha)^*). \tag{8.2}$$

Let $(N_\alpha)_\gamma^\beta$ denote the multiplicity of ρ_β in $\rho_\alpha \rho_\gamma$, and let $(Y_\alpha)_\beta = Y_{\alpha\beta}$. Then

8.1 Lemma [34]:

$$(i) \quad N_\alpha Y_\beta = \frac{1}{d_\beta} Y_{\alpha\beta} Y_\beta \tag{8.3}$$

i.e. Y_β is a joint eigenvector of the fusion matrices N_α .

$$(ii) \quad \text{Either } (Y_\alpha, Y_\beta) = 0 \text{ or } d_\alpha Y_\beta = d_\beta Y_\alpha. \tag{8.4}$$

Let $\rho_o = id$ denote the vacuum. A sector α is called degenerate if $(Y_\alpha, Y_o) \neq 0$, i.e. $Y_{\alpha\beta} = d_\alpha d_\beta$. Degenerate sectors have trivial monodromy with all other sectors. We assume now that there are no degenerate sectors. Then one obtains Rehren's Theorem.

8.2 Theorem [34]:
(i) The matrix Y is invertible.
(ii) Let $\sigma = \sum d_\alpha^2 \omega_\alpha^{-1}$. Then

$$|\sigma|^2 = \sum d_\alpha^2. \tag{8.5}$$

(iii) Let $S = |\sigma|^{-1}|Y, T = (\frac{\sigma}{|\sigma|})^{1/3} diag(\omega_\alpha)$. Then one obtains the Verlinde relations

$$\begin{aligned} SS^* &= TT^* = 1 \\ TSTST &= S \\ S^2 &= C, TC = T \end{aligned} \tag{8.6}$$

where C means the matrix $C_{\alpha\beta} = \delta_{\bar{\alpha}\beta}$ where $\rho_{\bar{\alpha}}$ is conjugate to ρ_α.
(iv) The multiplicities are given by

$$(N_\alpha)_\gamma^\beta = \sum_\delta \frac{S_{\alpha\delta} S_{\gamma\delta} S_{\beta\delta}^*}{S_{o\delta}}. \tag{8.7}$$

Verlinde found his matrices S and T by a description of the action of modular transformations $\tau \to -\tau^{-1}, \tau \to \tau + 1$, respectively, on the Virasoro characters

$$\chi_\alpha(\tau) = q^{-\frac{c}{24}} Tr_{\mathcal{H}_\alpha} q^{L_0} \tag{8.8}$$

with $q = e^{2\pi i \tau}$

$$\begin{aligned} \chi_\alpha(-\tau^{-1}) &= \sum_\beta S_{\alpha\beta} \chi_\beta(\tau) \\ \chi_\alpha(\tau + 1) &= \sum_\beta T_{\alpha\beta} \chi_\beta(\tau) \end{aligned} \tag{8.9}$$

Whether Rehren's matrices S and T which exist in generic 2 and 3 d models also have a geometric interpretation, is unknown up to now. In any case, the observed structure is a strong restriction on the possible symmetry of the theory.

In the presence of degenerate sectors one may proceed as follows: First one may consider only the degenerate sectors. For them the Doplicher-Roberts reconstruction of the field algebra and a compact symmetry group may be carried out. The enlarged algebra presumably has only nondegenerate sectors, hence Rehren's analysis applies provided there are only finitely many sectors. The structure in the case of infinitely many sectors is unknown.

9. Braid Group Statistics in 3 Dimensions

As pointed out before, sectors which satisfy the DHR selection criterion have permutation group statistics in $D \geq 3$ dimensions. There are, however, PER's which violate this criterion, e.g. charged sectors in gauge theories or soliton sectors in $D = 2$ dimensions, and, at least on the lattice, charged states may occur even in purely massive theories [35]. One may ask whether a PER is always related to a vacuum representation, and how this relation looks like. In the presence of massless particles, experience from quantum electrodynamics suggests a rather strong deviation from the DHR property; methods to deal with this situation have been developed by Buchholz [36]. Here we restrict ourselves to situations without massless particles.

A convenient starting point for the analysis is a PER where the energy momentum spectrum contains an isolated mass shell, thereby describing states of a single massive particle in the absence of massless excitations. Extending an earlier result of Swieca [37], Buchholz and I proved the following theorem:

9.1 Theorem [38]: Let π be an irreducible PER with an isolated mass shell in the energy momentum spectrum. Then there is a dense subspace $\mathcal{D} \subset \mathcal{H}_\pi$ such that

$$\partial_\mu(\phi, \pi\alpha_x(A)\phi), \quad \phi\epsilon\mathcal{D}, A\epsilon\mathcal{A}(0), 0\epsilon\mathcal{K} \tag{9.1}$$

is strongly decreasing as $|\underline{x}| - |x^0|$ tends to $+\infty$.

(A simplified proof appears in [39]).

An immediate consequence of the theorem is the convergence of expectation values of local observables as x tends to spacelike infinity,

$$\|\phi\|^{-2}(\phi, \pi\alpha_x(A)\phi) \to \omega_0(A)\,, x \to \text{spacelike infinity}. \tag{9.2}$$

From locality and the fact that π is irreducible, the limit does not depend on the choice of ϕ. ω_0 is a translation invariant state, and its restriction to each local algebra $\mathcal{A}(0)$ is normal, since the set of normal states is sequentially complete in the weak topology. Let $(\mathcal{H}_0, \pi_0, \Omega)$ denote the GNS-triple associated to ω_0 such that

$$\begin{aligned} (\Omega, \pi_0(A)\Omega) &= \omega_0(A) \\ \pi_0(\mathcal{A})\Omega & \text{ dense in } \mathcal{H}. \end{aligned} \tag{9.3}$$

Since ω_0 is translation invariant,

$$\omega_0\alpha_x = \omega_0, \tag{9.4}$$

there is a unitary implementation of the translations by

$$U_0(x)\pi_0(A)\Omega = \pi_0\alpha_x(A)\Omega\,, A\epsilon\mathcal{A}. \tag{9.5}$$

As ω_0 is locally normal (i.e. $\omega_0|\mathcal{A}(0)$ is normal for all $0\epsilon\mathcal{K}$) one can show that $x \to U_0(x)$ is strongly continuous. It then automatically satisfies the spectrum condition, and π_0 is a vacuum representation with a unique vacuum.

In $D \geq 3$ dimensions the limit in (9.2) is independent of the direction, due to the strong decrease in (9.1), hence there is a unique vacuum representation π_0 associated to π. In $D = 2$ dimensions the limits for right and left spacelike infinity may differ, and one obtains a right vacuum π^+ and a left vacuum π^- associated to π. This possibility actually occurs for soliton sectors; the arising structure will be discussed in Section 11.

In $D \geq 3$ dimensions one may ask whether there are fields which interpolate between the vacuum and the single particle sector, and what their localization properties are. As the basic estimate in (9.1) is an estimate on derivatives, one gets estimates on expectation values by integrating (9.1) along some path to spacelike infinity; observables which are localized sufficiently far spacelike separated from this path have expectation values which differ not much from the vacuum expectation values.

For a precise discussion, it is convenient to consider sets of the form

$$S = a + \bigcup_{\lambda > 0} \lambda \mathcal{O} \, , \mathcal{O} \epsilon \mathcal{K}, \bar{\mathcal{O}} \epsilon \{o\}', a \epsilon \mathbb{R}^D. \tag{9.6}$$

S may be called a spacelike cone [28]. Then one obtains the result:

<u>9.2 Theorem</u> [38]: For all spacelike cones S

$$\pi | \mathcal{A}(S') \simeq \pi_0 | \mathcal{A}(S'). \tag{9.7}$$

The result seems plausible for gauge theories where a gauge invariant charged field is heuristically given by Mandelstam's formula

$$\psi_C = \psi(x) P e^{\, i \int_C A} \tag{9.8}$$

where C is a path from x to spacelike infinity. The spacelike cone S in the theorem is a neighborhood of a certain path C. Note however, that the derivation of the theorem did not use any property which is special for gauge theories.

We now want to analyse the class of representations π satisfying (9.7) by generalizing the methods of DHR.

Let $\mathcal{S}$ denote the set of spacelike cones S defined in (9.6). Analogous to the DHR case we choose a fixed $S_0 \epsilon \mathcal{S}$. From (9.7) there is a unitary $V : \mathcal{H}_{\pi_0} \to \mathcal{H}_\pi$ such that

$$\pi(A) V = V \pi_0(A) \, , A \epsilon \mathcal{A}(S_0'). \tag{9.9}$$

We set $\pi_V = AdV^{-1} \circ \pi$ and obtain an equivalent representation in the vacuum Hilbert space $\mathcal{H}_{\pi_0} \equiv \mathcal{H}_0$ such that

$$\pi_V | \mathcal{A}(S_0') = \pi_0 | \mathcal{A}(S_0'). \tag{9.10}$$

As in the DHR case one finds that

$$\rho : \pi_0(A) \to \pi_V(A) \, , A \epsilon \mathcal{A} \tag{9.11}$$

is an isometric mapping, so we may again identify $\mathcal{A}$ with $\pi_0(\mathcal{A})$ by dropping the symbol π_0. In order to get information on the image of ρ we use locality together with (9.10). Let $0\epsilon\mathcal{K}$ and $S_1\epsilon\mathcal{S}$ with $S_1 \supset 0 \cup S_0$. Then for $B\epsilon\mathcal{A}(S_1')$ and $A\epsilon\mathcal{A}(0)$

$$[\pi_V(A), \pi_0(B)] = [\pi_V(A), \pi_V(B)] = \pi_V([A, B]) = 0, \qquad (9.12)$$

hence $\pi_V(A)\epsilon\pi_0(\mathcal{A}(S_1'))'$. One may now require Haag duality for spacelike cones. It is however sufficient to require essential duality, which means (cf. the discussion in Sect. 3) that for $S_1 \subset S_2', S_1, S_2\epsilon\mathcal{S}$

$$\pi_0(\mathcal{A}(S_1'))' \subset \pi_0(\mathcal{A}(S_2'))''. \qquad (9.13)$$

By (9.13) we conclude that (omitting the symbol π_0)

$$\rho(\mathcal{A}(0)) \subset \mathcal{A}(S_2')'' \qquad (9.14)$$

provided $0 \cup S_0 \subset S_2'$. The problem is that (9.14) does not imply that ρ is an endomorphism of $\mathcal{A}$, since the weak closure of the algebra of an unbounded region contains nonlocal operators. It would not help when Haag duality would hold for $S\epsilon\mathcal{S}$, since then we would find

$$\rho(\mathcal{A}(0)) \subset \mathcal{A}(S_1)'' \,, \; S_1 \supset 0. \qquad (9.15)$$

The way out chosen in [38] was an extension of ρ to larger algebras such that it became an endomorphism. For this purpose we choose a spacelike vector r with $r^2 = -1$ and define the set of spacelike cones containing the direction of r in their interior,

$$\mathcal{S}(r) = \{S\epsilon\mathcal{S} \mid \bar{S} + r \subset S\} \,. \qquad (9.16)$$

The set $\mathcal{S}(r)$ is directed with respect to inclusion. Therefore

$$B^r = \overline{\cup_{S\epsilon\mathcal{S}(r)}\mathcal{A}(S')''} \qquad (9.17)$$

is a C^*-algebra. One finds the following result [38]:

9.3 Theorem:

(i) ρ is weakly continuous on $\mathcal{A}(S')$ for all $S\epsilon\mathcal{S}$. Thus there exists a unique extension ρ^r of ρ to B^r which is weakly continuous on $\mathcal{A}(S')''$ for all $S\epsilon\mathcal{S}(r)$.

(ii) Let the direction of r be spacelike to the localization region S_0 of ρ (i.e. there is some $S_1\epsilon\mathcal{S}(r)$ with $S_1 \subset S_0'$). Then ρ^r is an endomorphism of B^r.

A composition law for sectors can now be defined as follows: let ρ_1, ρ_2 be representations of $\mathcal{A}$ in the vacuum Hilbert space $\mathcal{H}_0$ which satisfy (9.7) and are localized in $S_1, S_2\epsilon\mathcal{S}$, respectively. Then the composed sector is defined by

$$[\rho_1] \times [\rho_2] = [\rho_1\rho_2] \qquad (9.18)$$

where

$$\rho_1\rho_2 := \rho_1^r\rho_2 \,, r \subset S_2'. \qquad (9.19)$$

One checks that $\rho_1\rho_2$ does not depend on the choice of r, and that $[\rho_1\rho_2]$ does not depend on the choice of $\rho_1\epsilon[\rho_1]$ and $\rho_2\epsilon[\rho_2]$ [38].

One may avoid the extension of ρ to enlarged algebras and define the product $\rho_1\rho_2$ directly. For $D \geq 4$ dimensions this has been done in [1]. In any case, however, one needs an extension of ρ to intertwiners U which move the localization cone. It is at this place that $D = 3$ dimensions are special.

Let $S_1, S_2 \epsilon \mathcal{S}$, $S_1 \subset S_2'$ and let $\rho_1 = AdU \circ \rho_2$ with ρ_i localized in S_i, $i = 1,2$. The intertwiner U is in B^r for all $r \subset S_1' \cap S_2'$. Now $\rho^r(U)$ is locally constant in r. Since, however, the set of spacelike directions in the spacelike complement of $S_1 \cup S_2$ has two connected components, there may be two different extensions of ρ to U. This fact turns out to be responsible for the possible occurrence of braid group statistics in $D = 3$ dimensional quantum field theory.

We may proceed as in the DHR theory and show that the product (9.19) is commutative if ρ_1 and ρ_2 are localized in spacelike separated cones. This again proves, that the composition law for sectors is commutative, and we can compute the statistics operator ε in

$$\rho_2\rho_1 = Ad\varepsilon \circ \rho_1\rho_2 \tag{9.20}$$

for ρ_1, ρ_2 localized in $S \epsilon \mathcal{S}$. One finds the formula

$$\varepsilon(\rho_1, \rho_2) = U_2^{-1}\rho_1^r(U_2) \tag{9.21}$$

where $\hat{\rho}_2 = AdU_2 \circ \rho_2$ is localized in $\hat{S} \subset S'$ and $r \subset S' \cap \hat{S}'$. But due to the nonuniqueness of the extension of ρ_1 to U_2 we may get two different statistics operators. We reserve the symbol $\varepsilon(\rho_1, \rho_2)$ for the case where S, r and $\hat{S}$ are ordered in the sense of positively oriented rotation. For $\hat{S}$ between S and r' one finds

$$\varepsilon(\rho_2, \rho_1)^{-1} = U_2^{-1}\rho_1^{r'}(U_2)\ . \tag{9.22}$$

Therefore the monodromy operator $\varepsilon_M(\rho_1, \rho_2)$ directly measures the difference between possible extensions of ρ_1 to U_2,

$$\varepsilon_M(\rho_1, \rho_2) := \varepsilon(\rho_2, \rho_1)\varepsilon(\rho_1, \rho_2) = \rho_1^{r'}(U_2)^{-1}\rho_1^r(U_2). \tag{9.23}$$

There is an interesting connection of this fact to Poincaré covariance in composed sectors [40,42,48,51,52]. Namely, let $\rho_i, i = 1,2$ be Poincaré covariant, i.e. there are unitary representations U_i of the covering group P_c of the identity component of the Poincaré group in 3 dimensional Minkowski space such that

$$AdU_i(L) \circ \rho_i = \rho_i\alpha_L\ , L\epsilon P_c, \tag{9.24}$$

and let U_0 be the unitary implementation of $P_+^\uparrow(3)$ in the vacuum sector. Then α_L has an extension to $B(\mathcal{H}_0)$, and

$$\begin{aligned} \alpha_L(B^r) &= B^{Lr} \\ AdU_i(L) \circ \rho_i^r &= \rho_i^{Lr}\alpha_L|B^r. \end{aligned} \tag{9.25}$$

Now let φ denote rotation by the angle φ (in a fixed Lorentz system). Then

$$U_i(2\pi) = e^{2\pi s_i}1\ , s_i\epsilon\mathbb{R}\ , i = 1,2 \tag{9.26}$$

if ρ_1, ρ_2 are irreducible, and particles in the sector of ρ_i have spins $s_i + n, n\epsilon\mathbb{Z}$.

The composed sector $\rho_1\rho_2$ is again Poincaré covariant, and for small L one finds

$$U_{12}(L) = U_1(L)\rho_1^r(U_0(L)^{-1}U_2(L)). \tag{9.27}$$

Here we used the fact that

$$AdU_0(L)^{-1}U_2(L) \circ \rho_2 = \alpha_L^{-1}\rho_2\alpha_L =: \rho_{2,L}\ , \tag{9.28}$$

hence $U_0(L)^{-1}U_2(L)$ intertwines from ρ_2 (localized in $S\epsilon\mathcal{S}$) to $\rho_{2,L}$ (localized in $L^{-1}S$). Thus $U_0(L)^{-1}U_2(L)\epsilon\mathcal{B}^r$ for all $r\epsilon S' \cap (L^{-1}S)'$. In order to satisfy the condition that U_{12} is a representation of P_c for small L we choose a path $L(t)$ in $P_+^\uparrow(3)$ starting at the identity which belongs to L (considered as a homotopy class in $P_+^\uparrow(3)$). Then the set of spacelike directions in the spacelike complement of $\cup_t L(t)^{-1}S$ is connected, hence the definition in (9.27) is unique, and it extends to all of P_c by the requirement that U_{12} is a representation.

We now compute the possible spin quantum numbers in U_{12}. We have

$$\begin{aligned} U_{12}(2\pi) =& U_{12}(\pi)^2 = [U_1(\pi)\rho_1^r(U_0(\pi)^{-1}U_2(\pi))]^2 \\ =& U_1(2\pi)\rho_1^{\pi r}(U_0(2\pi)^{-1}U_2(\pi)U_0(\pi))\rho_1^r(U_0(\pi)^{-1}U_2(\pi)), \\ =& U_1(2\pi)\rho_1^{\pi r}(U_0(2\pi)^{-1}U_2(2\pi)) \\ & \times \rho_1^{\pi r}(U_2(\pi)^{-1}U_0(\pi))\rho_1^r(U_0(\pi)^{-1}U_2(\pi)) \\ =& e^{2\pi i(s_1+s_2)}\varepsilon_M(\rho_1,\rho_2) \end{aligned} \tag{9.29}$$

where the last equation follows from (9.23). The eigenvalues of the monodromy operators ε_M are given in terms of statistics phases. If ρ_α is equivalent to a subrepresentation of $\rho_1\rho_2$, then $\omega_\alpha\omega_1^{-1}\omega_2^{-1}$ is the associated eigenvalue of ε_M. Hence (9.29) leads to the following sum rule for spins

$$\frac{e^{2\pi i s_\alpha}}{e^{2\pi i(s_1+s_2)}} = \frac{\omega_\alpha}{\omega_1\omega_2}. \tag{9.30}$$

One may apply this rule to the case $\rho_1 = \rho, \rho_2 = \bar{\rho}$ and $\rho_\alpha = id$. Since $\omega_\rho = \omega_{\bar{\rho}}, s_\rho = s_{\bar{\rho}}$ and $\omega_{id} = 1, s_{id} = 0$ one finds the spin-statistics connection

$$e^{2\pi i s_\rho} = \pm\omega_\rho \tag{9.31}$$

Note that the usual spin-statistics theorem in $D = 4$ dimensions whose strongest version is due to Buchholz and Epstein [43] determines the sign in (9.31). An extension to the 3 dimensional case has still to be worked out.

Similar formulas hold in chiral conformal field theories in 2 dimensions. There the spin is replaced by the conformal dimension, and the sign in (9.31) has been determined by Rehren [40].

Relations (9.30) and (9.31) have been observed before in models in the case of abelian braid group statistics [41]. A general proof has been given also by Fröhlich, Gabbiani and Marchetti [42].

The possibility of statistics beyond the Bose-Fermi alternative in two spatial dimensions has first been observed by Leinaas and Myrheim [44] in the framework of nonrelativistic quantum mechanics on configuration spaces. Its possible occurrence in relativistic quantum field theory followed from the general analysis in [38]. The connection of these results with the Leinaas-Myrheim analysis has first been discussed in [41]. Wilczek [17] described the first models and mechanisms for anomalous statistics in two spatial dimensions.

Nowadays two mechanisms for the occurrence of braid group statistics in (2+1)-dimensional quantum field theory are known. One is the interplay between electrical charges and magnetic fluxes in gauge theories which have nontrivial monodromy due to the Aharonov-Bohm effect; dyons then may have nontrivial braid group statistics. The other one is the coupling of electrically charged particles to a gauge field with a so-called Chern-Simons term in the Lagrangian. These models will be discussed in more detail in the lectures of Professor Jackiw and in the seminar of Florian Nill. Up to now it is not clear whether one can find particles with nonabelian braid group statistics in these models. The long distance limit of theories with a Chern-Simons term are Witten's topological field theories where only the Chern-Simons term survives. The observation [45] that expectation values of Wilson loops in this limit are the knot invariants known from the Jones theory supports the conjecture [42] that the model described particles with nonabelian braid group statistics before the long distance limit was performed.

In the abelian case the statistics of particles is characterized by the statistics phase which can take any value, hence the name "Anyon" [17] seems to be appropriate. In the nonabelian case however, rather nontrivial quantization rules occur. We therefore proposed a new name for particles with nonabelian braid group statistics: "Plektons" from the Greek word πλεκτόs (braided).

Let me close this section by a remark on parity violation. Parity in 2 spatial dimensions means reflection at a line. Particles with braid group statistics cannot be eigenstates of a parity operator. Hence if parity is a symmetry each plekton has a mirror image which is different from the antiparticle. So braid group statistics leads either to parity violation or to a doubling in the spectrum of particles.

10. Conformal Light Cone Theories

In 2d conformal field theory a special role is played by the so-called chiral fields which are invariant under a 1-parameter group of light-like translations. They may be considered as fields on a one dimensional space which, due to conformal covariance, can be identified with the circle S^1. They generate a family of von Neumann algebras $(\mathcal{A}(I))_{I\epsilon\mathcal{J}}$ on a Hilbert space $\mathcal{H}_0$ where $\mathcal{J}$ denotes the set of nonempty open intervals $I \subset S^1$ which are not dense in S^1. The family $(\mathcal{A}(I))_{I\epsilon\mathcal{J}}$ may be assumed to satisfy the following version of the Haag-Kastler axioms:

$$\begin{array}{lll} (i) & I\subset J \Rightarrow \mathcal{A}(I)\subset\mathcal{A}(J) & \text{(isotony)} \\ (ii) & I\cap I=\emptyset \Rightarrow [\mathcal{A}(I),\mathcal{A}(J)]=0 & \text{(locality)} \end{array} \tag{10.1}$$

(iii) There is a unitary strongly continuous positive energy representation U of the real Moebius group $M \cong SL(2,\mathbb{R})/\mathbb{Z}_2$ on $\mathcal{H}_0$ with a unique ground state such that

$$U(\gamma)\mathcal{A}(I)U(\gamma)^{-1} = \mathcal{A}(\gamma I)\,, \gamma \epsilon M \qquad \text{(covariance)}. \tag{10.2}$$

From the Bisognano-Wichmann theorem [16] together with covariance one may sharpen (ii) to

$$\mathcal{A}(I)' = \mathcal{A}(I') \qquad \text{(Haag duality)} \tag{10.3}$$

where I' denotes the interior of the complement of I in S^1. For the vacuum representations of the Virasoro algebra with the admissable values of the central charge c this structure has been verified by Buchholz and Schulz-Mirbach [14].

A conformally covariant representation of $(\mathcal{A}(I))_{I\epsilon\mathcal{J}}$ may be defined as a family of representations

$$\pi^I : \mathcal{A}(I) \to B(\mathcal{H}_\pi) \tag{10.4}$$

such that

$$\pi^I|\mathcal{A}(J) = \pi^J \qquad \text{for } J \subset I, \tag{10.5}$$

together with a strongly continuous unitary representation U_π of the covering group $\tilde{M}$ of the Moebius group M such that

$$AdU_\pi(\tilde{\gamma}) \circ \pi^I = \pi^{\gamma I} \circ AdU(\gamma) \tag{10.6}$$

where $\tilde{\gamma} \mapsto \gamma$ denotes the covering homomorphism from $\tilde{M}$ onto M.

As in the case of local algebras on Minkowski space one may desire to introduce a global algebra. However, contrary to the set $\mathcal{K}$ of double cones in Minkowski space, the set $\mathcal{J}$ of intervals in S^1 is not directed, hence the union of the local algebras is not an algebra, and the global algebra cannot be defined as the inductive limit of the local ones as in (2.22). Instead of this one may use the following universal construction: the global algebra $\mathcal{A}$ is implicitly defined as a C^*-algebra containing all algebras $\mathcal{A}(I), I\epsilon\mathcal{J}$ as subalgebras such that for all families of representations $\pi^I : \mathcal{A}(I) \to B(\mathcal{H}_\pi), I\epsilon\mathcal{J}$, which satisfy the compatibility condition (10.5) there is a unique representation π of A in $\mathcal{H}_\pi$ such that

$$\pi|\mathcal{A}(I) = \pi^I \qquad , I\epsilon\mathcal{J}. \tag{10.7}$$

The global algebra $\mathcal{A}$ may be constructed as follows: let $\mathcal{A}_0$ denote the free *-algebra which is generated by the algebras $\mathcal{A}(I), I\epsilon\mathcal{J}$, disregarding all relations between algebras associated to different regions. On $\mathcal{A}_0$ we define the C^*-seminorm

$$\|A\| := \sup_\pi \|\pi(A)\| \qquad , A\epsilon\mathcal{A}_0 \tag{10.8}$$

where the supremum is taken over all representations π of $\mathcal{A}_0$ which satisfy

$$\pi^I|\mathcal{A}(J) = \pi^J \qquad , J \subset I \tag{10.9}$$

where π^I denotes the restriction of π to $\mathcal{A}(I)$. The set of elements with norm zero is a both sided ideal in $\mathcal{A}_0$, and the completion of the quotient is the desired C^*-algebra $\mathcal{A}$.

The abstract construction may be illustrated on the example of a $U(1)$ current $j(z), z\epsilon S^1$. Let

$$W(f) = \exp\{i \int \frac{dz}{2\pi i} f(z) j(z)\} \tag{10.10}$$

with a real valued smooth function f on S^1. The operators $W(f)$ are unitary, and the canonical commutation relations for the current can be formulated in Weyl form

$$W(f)W(g) = W(f+g)\exp\{-\frac{1}{2}\int \frac{dz}{2\pi i} f'(z)g(z)\}. \tag{10.11}$$

The operators $W(f)$ generate a unique C^*-algebra $\mathcal{A}$ with subalgebras $\mathcal{A}(I)$ which are generated by operators $W(f)$ with *supp* $f \subset I$. One easily checks that the algebra $\mathcal{A}$ is the unique C^*-algebra associated to the family $(\mathcal{A}(I))_{I\epsilon\mathcal{J}}$ by the universal construction described above [46].

We now return to the general discussion and consider a conformally covariant PER π of $\mathcal{A}$. According to [47], π is locally normal and

$$\pi^I \simeq \pi_0^I \quad , I\epsilon\mathcal{J}. \tag{10.12}$$

where π_0 is the (defining) vacuum representation of $\mathcal{A}$. Hence one may choose some $I_0\epsilon\mathcal{J}$ and $\hat{\pi} \simeq \pi$ such that $\mathcal{H}_{\hat{\pi}} = \mathcal{H}_0$ and

$$\hat{\pi}^{I_0'} = \pi_0^{I_0'}. \tag{10.13}$$

$\hat{\pi}$ may be called to be localized in I_0. In order to apply the fundamental idea of composition of sectors we search for an endomorphism ρ of $\mathcal{A}$ such that

$$\hat{\pi} = \pi_0 \circ \rho. \tag{10.14}$$

Because of Haag duality (10.3) the algebra $\hat{\pi}(\mathcal{A})$ is a subalgebra of $\pi_o(\mathcal{A})$; however π_0 is no longer faithful. Nevertheless one can find an endomorphism ρ of $\mathcal{A}$ which satisfies (10.14) and is localized in I_0; moreover, ρ is uniquely determined by the requirement to be "locally transportable" which means that to each $I_1\epsilon\mathcal{J}$ and each $I_{01}\epsilon\mathcal{J}$ with $I_{01} \subset I_0 \subset I_1$ there is a unitary $U\epsilon\mathcal{A}(I_{01})$ such that $AdU \circ \rho$ is localized in I_1 [46].

Hence any locally normal representation π of $\mathcal{A}$ is equivalent to a representation of the form $\pi_0 \circ \rho$ with a localized and locally transportable endomorphism ρ of $\mathcal{A}$, and the composition of endomorphisms induces a composition of sectors.

The formalism of the theory of superselection sectors may now be extended to field theories on S^1 with essentially the same results as for theories on 2d Minkowski space with DHR sectors. A similar formalism may be applied to sectors with localization in spacelike cones.

Apart from the conceptual interest on field theories on multiply connected spaces the described extension of the standard formalism of the theory of superselection sectors may have practical advantages since there may be endomorphisms of the global algebra which are related by an inner automorphism to a localized, locally transportable endomorphism an are explicitly given. This possibility has been exploited by Mack and Schomerus [26] in their treatment of the conformal Ising field theory; they describe the PER's of this model by endomorphisms of the global algebra which map the vacuum state onto the lowest weight states in the charged representations.

11. Soliton Sectors in 2d Minkowski Space

As discussed in Section 9, the general theorem on the localizability of particles in massive theories [38] leads in D=2 dimensions to two possibly different vacua $\pi^{\pm}$ associated to an irreducible PER with an isolated mass shell. The analogue of Theorem 9.2 for D=2 is

11.1. Theorem: Let π be an irreducible PER with an isolated mass shell in the energy momentum spectrum. Then there are two vacuum representations $\pi^{\pm}$ such that for all x

$$\pi|\mathcal{A}(W_{\pm}+x) \simeq \pi^{\pm}|\mathcal{A}(W_{\pm}+x). \tag{11.1}$$

Here $W_{\pm} = \{x \epsilon \mathrm{R}^2 |\, |x^0| < \pm x^1\}$ denote the right resp. the left wedge.

We call π a soliton sector if π^+ and π^- are inequivalent. In general, there seems to be no reason why soliton sectors should be related by endomorphisms. If the different vacua are related by a group of internal symmetries there might be endomorphisms ρ which act like the identity on a right wedge and like an internal symmetry transformation on a left wedge or vice versa. Such a structure has been proposed by Fröhlich [50].

In the case of a theory with different vacua which are not related by a symmetry group other methods are needed. Let (W) denote the class of all representations π which satisfy (11.1) for suitable vacuum representations $\pi^{\pm}$. We assume that all arising vacuum representations satisfy Haag duality for wedges. Then we find the following groupoid like composition rule. Let $\pi_1, \pi_2 \epsilon (W)$ such that $\pi_1^+ \cong \pi_2^-$. We may assume that all representations act on the same Hilbert space $\mathcal{H}$ such that $\pi_1^+ = \pi_2^-$ and

$$\begin{aligned} \pi_1|\mathcal{A}(W_+) &= \pi_1^+|\mathcal{A}(W_+) \\ \pi_2|\mathcal{A}(W_-) &= \pi_2^-|\mathcal{A}(W_-). \end{aligned} \tag{11.2}$$

Now let $0 \epsilon \mathcal{K}$. There is some $x \epsilon W_+$ such that $0 \subset W_+ + x$. Choose a unitary $U \epsilon B(\mathcal{H})$ such that

$$AdU \circ \pi_1|\mathcal{A}(W_+ + x) = \pi_1^+|\mathcal{A}(W_+ + x) \tag{11.3}$$

Then one defines

$$\pi_1\pi_2|\mathcal{A}(0) = AdU^{-1} \circ \pi_2|\mathcal{A}(0). \tag{11.4}$$

It is easily seen that the definition (11.4) of the composed representation $\pi_1\pi_2$ on $\mathcal{A}(0)$ does not depend on the choice of U and x, and leads to a representation

$\pi_1\pi_2$ of $\mathcal{A}$. Moreover, $\pi_1\pi_2$ is again in (W) and $(\pi_1\pi_2)^+ = \pi_2^+, (\pi_1\pi_2)^- = \pi_1^-$ [46].

The composition law (11.4) for soliton sectors may now be used for a general analysis of the superselection structure of 2d quantum field theories. E.g., using the methods of [19] one finds the antisoliton sector $\bar{\pi}\epsilon(W)$ with the properties

$$\bar{\pi}^{\pm} \simeq \pi^{\mp}, \pi\bar{\pi} \supset \pi^-, \bar{\pi}\pi \supset \pi^+. \tag{11.5}$$

One may also construct multisoliton scattering states. Here only states with well ordered velocities (in the sense of the order in the composition law (11.4)) can be expected. Whether there are structures similar to statistics is, at the present, unknown; from the analogy with quantum mechanics on configuration spaces [44] one expects no intrinsic notion of statistics.

References

1 S. Doplicher, J.E. Roberts: *Why there is a field algebra with a compact gauge group describing the superselection structure in particle physics*, Comm. Math. Phys. (to appear)

2 G.C. Wick. E.P. Wigner, A.S. Wightman: *Intrinsic parity of elementary particles*, Phys. Rev. **88**, 101 (1952)

3 R. Haag, D. Kastler: *An algebraic approach to field theory*, J. Math. Phys. **5**, 848 (1964)

4 N.P. Landsman: *Quantization and superselection sectors I. Tranformation group C^*-algebras*. DAMTP-89-44 (preprint); II. *Dirac monopole and Aharonov-Bohm effect*. DAMTP-90-7 (preprint)

5 H.-J. Borchers: *On the vacuum state in quantum field theory II*, Commun. Math. Phys. **1**, 57 (1965)

6 H.-J. Borchers: *Energy and Momentum as Observables in Quantum Field Theory*, Commun. Math. Phys. **2**, 49 (1966)

7 J. Fröhlich, G. Morchio, F. Strocchi: *Infrared problem and spontaneous breaking of the Lorentz group in QED*, Phys. Lett. **89B**, 61 (1979)

8 R.R. Streater, A.S. Wightman: *PCT, Spin and Statistics, and All That* (W.A. Benjamin Inc. , New York 1963)

9 S. Doplicher, R. Haag, J.E. Roberts: *Local observables and particle statistics*, Commun. Math. Phys. **23**, 199 (1971) and **35**, 49 (1974)

10 H.-J. Borchers: *Local rings and the connection of spin with statistics*, Commun. Math. Phys. **1**, 281 (1965)

11 H. Araki: *A Lattice of von Neumann Algebras associated with the Quantum Theory of a Free Bose Field*, J. Math. Phys. **4**, 1343 (1963); *Von Neumann Algebras of Local Observables for the Free Scalar Field*, J. Math. Phys. **5**, 1 (1964)

12 W. Driessler: *Duality and absence of locally generated superselection sectors for CCR-type algebras*, Comm. Math. Phys. **70**, 213 (1979)

13 S.J. Summers: *Normal product states for fermions and twisted duality for CCR- and CAR-type algebras with application to the Yukawa quantum field model*, Comm. Math. Phys. **86**, 111 (1982)

14 D. Buchholz, H. Schulz-Mirbach: *Haag duality in conformal quantum field theory* (to be published)

15 J.E. Roberts: *Spontaneously broken gauge symmetries and superselection rules*. Proceedings of the International School of Mathematical Physics, University of Camerino, 1974, ed. G. Gallavotti (1976)

16 J.J. Bisognano, E.H. Wichmann: *On the duality condition for a hermitean scalar field*, Journ. Math. Phys. **16**, 985 (1975)

17 F. Wilczek: *Quantum Mechanics of Fractional-Spin Particles*, Phys. Rev. Lett. **49**, 957 (1982)
18 K. Fredenhagen, K.H. Rehren, B. Schroer: *Superselection Sectors with Braid Group Statistics and Exchange Algebras*, Commun. Math. Phys. **125**, 201 (1989)
19 K. Fredenhagen: *On the Existence of Antiparticles*, Comm. Math. Phys. **79**, 141 (1981)
20 M.V. Pimsner: *A Class of Markov Traces*, preprint 1989
21 J. Birman: *Braids, Links and Mapping Class Groups*, Ann. Math. Studies **82** (Princeton University Press 1974)
22 W. Wenzl: *Hecke algebras of type A_n and subfactors*, Inv. Math. **92**, 349 (1988)
23 V. Jones: *Index for subfactors*, Invent. Math. **72**, 1 (1982)
24 R. Longo: *Index of Subfactors and statistics of quantum fields*, Comm. Math. Phys. **126**, 217 (1989); II (to appear)
25 P. Freyd, D. Yetter, J. Hoste, W.B.R. Lickorish, K. Millet, A. Ocneanu: *A new polynomial invariant of knots and links*, Bull. Am. Math. Soc. **12**, 239 (1985)
26 G. Mack, V. Schomerus: *Conformal field algebras with quantum symmetry from the theory of superselection sectors*, Universität Hamburg (1989) (preprint)
27 K.-H. Rehren: *Locality of conformal fields in two dimensions: Exchange algebra on the light-cone*, Comm. Math. Phys. **116**, 675 (1988)
28 K.-H. Rehren, B. Schroer: *Einstein Causality and Artin Braids*, Nucl. Phys. **B312**, 715 (1989)
29 A.A. Belavin, A.M. Polyakov, A.B. Zamolodchikov, Nucl. Phys. **B241**, 333 (1984)
30 J. Fröhlich: *Statistics of fields, the Yang Baxter equation and the theory of knots and links*, Proceedings of the 1987 Cargèse School
31 S. Doplicher, J.E. Roberts: *A new duality theory for compact groups*, Invent. Math. **98**, 157 (1989)
32 G. Moore, N. Seiberg: *Polynomial equations for rational conformal field theories*, Phys. Lett. **212B**, 451 (1988)
33 E. Verlinde: *Fusion rules and modular transformations in 2D conformal field theory*, Nucl. Phys. **B300** [FS 22], 360 (1988)
34 K.-H. Rehren: *Braid Group Statistics and their Superselection Rules*, University of Utrecht preprint (1989)
35 K. Fredenhagen, M. Marcu: *Charged states in Z(2) gauge theories*, Comm. Math. Phys. **92**, 81 (1983)
36 D. Buchholz: *On particles, infraparticles, and the problem of asymptotic completeness*, IAMP 1986 Marseille, eds. M. Mebkhout, R. Sénéor (World Scientific, Singapore 1987)
37 J.A. Swieca: *Charge screening and mass spectrum*, Phys. Rev. **D13**, 312 (1976)
38 D. Buchholz, K. Fredenhagen: *Locality and the Structure of Particle States*, Comm. Math. Phys. **84**, 1 (1982)
39 K. Fredenhagen: *Localizability of Particle States*, in "Algebraic Theory of Superselection Sectors and Field Theory" Istituto Guccia, Palermo 1989 ,ed. by D. Kastler
40 K. Fredenhagen, K.-H. Rehren, B. Schroer: *Superselection Sectors with Braid Group Statistics and Exchange Algebras II* (in preparation)
41 J. Fröhlich, P.A. Marchetti: *Quantum Field Theory of Vortices and Anyons*, Comm. Math. Phys. **121**, 177 (1989)
42 J. Fröhlich, F. Gabbiani, P.-A. Marchetti: *Superselection Structure and Statistics in Three-Dimensional Local Quantum Theory*, Zürich ETH-TH/89-22 (preprint)
43 D. Buchholz, H. Epstein: *Spins and Statistics of Quantum Topological Charges*, Fizika **17**, 329 (1985)
44 J.M. Leinaas, J. Myrheim: *On the Theory of Identical Particles*, Il Nuovo Cimento **37 B**, 1 (1977)
45 E. Witten: *Quantum Field Theory and the Jones polynomical*, Comm. Math. Phys. **121** (1989) 351
46 K. Fredenhagen: *Generalizations of the Theory of Superselection Sectors*, Istituto Guccia, Palermo 1989 ,ed. by D. Kastler
47 D. Buchholz, G. Mack, I.T. Todorov: *The current algebra on the circle as a germ of local field theories*, Nucl. Phys. B (Proc. Suppl.) **5B**, 20 (1988)
48 J. Fröhlich, F. Gabbiani, P.-A. Marchetti: *Braid Statistics in Three-Dimensional Local Quantum Theory*, Zürich ETH-TH/89-36
49 O. Bratteli, D. Robinson: *Operator algebras and statistical mechanics I, II* (Berlin, Heidelberg, New York: Springer 1981)

50 J. Fröhlich: *New Super-Selection Sectors ("Soliton States") in Two Dimensional Bose Quantum Field Models*, Comm. Math. Phys. **47**, 269 (1976)
51 K. Fredenhagen: *Sum Rules for Spins in (2+1)-Dimensional Quantum Field Theory*, Quantum Group Workshop Clausthal 1989, eds. H. Doebner, J. Henning
52 K. Fredenhagen: *Structure of Superselection Sectors in Low-Dimensional Quantum Field Theory*, Tahoe City 1989, eds. L.L. Chan, W. Nahm
53 D. Kastler, M. Mebkhout, K.-H. Rehren: *Introduction to the algebraic theory of superselection sectors*, Marseille 1989 (preprint)

From Integrable Models to Quantum Groups

L. Faddeev

Steklov Mathematical Institute, Leningrad, USSR

The history of quantum groups is quite exciting and instructive. In this notion a purely algebraic synthesis of the technical developments in mathematical physics is found. Its sources are:

1. Solution of the quantum magnetic chain, originated by H. Bethe.
2. Solution of two-dimensional lattice models in statistical physics, originated by L. Onsager in the case of the Ising model and developed by E. Lieb, R. Baxter and many others.
3. Solution of the many body scattering problem for a system of one-dimensional particles with δ-function interaction (McGuire, F. Beresin, C.N. Yang, E. Bresin and J. Zinn-Justin and many others).
4. The quantum inverse scattering method which is a long shot from the original proposal of Gardner-Green-Kruskal-Miura for the solution of the KdV equation. The development of this method in its algebraic form, performed by my colleagues and me in Leningrad at the end of the seventies embraced the sources [1-3] and envisaged the structures which eventually led to the introduction of quantum groups (the terminology was invented by V. Drinfeld).

In these lectures I shall present a short introduction to quantum groups in this historical way and then discuss their new applications to Conformal Field Theory. Chapter 1 contains a reminder of the main technical tools of the quantum inverse scattering method. The abstraction of this material will lead to the notion of a quantum group in Chapter 2. In Chapter 3 and 4 two main models of Conformal Field Theory, namely the Liouville model and the WZNW model, will be considered. It will be shown how the action of a quantum group enters naturally in the Lagrangian approach to these models.

In the course of preparing these lectures the discussions with A. Alexeev, O. Babelon, A. Kirillov, N. Reshetikhin, S. Shatashvili, F. Smirnoff, L. Takhtajan and A. Volkoff were most useful and I am very thankful to all of them.

1. Introduction to the Quantum Inverse Scattering Method

The method of the title (QISM) was introduced at the end of the seventies after almost 10 years of development stemming from my paper with Zakharov

Fields and Particles Editors: H. Mitter · W. Schweiger

[1] on the integrability of the KdV equation, going via a quasiclassical route, reviewed in [2], and leading to an understanding of the connection between the quantized Nonlinear Schrödinger Equation and the Bethe Ansatz method in the N-Body problem of one-dimensional particles with δ-function interaction in [3], [4]. The abstraction of formulas appearing in [3], [4] made in [5] led very soon to realizing the main algebraic content of the method in [6], [7]. The early review of this development was done by me in [8], and by Thaker in [9].

During the last decade the method was worked intensively on in Leningrad and elsewhere. For some reviews see [10], [11], [12]. Here we shall recall only some of the main formulas which will be needed for the introduction of quantum groups. This chapter can be considered as an accumulation of concrete material to be used in the following abstraction.

For definiteness we shall treat the concrete and simplest example of a magnetic chain – spin 1/2 XXX model. The dynamical variables – spin 1/2 operators s_n^a, $a = 1, 2, 3$ are attached to the sites n, $n = 1, \dots, N$ along the chain of length N. The full Hilbert space $\mathcal{H}$ of the model is a tensor product of the local spaces $\mathbb{C}^2$, in each of which a representation of spin operators $s^a = \frac{1}{2}\sigma^a$ is given, σ^a being Pauli matrices. In other words

$$\mathcal{H} = \prod^{n} \otimes \mathbb{C}^2$$

and

$$s_n^a = I \otimes \dots \underbrace{\otimes \frac{1}{2}\sigma^a \otimes}_{\text{site n}} \dots \otimes I \, .$$

These operators satisfy the usual commutation relations

$$[s_n^a, s_m^b] = i\epsilon^{abc} s_n^c \delta_{nm} \, .$$

We suppose the chain to be closed, so that the periodicity condition holds

$$s_n^a = s_{n+N}^a \, .$$

The Hamiltonian of the chain is given by

$$H = J \sum_{n,a} \left(s_n^a s_{n+1}^a - \frac{1}{4} \right) ,$$

where J is a coupling constant.

The main trick of the QISM is to associate with these data an auxiliary 2×2 matrix $L_n(\lambda)$ expressed in terms of the local dynamical variables s_n^a and a complex parameter λ as follows

$$L_n(\lambda) = \begin{pmatrix} \lambda + i s_n^3 & i s_n^- \\ i s_n^+ & \lambda - i s_n^3 \end{pmatrix} =$$

$$= \lambda I_n \otimes \sigma^0 + i s_n^a \otimes \sigma^a \, ,$$

where in the second line we used the basis of matrices $\sigma^0 = I$ and σ^a, $a = 1, 2, 3$, in the auxiliary space $V = \mathbb{C}^2$; moreover, $s_n^- = s_n^1 - i s_n^2$; $s_n^+ = s_n^1 + i s_n^2$.

The chief property of $L_n(\lambda)$ is the main commutation relation (MCR)

$$R(\lambda-\mu)L_n^1(\lambda)L_n^2(\mu) = L_n^2(\mu)L_n^1(\lambda)R(\lambda-\mu) .$$

Here the following notations are used: $L_n^1(\lambda), L_n^2(\mu)$ are 4×4 matrices attached to the space $V\otimes V$ as

$$L_n^1(\lambda) = L_n(\lambda)\otimes I ; \qquad L_n^2(\mu) = I\otimes L_n(\mu) ;$$

furthermore, $R(\lambda-\mu)$ is a structure constants matrix, independent of dynamical variables

$$R(\lambda) = \frac{1}{\lambda+i}(\lambda I + iP),$$

where P is a permutation in $V\otimes V$

$$P(a\otimes b) = b\otimes a$$

and can be written explicitly as

$$P = \frac{1}{2}(I\otimes I + \sigma^a\otimes\sigma^a).$$

In a natural basis $++, +-, -+, --$ in $\mathbb{C}^2\otimes\mathbb{C}^2$ it can be expressed as

$$R(\lambda) = \begin{pmatrix} 1 & 0 & 0 & 0 \\ 0 & b(\lambda) & c(\lambda) & 0 \\ 0 & c(\lambda) & b(\lambda) & 0 \\ 0 & 0 & 0 & 1 \end{pmatrix} ,$$

where

$$b(\lambda) = \frac{\lambda}{\lambda+i} ; \qquad c(\lambda) = \frac{i}{\lambda+i} .$$

Comparing the expressions for $L_n(\lambda)$ and $R(\lambda)$, after formally identifying the local quantum and auxiliary spaces, we see that

$$R(\lambda) = \frac{1}{\lambda+i}\, L(\lambda+\frac{i}{2}) .$$

This means that $R(\lambda)$ itself satisfies the relation

$$\begin{aligned} &R_{12}(\lambda-\mu)R_{13}(\lambda-\sigma)R_{23}(\mu-\sigma) = \\ &\qquad = R_{23}(\mu-\sigma)R_{13}(\lambda-\sigma)R_{12}(\lambda-\mu) , \end{aligned}$$

which holds in $V\otimes V\otimes V$ with the notations $R_{12} = R\otimes I$ etc. In [7] we proposed to call this formula the Yang-Baxter relation (YBR). To connect YBR and MCR it is sufficient to make a shift $\lambda\to\lambda-\sigma$, $\mu\to\mu-\sigma$ and identify the local quantum space with the third auxiliary space.

The advantage of the MCR consists in the following: introduce the monodromy matrix

$$T_N(\lambda) = L_N(\lambda)\dots L_1(\lambda) ,$$

associated with the auxiliary linear problem

$$\phi_{n+1} = L_n \phi_n \,.$$

The matrix elements of the 2×2 matrix $T_N(\lambda)$ are operators in the full quantum space $\mathcal{H}$. It follows from local MCR that $T_N(\lambda)$ also satisfies them

$$R(\lambda-\mu)T_N^1(\lambda)T_N^2(\mu) = T_N^2(\mu)T_N^1(\lambda)R(\lambda-\mu)\,.$$

To check it, it is sufficient to show that MCR are conserved after the multiplication of just two local matrices $L_n(\lambda)$. We have

$R(L_2^1 L_1^1)(L_2^2 L_1^2) =$ (using the commutativity of L_1^1 and L_2^2 due to the fact that both the quantum and auxiliary spaces are different)

$= RL_2^1 L_2^2 L_1^1 L_1^2 =$ (using the MCR twice)

$= L_2^2 L_2^1 L_1^2 L_1^1 R =$ (using the repetition of the first reasoning)

$= (L_2^2 L_1^2)(L_2^1 L_1^1)R\,.$

Thus the MCR for two L-matrices associated with the two different quantum spaces is conserved after their multiplication in the auxiliary space. It is this property which leads to the notion of a quantum group after a proper abstraction.

Returning to our model let us mention that it is the change of variables

$$s_n \to T_N(\lambda),$$

which is used in the exact solution of the model. In particular it follows from MCR that the family of operators

$$F_N(\lambda) = \operatorname{tr} T_N(\lambda)$$

is commutative

$$[F_N(\lambda), F_N(\mu)] = 0\,.$$

The Hamiltonian H is contained in this family

$$H = \text{const} + \frac{iJ}{2}\frac{d}{d\lambda}\ln F_N(\lambda)|_{\lambda=\frac{i}{2}}\,,$$

so that $F_N(\lambda)$ is an infinite set of commuting conservation laws. The off-diagonal elements of $T_N(\lambda)$ play the role of spectrum-raising operators. All this is explained in the literature already mentioned above (for a very explicit treatment of the XXX model, see [13]). However, we shall not need this for the goal of these lectures. Rather it will be instructive to describe the illustrative list of examples showing the generality of the QISM.

1.1 The Higher Spin Chain

MCR are satisfied for $L_n(\lambda)$ when for s_n^a we take any representation of the spin commutation relation. The quantum space $\mathbb{C}^{2j+1}$ is now different from the auxiliary space $\mathbb{C}^2$. The integrable Hamiltonian H can be shown to have the form (see [10])

$$H = \sum_n f_j(s_n^a s_{n+1}^a) ,$$

where f_j is some polynomial of degree $2j$.

1.2 Complex Spin

If ψ_n, ψ_n^* are oscillator variables

$$[\psi_n, \psi_m^*] = \delta_{nm} ,$$

the combinations

$$\begin{aligned} s_n^+ &= \psi_n^*(2S - \psi_n^*\psi_n)^{1/2}; \\ s_n^- &= (2S - \psi_n^*\psi_n)^{1/2}\psi_n ; \\ s_n^3 &= \psi_n^*\psi_n - S \end{aligned}$$

satisfy the spin commutation relations for any complex number S. The generalization of the polynomial f_j to be used to define an integrable Hamiltonian is discussed in [14], [15]. In particular it is shown in [15] how to get the Nonlinear Schrödinger model as a scaling limit with $S \to \infty$.

1.3 The Spin 1/2 XXZ Model

With the Hamiltonian

$$H = J\sum_n \left(s_n^1 s_{n+1}^1 + s_n^2 s_{n+1}^2 + \cos\gamma\; s_n^3 s_{n+1}^3\right)$$

one can associate a trigonometric generalization of the L_n-matrix

$$L_n(\lambda) = \begin{pmatrix} \sinh\lambda\cot\gamma + i\cosh\lambda s_n^3 & is_n^- \\ is_n^+ & \sinh\lambda\cot\gamma - i\cosh\lambda s_n^3 \end{pmatrix} .$$

It satisfies MCR with an R-matrix where the coefficients b and c are given by

$$b(\lambda) = \frac{\sinh\lambda}{\sinh(\lambda + i\gamma)} ; \qquad c(\lambda) = \frac{i\sin\gamma}{\sinh(\lambda + i\gamma)} .$$

1.4 The Higher Spin XXZ Model

The same R-matrix can be used for the higher spin case. However, some changes must be introduced in the L-matrix. It can be written in the form

$$L_n(\lambda) = \frac{1}{\sin\gamma}\begin{pmatrix} \sinh(\lambda + i\gamma s_n^3) & i\sin\gamma s_n^- \\ i\sin\gamma s_n^+ & \sinh(\lambda - i\gamma s_n^3) \end{pmatrix} ,$$

but the local dynamical variables s_n^3, $s_n^\pm$ have to satisfy the commutation relations

$$\begin{aligned} [s_n^3, s_n^\pm] &= \pm s_n^\pm ; \\ [s_n^+, s_n^-] &= \frac{\sin(2\gamma s_n^3)}{\sin\gamma} , \end{aligned}$$

which differ from the ordinary spin CR. The introduction of this algebra by Kulish and Reshetikhin in [16] is the second important contribution leading to the notion of quantum Lie algebra.

1.5 The Complex Spin XXZ Model

In terms of dynamical variables φ_n, π_n with the canonical commutation relations

$$[\pi_n, \varphi_m] = -i\gamma\delta_{nm}$$

we can realize the Kulish-Reshetikhin relations as follows:

$$\begin{aligned} s_n^\pm &= \frac{1}{2S\sin\gamma} e^{\pm i\frac{\pi_n}{2}} (1 + 2S^2\cos 2\varphi_n)^{1/2} e^{\pm i\frac{\pi_n}{2}} ; \\ s_n^3 &= \frac{\varphi_n}{\gamma} . \end{aligned}$$

The continuum limit of the corresponding Hamiltonian leads to the quantized Sine-Gordon Model (cf. [6], [15]).

1.6 The Liouville Limit

In the limit $i\varphi \to \infty$ and after the elementary conjugation

$$L_n(\lambda)\sigma_1 \to Q(\lambda) L_n^{\mathcal{L}} Q^{-1}(\lambda) ,$$

where

$$Q(\lambda) = \exp\frac{\lambda}{2}\sigma_3 ,$$

we get the L-matrix

$$L_n^{\mathcal{L}} = \begin{pmatrix} e^{i\pi_n/2}(1 + e^{2\varphi_n})^{1/2} e^{i\pi_n/2} & e^{\varphi_n} \\ e^{\varphi_n} & e^{-i\pi_n/2}(1 + e^{2\varphi_n})^{1/2} e^{-i\pi_n/2} \end{pmatrix} ,$$

which was shown in [17] to be suitable for the investigation of the Liouville model (see also Chapter 3). It satisfies a λ-independent form of the MCR

$$RL_n^1 L_n^2 = L_n^2 L_n^1 R \ ,$$

where

$$R = \begin{pmatrix} q & 0 & 0 & 0 \\ 0 & 1 & 0 & 0 \\ 0 & q - \frac{1}{q} & 1 & 0 \\ 0 & 0 & 0 & q \end{pmatrix}$$

and $q = e^{i\gamma}$. It can be obtained in the limit $\lambda \to \infty$ from the trigonometric R-matrix after the conjugation

$$R(\lambda - \mu) \to Q^1(\lambda)Q^2(\mu)R(\lambda - \mu)(Q^1(\lambda))^{-1}(Q^2)(\mu)^{-1}$$

and multiplication by $e^{i\gamma}$. This matrix satisfies the λ-independent Yang-Baxter relation

$$R_{12}R_{13}R_{23} = R_{23}R_{13}R_{12} \ .$$

The appearance of this algebra is the third important contribution to the definition of a quantum group.

The presented examples were associated with the simple classical Lie-group, namely $sl(2)$. The higher rank examples were also worked out in the eighties. The names of Kulish and Sklyanin, Belavin and Zamolodchikov, Fateev, Cherednic, De-Vega, Viallet and Babelon, Bazhanov, Izegring and Korepin, Karowsky, Jimbo and some others should be mentioned here. However, in our elementary course this development will not be discussed.

2. Quantum Groups

The key word in connection with the introduction of quantum groups is noncommutative geometry. Quite fashionable nowadays, this notion was implicitly used in physics long ago. Indeed, the correspondence principle, appropriately stated, is a very representative example of noncommutative geometry.

The main object entering the description of classical mechanics is the phase space Γ with the canonical variables p, q and the main Poisson bracket $\{p_i, q_k\} = \delta_{ik}$. The algebra of observables $\mathcal{A}$ consisting of the real valued functions $f(p, q)$ on the phase space has two bilinear operations: multiplication fg and the Poisson bracket $\{f, g\}$. It is a commutative associative algebra with respect to the former and a Lie algebra with respect to the latter. Phase space – a geometric object – appears in this algebraic language as the spectrum of $\mathcal{A}$.

In quantum mechanics we substitute the coordinates in Γ by operators P, Q with commutation relations

$$[P_i, Q_k] = \frac{\hbar}{i} I \delta_{ik} \ ,$$

where $\hbar$ is the Planck constant, and define the algebra of observables $\mathcal{A}_\hbar$ as set of functions $f(P,Q)$ where the operator ordering is properly fixed (i.e. à la H.Weyl). Now the algebra ceases to be commutative and can not be represented as an algebra of functions on some manifold. Rather the set of generators can be called coordinates on a noncommutative space, hence we can speak of noncommutative geometry.

In more detail, the correspondence between the symbols $f(p,q)$ and operators $f(P,Q)$ induces an associative product $f * g$ of symbols which can be represented as a formal series in $\hbar$

$$f * g = fg + \hbar\{f,g\}^{(1)} + \hbar^2\{f,g\}^{(2)} + \dots$$

(in the Weyl correspondence $\{f,g\}^{(1)} = \frac{1}{2i}\{f,g\}$). This means that quantization defines a deformation of the classical algebra of observables with $\hbar$ playing the role of a deformation parameter and $\{f,g\}$ being the direction of deformation. The Lie algebra structure

$$\begin{aligned}\{f,g\}_\hbar &= \frac{i}{\hbar}(f * g - g * f) = \\ &= \{f,g\} + \hbar\widetilde{\{f,g\}}^{(1)} + \dots\end{aligned}$$

is also deformed, but now it is not independent, but rather defined in terms of a multiplication $f * g$, which means that some degeneracy is lifted. Indeed, as was shown in [18], the deformed algebra $\mathcal{A}_\hbar$ is stable in a natural sense.

A quantum group gives us another example of a similar deformation. The quantized manifold is that of a Lie group. The structures which survive the deformation correspond to the group multiplication and the notion of an inverse element.

Let us forget for the time being about the latter and stick to the former. As a group manifold we shall take the matrix algebra of complex $n \times n$ matrices. The matrix elements T_{ij} of a matrix $T = \| T_{ij} \|$ can be considered as coordinates of a (commutative) manifold. In the commutative algebra $\mathcal{A}$ of functions $f(T)$ we have the structure Δ of co-multiplications

$$\Delta : \mathcal{A} \to \mathcal{A} \otimes \mathcal{A} ,$$

defined by

$$\Delta f(T) = f(T'T'') ,$$

where $T'T''$ is the product of matrices T' and T''

$$(T'T'')_{ij} = \sum_k T'_{ik} T''_{kj} .$$

In other words, Δ attaches a function of two variables T', T'' to any function of one variable T in such a way that

$$\Delta f \Delta g = \Delta(fg) .$$

In fact, one can reconstruct the matrix multiplication from this law of co-multiplication of functions.

Now we deform (quantize, make non commutative) the multiplication in $\mathcal{A}$ in such a way that co-multiplication survives. This means that the manifold of matrix elements T_{ij} we begin with becomes noncommutative, but matrix multiplication remains intact. Hence the new object can be called quantum matrix algebra (and later quantum group).

To describe the deformation we are looking for we have to impose the commutation relations on the matrix elements T_{ij} in such a way that if two sets T'_{ij} and T''_{ij} satisfy them and commute among each other, then

$$T_{ij} = \sum_k T'_{ik} T''_{kj}$$

also satisfy these relations.

In Chapter 1 we encounter exactly such relations – namely the MCR of local matrices L_n. Following this experience we introduce the commutation relations in a quantum matrix algebra in the form

$$RT^1T^2 = T^2T^1R ,$$

where T^1 and T^2 are $n^2 \times n^2$ matrices constructed in terms of T as follows

$$T^1 = T \otimes I , \qquad T^2 = I \otimes T$$

and R is a $n^2 \times n^2$ matrix of structure constants – an invertible complex matrix. The requirement on the commutation relations is satisfied as was already checked in Chapter 1.

There arises a natural question if the MCR are not too restrictive. In principle the quadratic MCR could produce new higher order relations on the generators T_{ij}. Indeed, let us transform the product $T^1T^2T^3$ into $T^3T^2T^1$ using the MCR. Notations here are evident, we deal with the product $V \otimes V \otimes V$ of three auxiliary spaces $V = \mathbb{C}^n$ and

$$T^1 = T \otimes I \otimes I; \quad T^2 = I \otimes T \otimes I; \quad T^3 = I \otimes I \otimes T.$$

We have two ways to do the rearrangement

$$\begin{array}{ccc} & \nearrow T^1T^3T^2 \ \rightarrow T^3T^1T^2 \searrow & \\ T^1T^2T^3 & & T^3T^2T^1 \\ & \searrow T^2T^1T^3 \ \rightarrow T^2T^3T^1 \nearrow & \end{array}$$

and so get the results

$$\begin{aligned} T^1T^2T^3 &= R^{123}T^3T^2T^1(R^{123})^{-1} = \\ &= R^{321}T^3T^2T^1(R^{321})^{-1} , \end{aligned}$$

where

$$R^{123} = R^{12}R^{13}R^{23}; \quad R^{321} = R^{23}R^{13}R^{12}.$$

So a new cubic relation appears unless we require that

$$R^{123} = R^{321} ,$$

i.e. that R satisfies a (λ independent) Yang-Baxter relation. If we make the YBR to hold, then no new relations will appear in higher orders. Thus it is natural to add the YBR to the definition of the quantum matrix algebra.

We see that the two important formulas of the QISM – namely the MCR and the YBR – get a new interpretation as defining the noncommutative deformation of a matrix algebra. The concrete examples require the solution of the YBR. But the QISM gives them as well. Indeed, at the end of Chapter 1 an explicit one-parameter family of 4×4 R-matrices was provided

$$R_q = \begin{pmatrix} q & 0 & 0 & 0 \\ 0 & 1 & 0 & 0 \\ 0 & q - \frac{1}{q} & 1 & 0 \\ 0 & 0 & 0 & q \end{pmatrix} .$$

In the limit $q \to \infty$ we have

$$R_q \Big|_{q=1} = I$$

and the MCR reduces to the commutativity of T_{ij}. Hence $q = 1$ is the "classical" limit and $q - 1$ is the deformation parameter of this example.

Let us consider this example in more detail. Writing the 2×2 matrix T as

$$T = \begin{pmatrix} a & b \\ c & d \end{pmatrix}$$

we can readily show that the MCR gives 6 independent relations

$$ab = qba \; ; \qquad db = \frac{1}{q} bd \; ;$$

$$ac = qca \; ; \qquad dc = \frac{1}{q} cd \; ;$$

$$bc = cb \; ; \qquad ad - da = (q - \frac{1}{q}) bc \, .$$

Now it is easy to check that the expression

$$\det{}_q T = ad - qbc$$

defines a central element (commuting with a, b, c, d). Putting

$$\det{}_q T = 1$$

we can define

$$T^{-1} = \begin{pmatrix} d & -\frac{1}{q} b \\ -qc & a \end{pmatrix}$$

and check that

$$TT^{-1} = T^{-1} T = I.$$

Hence the quantum group $sl_q(2)$ is defined.

I hope that the main idea of the quantum group is now clear. The natural question of generalization to any classical group is now resolved; I have no time to discuss it and refer to an original article [19]. There it is explained how one can associated a one-parameter family of R-matrices with all A, B, C, D series of classical groups. For the exceptional E, G, F cases information is less detailed. It is instructive to mention that these R-matrices systematically appear as an outcome of investigations of the Toda-lattice models [20], [21] – another example of integrable models, generalizing the Liouville model to be discussed in the following lecture. In fact the example just described appeared for the first time in connection with this last model [17].

Now let us turn to the interpretation of the relations mentioned in Chapter 1 and generalize those of the $sl(2)$ Lie algebra. They can also be put into our general scheme.

Consider two triangular matrices

$$L_+ = \begin{pmatrix} q^{H_+/2} & (q - \frac{1}{q})X_+ \\ 0 & q^{-H_+/2} \end{pmatrix} ;$$

$$L_- = \begin{pmatrix} q^{-H_-/2} & 0 \\ (\frac{1}{q} - q)X_- & q^{H_-/2} \end{pmatrix} ,$$

where for convenience we introduce a composite notation for the generators

$$k_\pm = q^{H_\pm/2}; \qquad k_\pm^{-1} = q^{-H_\pm/2}$$

with evident relations

$$k_\pm k_\pm^{-1} = I .$$

The main commutation relations, imposed separately on L_+ and L_-

$$R^+ L_\pm^1 L_\pm^2 = L_\pm^2 L_\pm^1 R^+ ,$$

where

$$R^+ = PRP ,$$

lead to Weyl-type relations for the generators

$$q^{H_+/2} X_+ = q X_+ q^{H_+/2} ;$$
$$q^{H_-/2} X_- = q^{-1} X_- q^{H_-/2} ,$$

or after (formally) taking the logarithm

$$[H_\pm, X_\pm] = \pm 2 X_\pm .$$

However, if we identify

$$H_+ = H_- = H$$

and impose one more relation

$$R^+ L_+^1 L_-^2 = L_-^2 L_+^1 R^+$$

then the last relation of Chapter 1

$$[X_+, X_-] = \frac{q^H - q^{-H}}{q - \frac{1}{q}} = \frac{\sin \gamma H}{\sin \gamma},$$

where $q = e^{i\gamma}$, will appear.

The algebra $\mathcal{B}$ generated by X_+, X_- and H has a natural co-multiplication Δ, defined on the generators by

$$\Delta(L_\pm) = L_\pm \dot{\otimes} L_\pm \ ,$$

where $\dot{\otimes}$ means matrix multiplication together with tensorizing of two algebras $\mathcal{B}$. If the algebra $\mathcal{A}$ could be considered as a q-deformation of a (commutative) algebra of functions on the Lie group $\mathcal{G}$, then the algebra $\mathcal{B}$ is a deformation of a (noncommutative) algebra of functions on the Lie algebra, in other words, of the universal enveloping algebra $U(\mathcal{G})$.

In the classical limit $q \to 1$ these algebras are dual to each other. Indeed, if we realize the elements of $U(\mathcal{G})$ as differential operators $P(X)$ on $\mathcal{G}$ then the natural pairing looks like

$$< P(X), f(g) >= P(X) f(g)|_{g=e} \ ,$$

where e is the unit element in $\mathcal{G}$.

This duality is preserved by quantization. The deformed pairing on the generators is given by

$$< L^1_\pm, T^2 >= R^\pm \ ,$$

where

$$R^+ = PRP \ ; \qquad R^- = R^{-1}$$

(see [19] for details).

The generalization of this construction to higher rank Lie algebras is also done in [19]. It is clear that the Borel subalgebras play the essential role.

With this formula we terminate the elementary introduction to quantum groups. It is clear that irrespective of its origin in the theory of integrable models, the notion of a quantum group is quite general and must have independent applications. The next two chapters will treat one of them. The most exciting and purely mathematical application already developed is the Theory of Knots (see i.e. [22]).

It is also clear that as soon as this notion appeared, all the mathematical development connected with classical Lie groups and algebras – differential geometry, representation theory etc. could be generalized to its q-analogues. Quite a wealth of papers on this subject already exists. But evidently this activity is still in the beginning.

To finish this chapter let us comment on a closer analogy of the quantization of symplectic manifolds and Lie groups. Let us put

$$q = 1 + i\gamma$$

and consider the main commutation relations up to the first nonvanishing term in γ (analogue of the Planck constant). With the usual identification

$$\{\,,\,\} = \frac{i}{\gamma}[\,,\,]$$

for the Poisson brackets we shall obtain such a bracket on the group manifold given on the coordinate functions (matrix elements T_{ij}) by

$$\{T^1, T^2\} = [r, T^1 T^2] \,,$$

where $[\,,\,]$ is just the matrix commutator and the "structure constants" matrix r is given by

$$R_q = I + i\gamma r + \mathcal{O}(\gamma^2) \,.$$

Thus the quantization of the group is a deformation in the direction of the Poisson bracket, existing on the group manifold – in full analogy with the case of mechanics.

In fact we have two Poisson brackets, the first produced by multiplication and the second by co-multiplication. For a discussion, see [23].

It is this Poisson structure which evidently was an important stimulus for Drinfeld to make his proposal [24], [25], [26]. Jimbo's approach [27] is nearer to the QISM. The exposition in this chapter follows the treatment of [28]. For a completely independent development see [29], [30].

3. Liouville Model

The field-theoretical model with one field $\varphi(x,t)$ in 1+1 dimensional space-time and the classical equation of motion

$$\frac{\partial^2\varphi}{\partial t^2} - \frac{\partial^2\varphi}{\partial x^2} + e^{2\varphi} = 0$$

got a new life recently in connection with string theory. As is well known, the constraints of this theory

$$T_{00} = \frac{1}{2}(\Pi + X')^2 \;; \qquad T_{01} = \Pi X'$$

acquire Schwinger terms after quantization turning them into second class. In the spirit of the general theory of anomalies (see e.g. my previous Schladming lectures [31]) this means that a new degree of freedom appears upon quantization. As was shown by Polyakov [32] this degree of freedom is exactly the Liouville field.

For a closed string space-time is closed with respect to the space-coordinate x, so that periodic boundary conditions are imposed

$$\varphi(x + 2\pi, t) = \varphi(x) \,.$$

In the Hamiltonian approach to the model we begin with the phase-space variables $\varphi(x)$, $\pi(x)$ with the Poisson bracket

$$\{\pi(x), \varphi(y)\} = \gamma\delta(x - y)$$

and the Hamiltonian functional

$$H = \frac{1}{2\gamma}\int_0^{2\pi}\left(\pi^2 + \varphi'^2 + e^{2\varphi}\right) dx ,$$

so that γ plays the role of the coupling constant. In Polyakov's quasiclassical proposal γ is connected with the dimension D of the physical space time as follows

$$\gamma = \frac{26 - D}{6\pi}.$$

The integrability of the model is realized by the change of variables

$$\begin{pmatrix}\varphi\\ \pi\end{pmatrix} \longrightarrow \begin{pmatrix}u\\ v\end{pmatrix}$$

to the new fields with trivial equations of motion

$$\frac{\partial u}{\partial t} = \frac{\partial u}{\partial x} ; \qquad \frac{\partial v}{\partial t} = -\frac{\partial v}{\partial x}$$

and some reasonably simple boundary conditions. One variant of this map can be based on the auxiliary linear problem (note the absence of the spectral parameter)

$$L\phi = \left(\frac{d}{dx} - Q(x)\right)\phi = 0 ,$$

where Q is a 2×2 matrix parametrized by the initial data

$$Q = \begin{pmatrix}\pi & e^{\varphi}\\ e^{\varphi} & -\pi\end{pmatrix} .$$

This matrix is easily seen to be the scaling limit of the Liouville L-matrix $L^{\mathcal{L}}$ presented at the end of Chapter 1. Let $T(x, y)$ be a fundamental matrix solution normalized by $T(x, x) = I$. Then $M = T(2\pi, 0)$ is called the monodromy matrix. The particular form of Q guarantees that $\det T = 1$ and $\operatorname{Tr} T > 2$, so that M is hyperbolic. Using the notation

$$T(x, 0) = T(x) = \begin{pmatrix}A(x) & B(x)\\ C(x) & D(x)\end{pmatrix}$$

we introduce functions $u(x)$ and $v(x)$

$$u(x) = \frac{A(x)}{B(x)} ; \qquad v(x) = \frac{C(x)}{D(x)}$$

with the following properties

$$u'(x) < 0 ; \qquad v'(x) > 0 ;$$
$$u(0) = \infty ; \qquad v(0) = 0 ; \qquad u > v$$

and

$$u(x+2\pi) = M(u(x)) \; ; \qquad v(x+2\pi) = M(v(x)) \; ,$$

where for any f

$$M(f) = \frac{Af+C}{Bf+D} \; ,$$

A, B, C, D being the matrix elements of the monodromy matrix. They are non-trivial observables so that the boundary conditions for u and v are not very transparent.

The Poisson brackets for φ and π lead to the fundamental Poisson bracket relations for the monodromy M

$$\{M^1, M^2\} = [r, M^1 M^2]$$

where the 4×4 matrix r is given by

$$r = \gamma \begin{pmatrix} 0 & 0 & 0 & 0 \\ 0 & -1 & 2 & 0 \\ 0 & 0 & -1 & 0 \\ 0 & 0 & 0 & 0 \end{pmatrix} .$$

We see that the monodromy matrix gives a quasiclassical realization of the quantum group – in the sense of the end of Chapter 2. In fact, the quantum version of this relation, given in [17], was the first example of the appearance of a quantum group in conformal field theory. It follows from an evident relation

$$T(x) = T(x,y)T(y) \; , \qquad x > y$$

and commutativity of the matrix elements of $T(y)$ and $T(x,y)$ (ultralocality) that the following relations are true

$$\begin{aligned}
\{u(x), u(y)\} &= \gamma\epsilon(x-y)(u(x)-u(y))^2 + \gamma(u^2(x) - u^2(y)) \; ; \\
\{v(x), v(y)\} &= -\gamma\epsilon(x-y)(v(x)-v(y))^2 + \gamma(v^2(x) - v^2(y)) \; ; \\
\{u(x), v(y)\} &= 2\gamma(u(x)v(y) - v^2(y)) \; ,
\end{aligned}$$

where $\epsilon(x)$ is a sign function and we confine ourselves to the fixed fundamental domain $0 \le x, y \le 2\pi$. The following chain ansatz

$$\begin{aligned}
u &\to \xi = u' \; ; \\
\xi &\to p = \frac{d}{dx} \ln \xi^{-1/2} = -\frac{1}{2}\frac{u''}{u'} \; ; \\
p &\to s = p^2 + p' = \frac{1}{2}\left(\frac{3}{2}\left(\frac{u''}{u'}\right)^2 - \frac{u'''}{u'}\right) ,
\end{aligned}$$

and analogously for v, is now introduced. The final object (the Schwartz derivative $s[u]$ of u) is known to be invariant under fractional-linear transformations of u and so the simple boundary conditions hold

$$s[u(x+2\pi)] \;=\; s[u(x)] \; ; \qquad s[v(x+2\pi)] \;=\; s[v(x)].$$

It is interesting that the variables introduced in the chain ansatz have beautiful Poisson brackets of their own. With the notations $\xi(x) = \xi[u(x)]$, $\hat{\xi}(x) = \xi[v(x)]$ and so on we have the following list of formulae

$$\{\xi(x), \xi(y)\} = -2\gamma\epsilon(x-y)\xi(x)\xi(y) \; ;$$
$$\{\hat{\xi}(x), \hat{\xi}(y)\} = 2\gamma\epsilon(x-y)\hat{\xi}(x)\hat{\xi}(y) \; ;$$
$$\{\hat{\xi}(x), \xi(y)\} = -2\gamma\hat{\xi}(x)\xi(y) \; ,$$

so that for $\ln \xi$ the Poisson brackets are field independent. Continuing the differentiation we get

$$\{p(x), p(y)\} = \gamma\delta'(x-y) \; ;$$
$$\{\hat{p}(x), \hat{p}(y)\} = -\gamma\delta'(x-y) \; ;$$
$$\{p(x), \hat{p}(y)\} = 0 \; ,$$

so that the p and $\hat{p}$ fields decouple. Finally for s we get the brackets

$$\{s(x), s(y)\} = 2\gamma(s(x) + s(y))\delta'(x-y) + \gamma\delta'''(x-y)$$

and analogous relations for $\hat{s}(x)$, which are characteristic of the Virasoro algebra. Thus the phase space of the Liouville model is essentially the product of two such algebras. This explains why this model is an example of CFT.

It can be shown that the Hamiltonian of our model has a simple expression in terms of $s[x]$ and $s[v]$

$$H = \frac{1}{\gamma}\int_0^{2\pi} (s(x) + \hat{s}(x))dx,$$

leading to the equation of motion

$$\dot{s} = \{H, s\} = s'$$

and similar ones for u and v. This allows us to call u, v the angle action variables.

The inverse map is given by the famous Liouville formula

$$e^{2\varphi} = -\frac{u'v'}{(u-v)^2}$$

with periodic $\varphi(x)$.

The solution of the equations of motion

$$u(x,t) = u(x-t) \; ; \qquad v(x,t) = v(x+t)$$

gives the final expression for the local field $\varphi(x,t)$. This expression can be simplified with the help of one more change of variables and using $e^{-\varphi}$ instead of $e^{2\varphi}$ as a local field. In terms of the functions

$$\psi_1 = \frac{1}{\sqrt{-u'}} \; ; \quad \psi_2 = \frac{u}{\sqrt{-u'}} \; ;$$
$$\chi_1 = \frac{1}{\sqrt{v'}} \; ; \qquad \chi_2 = \frac{v}{\sqrt{v'}}$$

we have

$$e^{-\varphi} = \psi_1\chi_2 - \psi_2\chi_1 = \psi\sigma\chi \ .$$

In the last expression we used the row vector

$$\psi = (\psi_1, \psi_2) \ ,$$

the column vector

$$\chi = \begin{pmatrix} \chi_1 \\ \chi_2 \end{pmatrix} \ ,$$

and the constant matrix

$$\sigma = \begin{pmatrix} 0 & -1 \\ 1 & 0 \end{pmatrix} \ .$$

The functions ψ_i, χ_i , $i = 1, 2$ are solutions of the Sturm-Liouville equations

$$\psi'' + s\psi = 0$$

and

$$\chi'' + \hat{s}\chi = 0 \ .$$

The boundary conditions are described by the monodromy matrix acting linearly

$$\psi(x + 2\pi) = \psi(x)M \ ;$$
$$\chi(x + 2\pi) = \sigma M^{-1}\sigma\chi(x) \ .$$

We can characterize the monodromy in natural geometrical terms: eigenvalues $e^{\pm p}$ and fixed points z_1, z_2. This is realized in the formula

$$M = \frac{1}{z_1 - z_2}\begin{pmatrix} 1 & 1 \\ -z_1 & -z_2 \end{pmatrix}\begin{pmatrix} e^p & 0 \\ 0 & e^{-p} \end{pmatrix}\begin{pmatrix} -z_2 & -1 \\ z_1 & 1 \end{pmatrix} \ .$$

The Poisson brackets for these entries can be calculated after the change of notation

$$z_1 = \kappa e^q \ ; \qquad z_2 = \kappa e^{-q}$$

to be

$$\{p, q\} = \gamma \ ; \qquad \{p, \kappa\} = 0 \ ; \qquad \{q, \kappa\} = 0 \ .$$

The canonical variables p, q are called zero modes to be distinguished from the oscillator type variables entering left and right fields u and v.

In fact these fields can be made "zero-modes free" in one fundamental domain $0 \le x \le 2\pi$. To do it we introduce the fractional-linear transformation [33]

$$(u, v) \ \rightarrow \ (\tilde{u}, \tilde{v})$$

as follows:

$$\tilde{u} = Bu - A; \quad \tilde{v} = \frac{v}{C - Dv} \ .$$

The new fields have the boundary values

$$\tilde{u}(0) = \infty \ ; \quad \tilde{v}(0) = 0 \ ;$$
$$\tilde{u}(2\pi) = 0 \ ; \quad \tilde{v}(2\pi) = \infty \ .$$

The monodromy is now given by

$$u(x+2\pi) = M_+[u(x)] ;$$
$$v(x+2\pi) = M_-[v(x)] ,$$

where

$$M_+ = \begin{pmatrix} 0 & 1 \\ -1 & 2\cosh p \end{pmatrix} ; \qquad M_- = \begin{pmatrix} 2\cosh p & -1 \\ 1 & 0 \end{pmatrix} .$$

In terms of solutions $\tilde{\psi}$ and $\tilde{\chi}$, corresponding to $\tilde{u}, \tilde{v}$ in the same fashion as ψ and χ to u and v we have the representation for the local field $e^{-\varphi}$

$$e^{-\varphi} = \tilde{\psi} S \tilde{\chi} ,$$

where now S depends on the zero-modes p and q;

$$S = \frac{1}{\sinh p} \begin{pmatrix} \cosh q & -\cosh(p+q) \\ \cosh(p-q) & -\cosh q \end{pmatrix} ;$$

on the other hand $\tilde{\psi}$ and $\tilde{\chi}$ do not depend on p and q in our fixed fundamental domain.

More exactly we have the following set of Poisson bracket relations

$$\{\tilde{\psi}^1(x), \tilde{\psi}^2(y)\} = \tilde{\psi}^1(x)\tilde{\psi}^2(y) r^{\pm} ;$$
$$\{\tilde{\chi}^1(x), \tilde{\chi}^2(y)\} = l^{\pm} \tilde{\chi}^1(x)\tilde{\chi}^2(y) ;$$
$$\{\psi^1(x), \tilde{\chi}^2(y)\} = 0 ;$$
$$\{\tilde{\psi}(x), p\} = 0 ; \qquad \{\tilde{\psi}(x), q\} = 0 ;$$
$$\{\tilde{\chi}(x), q\} = 0 ; \qquad \{\tilde{\chi}(x), p\} = 0 ;$$
$$\{\tilde{\psi}(x), \kappa\} = \tilde{\psi}(x) ; \qquad \{\tilde{\chi}(x), \kappa\} = -\tilde{\chi}(x) ;$$
$$\{S^1, S^2\} = -r\, S^1 S^2 + S^1 S^2 l .$$

We used the following notations: $+(-)$ correspond to $x > y$ ($x < y$); furthermore, $\psi^1\psi^2, \chi^1\chi^2, \psi^1\chi^2$ are the 4-row, 4-column and 2×2 matrices constructed in a natural way by building the tensor product, i.e.

$$\psi^1\psi^2 = (\psi_1\psi_1, \psi_1\psi_2, \psi_2\psi_1, \psi_2\psi_2)$$

with a similar convention for the Poisson brackets; finally, the matrices $r^{\pm}$ and $l^{\pm}$ are different guises for the quasi-classical matrix r introduced already

$$r^+ = r ; \qquad r^- = -PrP ; \qquad l^{\pm} = -r^{\mp} .$$

In the commutation relation for the matrix S one can use either r^+ and l^+, or r^- and l^-.

It is instructive to observe that a combination of the matrix r of structure constants of the quantum group enters these relations. The quantum group itself can be considered as a generalized symmetry of the model showing itself after the introduction of the chiral components of the local fields.

Indeed, let t be a generator matrix of the nondynamical quantum group with the quasiclassical relations

$$\{t^1, t^2\} = [r, t^1 t^2] .$$

It is then easy to show that the symmetry is given by the (co)action

$$\tilde{\psi} \to \tilde{\psi} t \; ; \qquad \tilde{\chi} \to t^{-1} \tilde{\chi} \; ;$$

$$S \to t^{-1} S t \; ,$$

which does not change the Poisson relation and the expression for the local field $e^{-\varphi}$.

This role of a quantum group has important implications. In a way one can argue that it substitutes for the dynamics in the model under consideration. We shall discuss it in more detail in the next chapter.

Direct quantization of the picture described is rather simple but it leads to unsatisfactory results in the string theory application. Indeed, the phase space for the zero modes p, q is noncompact: for the hyperbolic monodromy matrix we have values of p and q running through infinite intervals

$$0 \leq p \leq \infty; \quad -\infty \leq q \leq \infty$$

and so the quantum spectrum of p is continuous. The operator p^2 gives a contribution to the mass of string excitations and to have a particle interpretation we can not allow a continuous spectrum. An elliptic monodromy can appear only for a local field with singularities (see e.g. [34], [35]). But for singular solutions we get a negative energy and we are in trouble once more. These were the reasons why the Liouville model was abandoned in the mid-eighties.

Of course the change to a complex field

$$e^{\varphi} \to e^{i\varphi} \; , \qquad \pi \to i\pi$$

turns the monodromy into an elliptic one. The zero modes p, q now run through the (half) torus

$$0 \leq p \leq \pi \; ; \qquad 0 \leq q \leq 2\pi$$

and the semiclassical number of states N is finite

$$N = \frac{\pi \cdot 2\pi}{2\pi\gamma} = \frac{\pi}{\gamma} .$$

As is clear now, this model corresponds to the minimal model of CFT with $c < 1$, c being the central charge of the Virasoro algebra.

The WZNW model considered in the next chapter has an elliptic monodromy in a natural way. We shall discuss the quantization of zero modes in more detail there.

4. The Wess-Zumino-Novikov-Witten Model

The WZNW model is another example of conformal field theory which illustrates once more the main feature of this theory: its phase space is essentially the sum of two copies of some infinite-dimensional Lie algebra. This algebra was the Virasoro algebra in case of the Liouville model; it will be the Kac-Moody algebra in the WZNW case.

The local field of the model is a chiral field $g(x,t)$ with values in some compact finite dimensional group. Following the tradition of these lectures we shall take $SU(2)$ as this group.

The action functional

$$A = \frac{1}{8\gamma}\int t_2(\partial_\mu g g^{-1})^2 dx dt + \frac{1}{12\gamma}\int d^{-1} tr(dg\, g^{-1})^3$$

(the second term being the Wess-Zumino functional) leads to the equation of motion

$$\partial_-(g^{-1}\partial_+ g) = 0 \ .$$

The Hamilton treatment begins with the introduction of the phase space, constituted by fixed-time variables $g(x)$ and $L_0 = \partial_0 g g^{-1}|_{t=\text{fixed}}$, periodic in x

$$g(x+2\pi) = g(x) \ ; \qquad L_0(x+2\pi) = L_0(x) \ .$$

However, the Poisson brackets produced by the action above, can be written down more readily in terms of the variables

$$L_-(x) = \frac{1}{2}(L_0(x) + g'(x)g^{-1}(x)) \ ;$$
$$R_+(x) = \frac{1}{2}(L_0(x) - g^{-1}(x)g'(x)) \ ,$$

where $g' = \frac{d}{dx}g$. We get two Kac-Moody type relations

$$\{L_-^1(x), L_-^2(y)\} = \gamma K\delta'(x-y) + \frac{\gamma}{2}\left[L_-^1(x) - L_-^2(y), K\right]\delta(x-y) \ ;$$
$$\{R_+^1(x), R_+^2(y)\} = -\gamma K\delta'(x-y) + \frac{\gamma}{2}\left[R_+^1(x) - R_+^2(y), K\right]\delta(x-y) \ ,$$

and

$$\{L_-^1(x), R_+^2(y)\} = 0.$$

Here the matrix K is given by

$$K = \sum \sigma^a \otimes \sigma^a,$$

$\sigma^a, a = 1,2,3$ being Pauli matrices.

It is clear that the coupling constant γ defines the corresponding central charge, usually called level l; we have

$$l_{cl} = \frac{\pi}{\gamma} \ .$$

All this was nicely explained by Witten [36] (see also [37]).

The local field $g(x)$ is expressed in terms of the KM currents L_- and R_+ in the following way: introduce the solutions u and v of the equations

$$u' = L_- u \; ; \qquad v' = -vR_+$$

(playing a role analogous to the Sturm-Liouville equations of Chapter 3) with the initial conditions

$$u(0) = I \; ; \qquad v(0) = I \; .$$

Then

$$g(x) = u(x)g(0)v(x) \; ,$$

where $g(0)$ plays the role of the integration constant, lost in the transition

$$(g, L_0) \; \rightarrow \; (L_-, R_+) \; .$$

The solutions u and v are not periodic; let $M_\pm$ be their monodromies

$$u(x+2\pi) = u(x)M_- \; ; \qquad v(x+2\pi) = M_+ v(x) \; .$$

Periodicity of $g(x)$ and the initial conditions for u and v lead to the relation

$$g(0) = M_- g(0) M_+$$

so that M_- is conjugate to M_+^{-1}. Diagonalizing the monodromies

$$M_+ = Z_+ D^{-1} Z_+^{-1} \; ; \qquad M_- = Z_- D Z_-^{-1} \; ,$$

where D is diagonal

$$D = \begin{pmatrix} e^{ip} & 0 \\ 0 & e^{-ip} \end{pmatrix} \; , \qquad 0 \leq p \leq \pi \; ,$$

and introducing a new normalization for the solutions

$$\hat{u} = u Z_- \; ; \qquad \hat{v} = Z_+^{-1} v \; ,$$

we can rewrite the local field in the form

$$g(x) = \hat{u}(x) Q \hat{v}(x) \; ,$$

where

$$Q = Z_-^{-1} g(0) Z_+ \; .$$

The periodicity of $g(x)$ is now expressed by

$$DQD^{-1} = Q$$

so that Q is also diagonal and can be parametrized as

$$Q = \begin{pmatrix} e^{iq} & 0 \\ 0 & e^{-iq} \end{pmatrix} \; , \qquad 0 \leq q \leq 2\pi \; .$$

The variables p and q are the zero-modes of our model. They constitute a canonical pair

$$\{p, q\} = \gamma \ .$$

The variable p enters the fields (chiral components) $\tilde{u}$ and $\tilde{v}$ via the quasi-periodicity conditions

$$\hat{u}(x+2\pi) = \hat{u}(x)D \ ; \qquad \hat{v}(x+2\pi) = D^{-1}\hat{v}(x),$$

and these fields are independent of the variable q,

$$\{\hat{u}(x), p\} = 0 \ ; \qquad \{\hat{v}(x), p\} = 0.$$

The Poisson relations for $\hat{u}$ and $\hat{v}$ themselves are not so transparent. The difficulty in their derivation consists in the absence of ultralocality in the relations for L_- and R_+. This is in contrast to the ultralocality for the matrix-potential Q in the auxiliary linear problem for the Liouville case of Chapter 3. However, the connection of the Liouville and WZNW models via the so called Drinfeld-Sokolov construction [38], properly interpreted [39], can be used to write down the relations looked for and their quantum generalizations. Corresponding results are already published [40] with some indirect motivation. See also recent papers [41] - [44], where a related problem is treated in an alternative way. (Papers [43], [44] in fact treat the Liouville model and their origin is a pioneer paper [45].)

There are several variants of describing these relations. In one, as used in [40], the ingredients are the fields

$$u_F = \hat{u}Q \ ; \qquad v_F = Q^{-1}v$$

in terms of which the local field is given by the same formula

$$g = u_F Q v_F$$

as in the case of $\hat{u}$ and $\hat{v}$. The quantum commutation relations which we shall write from now on, referring to the general experience of the first two chapters, can be written as follows

$$u_F^1(x) u_F^2(y) = u_F^2(y) u_F^1(x) R_F^{\pm}(p) \ ,$$

where the unitary structure matrix $R_F^{\pm}(p)$ depends on the p-variable and is given by the expression

$$R_F^{+}(p) = \begin{pmatrix} e^{i\gamma/2} & 0 & 0 & 0 \\ 0 & a(p) & b(p) & 0 \\ 0 & c(p) & d(p) & 0 \\ 0 & 0 & 0 & e^{i\gamma/2} \end{pmatrix},$$

with

$$a(p) = e^{-i\gamma/2} \frac{(\sin(p+\gamma)\sin(p-\gamma))^{1/2}}{\sin p} \ ;$$

$$b(p) = \frac{\sin\gamma}{\sin p} e^{ip - i\gamma/2} \ ;$$

$$d(p) = a(-p) \ ; \qquad c(p) = b(-p) \ ;$$

furthermore,

$$R_F^- = PR_F^{-1}P,$$

where P is, as always, the permutation matrix.

Compactness of the phase-space for the zero-modes p, q leads to the quantization of the allowed values for p. For the representation of Kac-Moody algebras L_- and R_+ of level l we have

$$\gamma = \frac{\pi}{l+2}$$

(observe the famous quantum correction $l \to l+2$) and eigenvalues of p are given by

$$p = (2j+1)\gamma\ , \qquad j = 0, \frac{1}{2}, \ldots, \frac{l}{2}\ .$$

The eigenvalue $p = 0$ can not be used in R_F due to the singularity of $1/\sin p$.

The matrices $R_F^{\pm}(p)$ are intimately connected with the theory of quantum groups. As was shown first by Reshetikhin and Kirillov [46] they constitute the q-analogues of the 6-j symbols of $SU(2)$ for $q = e^{2i\gamma}$. So the quantized $SU_q(2)$ shows itself in the exchange relations for the quantized chiral components of the local fields.

The inconvenience of the chiral components u_F and v_F is due to their mutual noncommutativity. I shall not discuss the relation connecting $u^1(x)v^2(y)$ and $v^2(y)u^1(x)$. Rather I shall appeal to the experience in connection with the Liouville model to conjecture that there exists a more suitable normalization of the chiral components $\tilde{u}(x)$ and $\tilde{v}(x)$ which does not depend on zero-modes at all (of course, only in one fundamental domain $0 \le x \le 2\pi$)

$$[\tilde{u}, p] = 0\ ; \qquad [\tilde{v}, p] = 0\ ;$$
$$[\tilde{u}, q] = 0\ ; \qquad [\tilde{v}, q] = 0\ ;$$

moreover their exchange relations contain the usual R-matrix of the quantum $sl(2)$ group

$$\tilde{u}^1(x)\tilde{u}^2(y) = \tilde{u}^2(y)u^1(x)R^{\pm}\ ;$$
$$\tilde{v}^1(x)\tilde{v}^2(y) = L^{\pm}\tilde{v}^2(y)\tilde{v}^1(x)\ ;$$

where

$$R^+ = \begin{pmatrix} e^{i\gamma/2} & 0 & 0 & 0 \\ 0 & e^{-i\gamma/2} & e^{i\gamma/2} - e^{-3i\gamma/2} & 0 \\ 0 & 0 & e^{-i\gamma/2} & 0 \\ 0 & 0 & 0 & e^{i\gamma/2} \end{pmatrix},$$

and

$$R^- = P(R^+)^{-1}P\ ; \qquad L^{\pm} = R^{\mp}\ ;$$

moreover $\tilde{u}$ and $\tilde{v}$ commute

$$\tilde{u}^1(x)\tilde{v}^2(y) = \tilde{v}^2(y)\tilde{u}^1(x).$$

The local field $g(x)$ is given in terms of $\tilde{u}$ and $\tilde{v}$ by the formula

$$g(x) = \tilde{u}(x)S(p,q)\tilde{v}(x)\ ,$$

where the zero-modes dependent matrix S is given by

$$S = \left(e^{ip} - e^{-ip}\right)^{-\frac{1}{2}} \begin{pmatrix} e^{ip} & e^{-ip} \\ 1 & 1 \end{pmatrix} \begin{pmatrix} 0 & -e^{iq} \\ e^{-iq} & 0 \end{pmatrix} \begin{pmatrix} e^{ip} & -1 \\ e^{-ip} & 1 \end{pmatrix} \left(e^{ip} - e^{-ip}\right)^{\frac{1}{2}} .$$

Note that zero-modes enter only via the combinations

$$\alpha = e^{ip} ; \qquad \beta = e^{iq} .$$

The Weyl relation

$$\alpha\beta = e^{i\gamma}\beta\alpha$$

generalizing the classical relation between p and q can not be satisfied literally. This is due to the restriction $0 \leq p \leq \pi$ in the classical case.

In the quantum case this restriction can be realized in the following way[1]. The Weyl relation is invariant under the automorphism σ

$$\alpha \to \alpha^{\sigma} = \alpha^{-1}; \quad \beta \to \beta^{\sigma} = \beta^{-1}$$

and we have to introduce a superselection rule: observables are invariant under σ.

To make all this more explicit, let us mention that an irreducible representation of the Weyl relation for $\gamma = \pi/l + 2$ is given in $\mathbb{C}^{2(l+2)}$ by the matrices

$$\alpha = \begin{pmatrix} 1 & & & \\ & e^{i\gamma} & & \\ & & \ddots & \\ & & & e^{i(2l+3)\gamma} \end{pmatrix} ; \qquad \beta = \begin{pmatrix} & & 1 \\ & \iddots & \\ 1 & & \end{pmatrix} ,$$

where only nonzero matrix elements are indicated. The automorphism σ is inner and given by the matrix

$$Z = \begin{pmatrix} 1 & & \\ & & 1 \\ & 1 & \end{pmatrix} .$$

This matrix has $l+3$ eigenvalues 1 and $l+1$ eigenvalues -1. The factor algebra with respect to Z has $l+1$ states in which $\alpha = e^{ip}$ has effectively eigenvalues $\exp[(2j+1)i\gamma]$, $j = 0, 1/2, \ldots l/2$.

This correction does not influence the derivation of the commutation relation for the matrix S

$$RS^1 S^2 = S^2 S^1 L^{-1} ,$$

where R and L enter as R^+ and L^+ or R^- and L^-.

Now it is easy to check the commutativity for the local field $g(x)$. Taking into account commutativity of $\tilde{u}, \tilde{v}$ and S and omitting the x and y dependence we have

[1] Here I deviate from a somewhat premature exposition, presented at the School.

$$
\begin{aligned}
g^1 g^2 &= \tilde{u}^1 S^1 \tilde{v}^1 \tilde{u}^2 S^2 \tilde{v}^2 = \\
&= \tilde{u}^1 \tilde{u}^2 S^1 S^2 \tilde{v}^1 v^2 = \\
&= \tilde{u}^2 \tilde{u}^1 R S^1 S^2 L \tilde{v}^2 \tilde{v}^1 = \\
&= \tilde{u}^2 \tilde{u}^1 S^2 S^1 \tilde{v}^2 \tilde{v}^1 = g^2 g^1 \ .
\end{aligned}
$$

We also see the appearance of the action of a quantum group. If t is a matrix of generators

$$
R t^1 t^2 = t^2 t^1 R \ ,
$$

or

$$
L t^1 t^2 = t^2 t^1 L \ ,
$$

then the action

$$
\tilde{u} \to \tilde{u} t \ ; \qquad \tilde{v} \to t^{-1} \tilde{v} \ ; \qquad S \to t^{-1} S t
$$

does not change the commutation relations of the chiral components $\tilde{u}, \tilde{v}$ (exchange algebra) and the expression for the local field g. So we see that the quantum group invariance in the WZNW model is uncovered after the introduction of the chiral components and is hidden in the original variables.

Of course the main role of the chiral components consists in the simplicity of the solution of the equations of motion

$$
\tilde{u}(x,t) = \tilde{u}(x-t) \ ; \qquad \tilde{v}(x,t) = \tilde{v}(x+t) \ .
$$

So their introduction is indispensible for describing the dynamics and in particular for writing down the Green functions. The quantum-group symmetry is quite convenient in this connection.

The quantum picture already described can be recapitulated as the following scenario (using an irresponsible but fashionable terminology): The full Hilbert space of the quantized WZNW model has the structure

$$
\mathcal{H} = \sum_{j=0}^{l/2} \oplus \left(\mathcal{H}_j^L \otimes \mathcal{H}_j^R \right) .
$$

The currents L_- and R_+, realizing the representations of Kac-Moody algebras of level l, act in the spaces $\mathcal{H}_j^L$ and $\mathcal{H}_j^R$ in an irreducible way and realize the integrable representation of spin j. The operator e^{ip} leaves the components H_j invariant

$$
\begin{aligned}
e^{ip} H_j^L &= e^{(2j+1)\gamma} H_j^L \ ; \\
e^{ip} H_j^R &= e^{(2j+1)\gamma} H_j^R \ .
\end{aligned}
$$

The operator e^{iq} plays the role of an intertwiner

$$
e^{iq} H_j^{R,L} = H_{j+1/2}^{R,L} \ .
$$

We see that the full Hilbert space $\mathcal{H}$ looks like the space for a "regular representation", where each irreducible representation enters in its tensor square.

The Green functions are defined with respect to the state

$$|\mathrm{vac}>=|0>_R \otimes |0>_L \otimes e_0 \ ,$$

where $|0>_{R,L}$ are the vacuum vectors for the representation of the K-M algebras of spin $j=0$ and e_0 is the eigenvector of α

$$e^{ip}e_0 = e^{i\gamma}e_0 \ .$$

So that

$$G(x_1 t_1, x_2, t_2, \ldots, x_n, t_n) =< 0|g^1(x_1,t_1)\ldots g^n(x_n,t_n)|0> .$$

Taking into account the representation of the local field $g(x,t)$ in terms of the commuting chiral components $\tilde{u}(x-t)$ and $\tilde{v}(x+t)$ we have the decomposition of the Green functions

$$\begin{aligned} &G(x_1,t_1,\ldots,x_n,t_n) = \\ &=_L< 0|u^1(x_1-t_1)\ldots u^n(x_n-t_n)|0>_L< e_0|S^1\ldots S^n|e_0> \times \\ &\times_R < 0|v^1(x_1+t_1)\ldots v^n(x_n+t_n)|0>_R \ . \end{aligned}$$

In terms of the complex variables

$$z_n = e^{ix_n+t_n}$$

after the change to the Euclidean time $t \to it$ we get analytic and antianalytic conformal blocks $_L < 0|u^1(z_1)\ldots u^n(z_n)|0>_L$ and $_R < 0|v^1(\bar{z}_1)\ldots v^n(\bar{z}_n)|0>_R$. The matrix combining them into the full Green function is given by the average $< e_0|S^1\ldots S^n|e_0>$ which can be easily calculated. Indeed, due to the quantum group invariance it is just the singlet with respect to the tensorized action of this group. Such singlets are described in [47].

To get the expression for conformal blocks themselves the use of quantum-group related considerations is also quite convenient. Indeed, the exchange algebra relations lead to a particular Riemann-Hilbert problem for the conformal blocks in which one has to factorize the matrix, combined by the product of the R-matrices. A related more complicated problem for the rapidity dependent R-matrix was solved by F. Smirnoff in his study of the form factors in integrable models [48], [49]. Realization of this program is not finished yet and the work is in progress. But even the results already described show how the quantum group enters naturally the subject of CFT in their Hamiltonian formulation. This prospective application stimulates the development of the theory of quantum groups, its representations and construction of relevant invariants. It is clear that the major problem of classification of conformal field models is intimately connected with the theory of quantum groups. With this statement we finish our lectures.

References

1 V.E. Zakharov, L.D. Faddeev: Fun. Anal Appl. **5**, 280 (1971)
2 L.D. Faddeev, V.E. Korepin: Phys. Rev. **C42**, 3 (1978)
3 H.B. Thacker, D. Wilkinson: Phys. Rev. **D19**, 3660 (1979)
4 E.K. Sklyanin, L.D. Faddeev: Sov. Phys. Doklady **23**, 902 (1978)
5 E.K. Sklyanin: Sov. Phys. Doklady **24**, 107 (1979)
6 E.K. Sklyanin, L.A. Takhtajan, L.D. Faddeev: Theor. Math. Phys. **40**, 688 (1980)
7 L.A. Takhtajan, L.D. Faddeev: Russian Math. Surveys **34**, 11 (1979)
8 L.D. Faddeev: Sov. Sci. Rev. Sec. C: Math. Phys. **1**, 107 (1980)
9 H. Thacker: Rev. Mod. Phys. **53**, 253 (1981)
10 E.K. Sklyanin, P.P. Kulish: Lect. Notes Phys. **151**, 61 (1982)
11 A.G. Izergin, V.E. Korepin: Fisika Echaya, (JINR, Dubna) **13**, 501 (1982)
12 L.D. Faddeev: *"Les Houches Lectures 1982"* (Elsevier, Amsterdam 1984)
13 L.D. Takhtajan, L.D. Faddeev: Zap. Seminarov LOMI **109**, 134 (1981)
14 P.P. Kulish, N.Yu. Reshetikhin, E.K. Sklyanin: Lett. Math. Phys. **5**, 393 (1981)
15 V.O. Tarasov, L.A. Takhtajan, L.D. Faddeev: Theor. Math. Phys. **57**, 163 (1983) (in Russian)
16 P.P. Kulish, N.Yu. Reshetikhin: Zap. nauch. seminarov LOMI, **101** 101 (1981) (in Russian); Journ. Sov. Math. **23**, 2435 (1983)
17 L.D. Faddeev, L.A. Takhtajan: Lecture Notes Physics **246**, 166 (1986)
18 F. Bayen, M. Flato, C. Fronsdal, A. Lichnerowitz, D. Sternheimer: Ann. Phys. **111**, 61 (1978); **111**, 111 (1978)
19 N.Yu. Reshetkhin, L.A. Takhtajan, L.D. Faddeev: Algebra and Analysis **1**, 178 (1989)
20 M. Jimbo: Comm. Math. Phys. **102**, 537 (1986)
21 V. Bazhanov: Comm. Math. Phys. **113**, 471 (1987)
22 *Braid Group, Knot Theory and Statistical Mechanics*: ed. M.L. Ge, C.N. Yang (World Scientific 1989)
23 M.A. Semenov-Tian-Shansky: Publ. RIMS Kyoto Univ. **21**, 1237 (1985)
24 V.G. Drinfeld: Doklady AN SSSR **273**, 531 (1983)
25 V.G. Drinfeld: Doklady AN SSSR **283**, 1060 (1985)
26 V.G. Drinfeld: in *Proc. ICM Berkely*, 768 (1986)
27 M. Jimbo: Lett. Math. Phys. **11**, 247 (1986)
28 L.D. Faddeev, N.Yu. Reshetikhin, L.A. Takhtajan:*Algebraic Analysis vol.I* (Academic Press 1988) p.129
29 S. Woronowicz: Comm. Math. Phys. **111**, 613 (1987)
30 S. Woronowicz: Publ. RIMS Kyoto Univ. **23**, 117 (1987)
31 L.D. Faddeev: in *Recent Developments in Mathemtical Physics*: ed. H. Mitter (Proceedings of the XXVI Schladming Winter School, Springer 1987)
32 A.M. Polyakov: Phys. Lett. B **103**, 207 (1981)
33 A.Yu. Vokoff: to be published
34 A.K. Pogrebkov, M.K. Polivanov: Sov. Sci. Rev. Ser. C. Math. Phys. Vol. 3 (1985)
35 R. Marnelius: ITP Göteborg preprints (1982-1985)
36 E. Witten: Comm. Math. Phys. **92**, 455 (1984)
37 A.S. Budagov, N.M. Bogolyubov: Zap. nauch. Seminarov LOMI **146**, 3 (1985)
38 V.G. Drinfeld, V.V. Sokolow: Sovremennye problemy matematiki VINITI USSR **23**, 81 (1984)
39 L.D. Faddeev, A.Yu. Volkoff: to be published
40 L.D. Faddeev: Preprint HU-TFT-89-56; Comm. Math. Phys. (to be published)
41 B. Blok: Phys. Lett. **233B**, 359 (1989)
42 A. Alekseev, S.L. Shatashvili: Preprint EFI-89-67, (Comm. Math. Phys.) to be published
43 J.-L. Gervais: preprint LPTENS 89/14
44 O. Babelon: Phys. Lett. **B215**, 523 (1988)
45 J.-L. Gervais, A. Neveu: Nucl. Phys. B **238**, 125 (1984); **B224**, 329 (1983)
46 A.N. Kirillov, N.Yu. Reshetikhin: Adv. Series in Math. Phys. **7**, 285 (World Scientific 1989)
47 N.Yu. Reshetikhin, F.A. Smirnoff: Harvard preprint, 1989
48 F.A. Smirnoff: J. Phys. A **19**, L575 (1986)
49 F.A. Smirnoff: *Form-factors in Completely Integrable Models* (World Scientific, to be published)

Topics in Planar Physics[1]

R. Jackiw[2]

Department of Physics, Columbia University
538 West 120th Street, New York, NY 10027, U.S.A.

1. Overview

While the evident goal of physics is to explain phenomena in four-dimensional space-time, where physical Nature resides, and perhaps even to explain why Nature resides in four dimensions, the means that we have come to employ in reaching this goal are sufficiently intricate that it has proven useful to make a detour from the direct path, and to wander into lower-dimensional worlds, with the hope that in the simpler setting we can learn useful things about the agreed upon four-dimensional problem. This indeed has happened, initially in two dimensions, where we first encountered spontaneous gauge symmetry breaking, anomalies, the soliton phenomenon, to name three important examples. Moreover, when it was appreciated that there exist physical environments — not in particle physics but in condensed matter and statistical systems — which are properly described by two-dimensional field theories — *e.g.* linear chains whose time evolution gives rise to two-dimensional dynamics, or planar arrays in equilibrium whose static properties are governed by two-dimensional Euclidean field theory — a physical application of the pedagogical investigations could be made, for example to solitons and fractional charge in polyacetylene or to conformally invariant critical phenomena. Additionally, mathematical and speculative uses for two-dimensional field theories were found in the string program.

Thus the foray into two dimensions proved very useful and it was natural to seek a repetition of these successes in three dimensions. My research in this area began in the early 1980's, when conversations with G. 't Hooft during the 1980 Schladming school [1] persuaded me that three-dimensional gauge theories were very interesting and little understood. Three-dimensions provided an unexplored terrain where discoveries could be made, because most physicists were populating the vast expanses in dimensions greater than four.

In my lectures, I shall describe some of the interesting things that we have found in the intervening decade. The subject has become very large, because

1 This work is supported in part by funds provided by the U. S. Department of Energy (D.O.E.) under contract #DE-AC02-76ER03069.

2 On sabbatical leave from the Center for Theoretical Physics, Laboratory for Nuclear Science and Department of Physics, Massachusetts Institute of Technology, Cambridge, MA 02139 U.S.A.

Fields and Particles Editors: H. Mitter · W. Schweiger

many higher- and lower-dimensional colleagues have descended/ascended to three dimensions. Here there is time only for a selection of topics, drawn from the research by my collaborators and by me on geometrical planar models: gauge [2] and gravitational [3] theories. Regrettably I cannot acquaint you with the many interesting results in this area by the Princeton group, [4] nor with the investigations of the Texas group on non-geometrical planar field theories [5].

The reasons for studying planar theories in three-dimensional space-time are pretty much the same as those put forward above for studying two-dimensional models. First, there is the pedagogical motive: there is still much to learn about quantum field theory whose analysis is more accessible in three dimensions than in four; also there are interesting structures to explore that are peculiar to three dimensions [more generally to odd dimensions]. Second, there are possible physical applications: the high-temperature behavior of four-dimensional field theories is governed by their three-dimensional analogs; interesting condensed matter phenomena like the quantum Hall effect and high-T_c superconductivity appear to involve planar gauge theoretic dynamics; motion in the presence of cosmic strings is adequately described by planar gravity. Third, there are mathematical and speculative applications: field theoretic construction of mathematically interesting three-dimensional characteristics and invariants; a fresh perspective on conformal two-dimensional field theories; description of membranes, which for some represent the next step beyond strings.

2. Planar Gauge Theories

2.1 Topologically Massive Gauge Theories

Gauge theoretic dynamics in any dimension can be governed by the Maxwell/Yang–Mills Lagrange density.

$$\mathcal{L}_{\mathrm{YM}} = \frac{1}{2}\,\mathrm{tr}\, F^{\mu\nu}F_{\mu\nu} \tag{2.1.1}$$

$$F_{\mu\nu} = \partial_\mu A_\nu - \partial_\nu A_\mu + [A_\mu, A_\nu] \tag{2.1.2}$$

Here A_μ and $F_{\mu\nu}$, the gauge connection and curvature, are anti-Hermitian matrices that belong to the Lie algebra of the gauge group, which is generated by matrices T^a satisfying

$$[T^a, T^b] = f^{abc}T^c \tag{2.1.3a}$$

The T^a are normalized by

$$\mathrm{tr}\, T^aT^b = -\frac{1}{2}\delta^{ab} \tag{2.1.3b}$$

and provide a basis for expanding A_μ and $F_{\mu\nu}$ in components.

$$A_\mu = A^a_\mu T^a \tag{2.1.4a}$$

$$F_{\mu\nu} = F^a_{\mu\nu} T^a \tag{2.1.4b}$$

Our metric for flat three-dimensional space-time is $\eta_{\mu\nu} = \eta^{\mu\nu} = \mathrm{diag}(1, -1, -1)$.

However, in three dimensions another structure is available that can supplement/replace (2.1.1): the Chern–Simons term.

$$\Omega(A) = -\frac{1}{8\pi^2}\epsilon^{\alpha\beta\gamma}\,\mathrm{tr}\,\left(\partial_\alpha A_\beta A_\gamma + \frac{2}{3}A_\alpha A_\beta A_\gamma\right) \tag{2.1.5}$$

For dimensional balance with (2.1.1) the strength κ with which the Chern–Simons term enters dynamics must have dimensionality of mass [in our units where $\hbar$, c, and the gauge coupling are set to unity]. Thus we are led to consider the Lagrange density [2a]

$$\mathcal{L} = \frac{1}{2}\,\mathrm{tr}\,F^{\mu\nu}F_{\mu\nu} + 8\pi^2\kappa\Omega(A) = \mathcal{L}_{\mathrm{YM}} + \mathcal{L}_{\mathrm{CS}} \tag{2.1.6a}$$

$$\mathcal{L}_{\mathrm{CS}} = -\kappa\epsilon^{\alpha\beta\gamma}\,\mathrm{tr}\,\left(\partial_\alpha A_\beta A_\gamma + \frac{2}{3}A_\alpha A_\beta A_\gamma\right) \tag{2.1.6b}$$

The field equation that follows from (2.1.6) is

$$D_\mu F^{\mu\nu} + \frac{\kappa}{2}\epsilon^{\nu\alpha\beta}F_{\alpha\beta} = 0 \tag{2.1.7}$$

The covariant derivative D_μ acts by differentiation and commutation: $D_\mu \equiv \partial_\mu + [A_\mu,\ \ ]$.

$\mathcal{L}_{\mathrm{CS}}$ is not gauge invariant, but changes under the gauge transformation g.

$$A_\mu \to g^{-1}A_\mu g + g^{-1}\partial_\mu g \tag{2.1.8}$$

$$\mathcal{L}_{\mathrm{CS}} \to \mathcal{L}_{\mathrm{CS}} + \kappa\epsilon^{\alpha\beta\gamma}\partial_\alpha\,\mathrm{tr}\,\left(\partial_\beta g\, g^{-1}A_\gamma\right) + 8\pi^2\kappa w(g) \tag{2.1.9}$$

$$w(g) \equiv \frac{1}{24\pi^2}\epsilon^{\alpha\beta\gamma}\,\mathrm{tr}\,\left(g^{-1}\partial_\alpha g\, g^{-1}\partial_\beta g\, g^{-1}\partial_\gamma g\right) \tag{2.1.10}$$

Since the field equation (2.1.7) is gauge covariant, the change in the Lagrange density must be a total derivative. This is seen explicitly in the next-to-last term of (2.1.9). However, for the last term the identity

$$w(g) = \partial_\alpha w^\alpha(g) \tag{2.1.11}$$

can be established only locally in group space. For example, for the $SU(2)$ gauge group, with g parametrized as $g = e^\lambda$, where λ is in the Lie algebra, one verifies that

$$w^\alpha(g) = \frac{1}{4\pi^2}\epsilon^{\alpha\beta\gamma}\,\mathrm{tr}\,\lambda\partial_\beta\lambda\partial_\gamma\lambda\left(\frac{|\lambda| - \sin|\lambda|}{|\lambda|^3}\right) \tag{2.1.12}$$

$$|\lambda|^2 \equiv -2\,\mathrm{tr}\,\lambda^2 \ .$$

The Chern–Simons action $I_{\mathrm{CS}} = \int d^3x\mathcal{L}_{\mathrm{CS}}$ remains gauge non-invariant: although the next-to-last term in (2.1.9) integrates to zero [for g tending to the identity at infinity and for sufficiently well-behaved vector potentials], we recognize that the integral of the last term is proportional to the winding number $W(g)$ of g.

$$W(g) = \int d^3x\, w(g) = \frac{1}{24\pi^2}\int d^3x\, \epsilon^{\alpha\beta\gamma}\,\mathrm{tr}\,\left(g^{-1}\partial_\alpha g\, g^{-1}\partial_\beta g\, g^{-1}\partial_\gamma g\right) \tag{2.1.13}$$

When the gauge group is compact and non-Abelian, and thus has Π_3 equal to $\mathbb{Z}$, $W(g)$ takes integer values. Hence, gauge invariance of the quantum theory [defined by the functional integral of the phase exponential of the action] requires quantizing the coupling constant κ [2b].

$$\kappa = \frac{n}{4\pi} \;, \qquad n \in \mathbb{Z} \tag{2.1.14}$$

For a canonical, Hamiltonian description, we work in the Weyl $[A_0 = 0]$ gauge, where the canonical variables are $\mathbf{A}$, while the canonically conjugate momenta Π possess a contribution from the Chern-Simons term.

$$\Pi^i = \dot{A}^i + \frac{\kappa}{2}\epsilon^{ij}A^j \tag{2.1.15}$$

The Hamiltonian H, when expressed in terms of the electric and magnetic fields, $\mathbf{E} = -\dot{\mathbf{A}}$, and $B = -\frac{1}{2}\epsilon^{ij}F_{ij}$ respectively, does not see the Chern–Simons coupling.

$$H = -\int_{\mathbf{x}} \mathrm{tr}\,\left(E^2 + B^2\right) \tag{2.1.16}$$

This is a consequence of the topological nature of the Chern–Simons term: even in curved space-time, the generally covariant generalization of $\mathcal{L}_{\mathrm{CS}}$ does not make use of the space-time metric tensor $g_{\mu\nu}$ — in contrast to $\mathcal{L}_{\mathrm{YM}}$. Therefore, when $g_{\mu\nu}$ is varied to produce the energy-momentum tensor $T^{\mu\nu}$ no contribution arises from the Chern–Simons term and $T^{\mu\nu}$ as well as the energy density and the Hamiltonian retain their Yang–Mills form, when expressed in terms of configuration space variables: $\mathbf{A}$ and its derivatives. Of course the Chern–Simons term reappears when H is expressed in canonical variables $\mathbf{A}$ and Π from (2.1.15).

The Hamiltonian equations that follow from (2.1.16) must be supplemented by a subsidiary condition which coincides with the time component of the field equation (2.1.7) *i.e.* Gauss' law.

$$\mathbf{D}\cdot\mathbf{E} - \kappa B = 0 \tag{2.1.17a}$$

In terms of canonical variables this reads

$$\mathbf{D}\cdot\Pi + \frac{\kappa}{2}\nabla\times\mathbf{A} = 0 \tag{2.1.17b}$$

In the quantum theory, (2.1.17) is imposed as a condition on states. This is most transparent in a Schrödinger representation, [2g,h] where states are functionals of the dynamical variable $\mathbf{A}(\mathbf{x})$,

$$|\Psi\rangle \longleftrightarrow \Psi(\mathbf{A}) \tag{2.1.18a}$$

on which the operator $\mathbf{A}$ acts by multiplication,

$$\mathbf{A}^a(\mathbf{x})|\Psi\rangle \longleftrightarrow \mathbf{A}^a(\mathbf{x})\Psi(\mathbf{A}) \tag{2.1.18b}$$

and the canonical momentum operator, by functional differentiation.

$$\Pi^a(\mathbf{x})|\Psi\rangle \longleftrightarrow \frac{1}{i}\,\frac{\delta}{\delta \mathbf{A}^a(\mathbf{x})}\Psi(\mathbf{A}) \tag{2.1.18c}$$

[The Schrödinger representation is at fixed time, hence the time argument of all operators is omitted.] Then (2.1.17) is realized as a functional differential equation that is obeyed by all physical states.

$$\left\{\left(\mathbf{D}\cdot\frac{\delta}{\delta\mathbf{A}}\right)^a + i\frac{\kappa}{2}\nabla\times\mathbf{A}^a\right\}\Psi(\mathbf{A}) = 0 \tag{2.1.19}$$

Gauss' law expresses gauge invariance in a quantum theory: the quantity that must vanish — the left-hand side of (2.1.17) — forms the generator of fixed-time gauge transformations that remain invariances of the theory in the Weyl gauge, and the condition (2.1.19) demands that states be annihilated by this generator. Without the Chern–Simons term, (2.1.19) translates into the statement that physical states are gauge invariant: $\Psi(\mathbf{A}^g) = \Psi(\mathbf{A})$. However, the Chern–Simons term alters the result: states in the quantum theory of (2.1.6) and (2.1.16), satisfying (2.1.19), respond to a gauge transformation with a 1-cocycle [2g].

$$\Psi(\mathbf{A}^g) = e^{2\pi i\alpha_1(\mathbf{A};g)}\Psi(\mathbf{A}) \tag{2.1.20a}$$

$$\alpha_1(\mathbf{A};g) = -\frac{\kappa}{2\pi}\int_{\mathbf{x}}\epsilon^{ij}\,\mathrm{tr}\,\partial_i g\,g^{-1}A^j + 4\pi\kappa\int_{\mathbf{x}} w^0(g) \tag{2.1.20b}$$

Here $w^0(g)$ is given by (2.1.10) and (2.1.11); $\int d^2\mathbf{x}\, w^0(g)$ is globally ill-defined, owing to an integer ambiguity: it changes by $4\pi\kappa W(g) = 4\pi\kappa\times$ (integer) when g is taken through a smooth closed loop of group elements depending on the spatial two-vector $\mathbf{x}$ and on a homotopy parameter. This multivaluedness does not affect the exponentiated form in (2.1.20a), provided κ is quantized according to (2.1.14). Thus we obtain another perspective on (2.1.14): κ must be quantized so that physical states, which necessarily satisfy (2.1.20a), be single-valued.

Note also from (2.1.9) that $\frac{d}{dt}2\pi\alpha_1$ is precisely the change under a gauge transformation of our Lagrangian $L = \int_{\mathbf{x}}\mathcal{L}$; this examplifies a general relation between a 1-cocycle in the action of a symmetry transformation on states and the non-invariance of the Lagrangian against the symmetry transformation in question [2g].

The Abelian theory can be analyzed completely and it is established that κ provides a mass for the excitations. We are dealing with a massive "photon," which nevertheless respects gauge invariance. One may explicitly construct the states of this non-interacting, but nevertheless interesting model. For example, the vacuum state Ψ_0, *i.e.* the lowest eigenstate of

$$H = \frac{1}{2}\int_{\mathbf{x}}\left(\left(i\frac{\delta}{\delta A^i(\mathbf{x})} + \frac{\kappa}{2}\epsilon^{ij}A^j(\mathbf{x})\right)^2 + B^2(\mathbf{x})\right) \tag{2.1.21}$$

is

$$\Psi_0(\mathbf{A}) = \left(\exp i\frac{\kappa}{2}\int BA_L\right)\left(\exp -\frac{1}{2}\int A_T^i\sqrt{-\nabla^2+\kappa^2}\,A_T^i\right) \tag{2.1.22a}$$

where $\mathbf{A}$ has been decomposed into its transverse and longitudinal parts.

$$\mathbf{A} = \nabla A_L + \mathbf{A}_T \tag{2.1.22b}$$

The first factor in (2.1.22a) gives rise to the [Abelian] 1-cocycle after a gauge transformation is performed.

$$\Psi_0\,(\mathbf{A} - \nabla\lambda) = e^{-i\frac{\kappa}{2}\int B\lambda}\Psi_0(\mathbf{A}) \tag{2.1.23}$$

The kernel in the second factor,

$$\left(\sqrt{-\nabla^2 + \kappa^2}\right)(\mathbf{x}, \mathbf{y}) = \int \frac{d^2\mathbf{k}}{(2\pi)^2} e^{-i\mathbf{k}\cdot(\mathbf{x}-\mathbf{y})}\sqrt{k^2 + \kappa^2} \tag{2.1.24}$$

exhibits the massive nature of the excitations. [Higher states are obtained by multiplying Ψ_0 by polynomials in $\mathbf{A}$.]

The spin of the excitation is ± 1, the sign being correlated with the sign of κ.

While the non-Abelian theory cannot be similarly solved, its linear approximation coincides of course with the Abelian model discussed above, and there is no reason to doubt that here too the excitations are massive. For this reason we call the model (2.1.6) a ***topologically massive gauge theory***.

Let us further examine the Chern–Simons modified Gauss law (2.1.17). Specifically in the Abelian case, and also with matter couplings arising from a charge density ρ, (2.1.17a) reads

$$\nabla \cdot \mathbf{E} - \kappa B = \rho \tag{2.1.25}$$

Integrating this over all space gives zero for the integral of $\nabla \cdot \mathbf{E}$, because the gauge invariant electric field is short-range owing to long-distance damping caused by the mass κ; the integral of B is the flux of Φ through the plane and the integral of ρ is the total charge Q. Hence we get

$$\Phi = -\frac{1}{\kappa}Q \tag{2.1.26}$$

This means that particles carrying charge Q also carry magnetic flux $-Q/\kappa$. Since $B = \nabla \times \mathbf{A}$, we further see that the gauge-variant vector potential $\mathbf{A}$ is long-range, so that $\int d^2\mathbf{x}\, \nabla \times \mathbf{A} \neq 0$, while the gauge-invariant magnetic field is short-range, so that $\int d^2\mathbf{x}\, B$ converges. In other words, we are dealing with a vortex-like object [2a,b].

We conclude this discussion of topologically massive gauge theories with the following observations.

(a) The Chern–Simons term violates P and T, and conserves C and PT.

(b) When fermions couple to the gauge field, a Chern–Simons term is induced by fermion radiative corrections [2b,4a,6]. Fermi fields in three-dimensional space-time are described by two-component spinors, and the three "Dirac" matrices can be chosen to be the 2×2 Pauli matrices. A fermion mass term, constructed from these two-component spinors, also violates P and T; indeed it is the supersymmetric partner of the Chern–Simons mass term [7]. Thus it is natural that massive fermions radiatively generate the Chern–Simons term. However, also massless fermions do so, owing to the mass term that is present in Pauli–Villars regularization, which is needed to preserve gauge invariance against "small" gauge transformations with vanishing winding number. The technical mechanism that induces the Chern–Simons term through fermion loops relies on the trace of three Pauli matrices being non-zero, but proportional to the three-index Levi–Civita anti-symmetric epsilon tensor. Moreover, in the non-Abelian theory the coefficient of the Chern–Simons term, induced by a minimal set of fermions, is not properly quantized. To preserve gauge invariance against "large" gauge transformations with non-zero winding number, while retaining a minimal fermion content, it is necessary to include a "bare" Chern–Simons term so that the total coefficient is properly quantized. As a consequence there is no gauge invariant, parity preserving, non-Abelian gauge theory interacting with a minimal set of fermions. This is the "parity anomaly" of three-dimensions — the lower dimensional analog of the four-dimensional chiral anomaly.

(c) It is to be emphasized that there is no topological quantization of κ in the Abelian theory: Π_3 is trivial. This fact may also be established through a gauge invariant formulation of the Abelian model [2c]. Consider the Lagrange density

$$\mathcal{L} = \frac{1}{2}\epsilon^{\alpha\beta\gamma}\partial_\alpha F_\beta F_\gamma - \frac{\kappa}{2}F_\alpha F^\alpha \tag{2.1.27}$$

leading to the field equation,

$$\epsilon^{\mu\alpha\beta}\partial_\alpha F_\beta - \kappa F^\mu = 0 \tag{2.1.28a}$$

which also implies transversality of F^μ.

$$\partial_\mu F^\mu = 0 \tag{2.1.28b}$$

In three dimensions, a vector is dual to an antisymmetric tensor.

$$F^\mu = \frac{1}{2}\epsilon^{\mu\alpha\beta}F_{\alpha\beta} \quad , \quad F^{\alpha\beta} = \epsilon^{\alpha\beta\mu}F_\mu \tag{2.1.29}$$

Substituting (2.1.29) into (2.1.28a) shows that $F_{\mu\nu}$ satisfies [the Abelian version of] (2.1.7). We recognize that $F_{\alpha\beta}$ is just the gauge curvature, F^μ is its dual, both are gauge-invariant. In the absence of dynamical charged matter [external conserved matter currents can be coupled through $j_\mu F^\mu$], there is no need for a gauge-variant vector potential, which, as a consequence of (2.1.28b) and (2.1.29), can be introduced in topologically simple spaces, where a transverse vector can be written as a curl.

$$F^\mu = \epsilon^{\mu\alpha\beta}\partial_\alpha A_\beta \quad , \quad F_{\alpha\beta} = \partial_\alpha A_\beta - \partial_\beta A_\alpha \tag{2.1.30}$$

Of course A_μ is gauge-variant. However, the gauge symmetry acts trivially on (2.1.27) and does not constrain κ.

2.2 Non-Abelian Chern–Simons Gauge Theories

Because Eq. (2.1.26) encapsulates the physically novel and important consequences of the Chern–Simons term, it is natural to consider a truncation where (2.1.26) holds locally in space. This is achieved when the kinetic action for the gauge field is just the Chern–Simons term, with no Maxwell/Yang–Mills term. Such a model can be viewed as the $\kappa \to \infty$ limit of (2.1.6) and (2.1.7); it is a physically meaningful truncation at low energy, or at large distance, where the lower-derivative Chern–Simons term dominates the higher-derivative Maxwell term.

The *Chern–Simons theory* [without matter interactions] is governed by the Lagrange density [8]

$$\mathcal{L}_{\mathrm{CS}} = 8\pi^2 \kappa \Omega(A) = -\kappa \epsilon^{\alpha\beta\gamma} \operatorname{tr}\left(\partial_\alpha A_\beta A_\gamma + \frac{2}{3} A_\alpha A_\beta A_\gamma\right) \tag{2.2.1}$$

with field equation

$$\kappa \epsilon^{\alpha\beta\gamma} F_{\beta\gamma} = 0 \tag{2.2.2}$$

which implies that $F_{\alpha\beta}$ vanishes, and so A_α must be pure gauge, at least locally. Nevertheless, the quantum theory retains non-trivial, interesting features. Of course, the gauge properties of the Chern–Simons theory are the same as those of the topologically massive model; in particular for non-Abelian gauge groups κ is quantized as in (2.1.14).

The canonical formulation in the Weyl gauge begins with (2.2.1) at $A_0 = 0$.

$$\mathcal{L}_{\mathrm{CS}} = \frac{\kappa}{2} \epsilon^{ij} \dot{A}_i^a A_j^a \tag{2.2.3}$$

The Hamiltonian vanishes. The Euler–Lagrange equation which follows from (2.2.3)

$$\dot{A}_i^a = 0 \tag{2.2.4}$$

coincides with the spatial component of (2.2.2) when $A_0 = 0$, while the time component — the Chern–Simons Gauss law — is imposed as a constraint.

$$G^a \equiv -\frac{\kappa}{2} \epsilon^{ij} F_{ij}^a = 0 \tag{2.2.5}$$

In the Weyl gauge, the theory is invariant under static gauge transformations,

$$\delta A_i^a = \partial_i \lambda^a + f^{abc} A_i^b \lambda^c \quad , \tag{2.2.6}$$

which are generated by

$$G = \int_{\mathbf{x}} \lambda^a G^a \quad . \tag{2.2.7}$$

Thus the constraint sets the generator to zero.

$\mathcal{L}_{\mathrm{CS}}$ is first-order in time derivatives, and the quantization of the corresponding symplectic structure leads to non-trivial equal-time commutation relations between vector potentials [2c].

$$[A_i^a(\mathbf{x}), A_j^b(\mathbf{y})] = \frac{i}{\kappa} \epsilon_{ij} \delta^{ab} \delta^2(\mathbf{x} - \mathbf{y}) \tag{2.2.8}$$

Consequently, the generators follow the group's Lie algebra,

$$i\left[G^a(\mathbf{x}), G^b(\mathbf{y})\right] = f^{abc}G^c(\mathbf{x})\delta^2(\mathbf{x}-\mathbf{y}) \tag{2.2.9}$$

and there is no apparent obstruction to demanding that the constraint (2.2.5) be met by requiring that $G^a(\mathbf{x})$ annihilate physical states.

$$G^a(\mathbf{x})|\Psi\rangle = 0 \tag{2.2.10}$$

However, we show that (2.2.10) in fact cannot be satisfied unless κ is quantized. The Gauss law is all there is to this theory; since the Hamiltonian vanishes, (2.2.4) is trivially satisfied. Equation (2.2.10) is most readily analyzed on the Schrödinger representation.

In order to give a Schrödinger representation for the canonical algebra (2.2.8), we have to decide which operator is realized by multiplication, which by [functional] differentiation, and on what function(s) the state functionals depend. This procedure of dividing phase space into "coordinates" and "momenta" is called choosing a ***polarization***.

We choose a ***Cartesian polarization***: the state functionals depend on A_1^a, which we call φ^a, and so A_2^a is realized by functional differentiation with respect to φ^a.

$$|\Psi\rangle \longleftrightarrow \Psi(\varphi) \tag{2.2.11a}$$

$$A_1^a(\mathbf{x})|\Psi\rangle \longleftrightarrow \varphi^a(\mathbf{x})\Psi(\varphi) \ , \tag{2.2.11b}$$

$$A_2^a(\mathbf{x})|\Psi\rangle \longleftrightarrow \frac{1}{i\kappa}\,\frac{\delta}{\delta\varphi^a(\mathbf{x})}\Psi(\varphi) \ . \tag{2.2.11c}$$

Other polarizations are available. For example, A_1^a may be decomposed into longitudinal and transverse parts, which serve as coordinates and momenta. This choice has the advantage of being rotationally invariant, unlike the Cartesian polarization, and also permits a very simple treatment of the Abelian theory — a fact that we shall exploit later. However, non-Abelian gauge transformations are represented very awkwardly in this polarization, because they do not respect the longitudinal/transverse decomposition.

Another possible choice is a holomorphic polarization that uses the non-Hermitian pair: $A^a \equiv \frac{1}{\sqrt{2}}\left(A_1^a + iA_2^a\right)$ and $A^{a*} \equiv \frac{1}{\sqrt{2}}\left(A_1^a - iA_2^a\right)$. This too will be briefly described.

We now show that the action of the gauge group on states is realized with a 1-cocycle [2d]. The exponential of the generator G is the unitary operator $U(g)$ that implements a finite gauge transformation g,

$$\begin{aligned} U(g) &= \exp\left(i\int_{\mathbf{x}}\lambda^a G^a\right) \\ g &= e^{\lambda} \end{aligned} \tag{2.2.12}$$

and according to (2.2.9), the composition law follows that of the group.

$$U(g_1)U(g_2) = U(g_1g_2) \tag{2.2.13}$$

In the Cartesian polarization (2.2.11), the generator is a functional differential operator, acting on functionals of φ,

$$\int_{\mathbf{x}} \lambda^a G^a = - \int_{\mathbf{x}} \lambda^a(\mathbf{x}) \left(\partial_1 \frac{1}{i} \frac{\delta}{\delta\varphi^a(\mathbf{x})} + f^{abc} \varphi^b(\mathbf{x}) \frac{1}{i} \frac{\delta}{\delta\varphi^c(\mathbf{x})} \right)$$

$$- \kappa \int_{\mathbf{x}} \varphi^a(\mathbf{x}) \partial_2 \lambda^a(\mathbf{x}) \qquad (2.2.14)$$

$$\equiv G_\varphi + 2\kappa \int_{\mathbf{x}} \operatorname{tr} (\varphi \partial_2 \lambda) \ .$$

G_φ generates infinitesimal gauge transformations on $\varphi^a = A_1^a$; the last term, needed to generate the transformation on $\delta / i\kappa\delta\varphi^a = A_2^a$, is responsible for the 1-cocycle.

$$\begin{aligned} U(g)\Psi(\varphi) &= e^{iG} e^{-iG_\varphi} \Psi(\varphi^g) \\ \varphi^g &\equiv g^{-1}\varphi g + g^{-1}\partial_1 g \end{aligned} \qquad (2.2.15)$$

The prefactor $e^{iG}e^{-iG_\varphi}$ is evaluated by introducing a homotopy parameter τ, $\tau \in [0,1]$, and solving a differential equation in τ.

$$\begin{aligned} -i\frac{\partial}{\partial\tau} \left(e^{i\tau G} e^{-i\tau G_\varphi} \right) &= \left(e^{i\tau G} e^{-i\tau G_\varphi} \right) \left(2\kappa \int_{\mathbf{x}} \operatorname{tr} \left(\varphi^{g_\tau} \partial_2 \lambda \right) \right) \\ g_\tau &= e^{\tau\lambda} \end{aligned} \qquad (2.2.16)$$

The last factor in (2.2.16) reads

$$\begin{aligned} 2\kappa \int_{\mathbf{x}} \operatorname{tr} (\varphi^{g_\tau} \partial_2 \lambda) &= 2\kappa \int_{\mathbf{x}} \operatorname{tr} (g_\tau^{-1} \varphi g_\tau \partial_2 \lambda) + 2\kappa \int_{\mathbf{x}} \operatorname{tr} (g_\tau^{-1} \partial_1 g_\tau \partial_2 \lambda) \\ =& 2\kappa \int_{\mathbf{x}} \operatorname{tr} (\varphi g_\tau \partial_2 \lambda g_\tau^{-1}) + \kappa \int_{\mathbf{x}} \operatorname{tr} (g_\tau^{-1} \partial_1 g_\tau \partial_2 \lambda + g_\tau^{-1} \partial_2 g_\tau \partial_1 \lambda) \\ &+ \kappa \int_{\mathbf{x}} \epsilon^{ij} \operatorname{tr} (g_\tau^{-1} \partial_i g_\tau \partial_j \lambda) \end{aligned} \qquad (2.2.17)$$

Since $\partial_\tau g_\tau = g_\tau \lambda$, the first term in the last equality is recognized as the τ derivative of $2\kappa \int \operatorname{tr} (\varphi \partial_2 g_\tau g_\tau^{-1})$, and the second as the τ derivative of $\kappa \int_{\mathbf{x}} \operatorname{tr} (g_\tau^{-1} \partial_1 g_\tau g_\tau^{-1} \partial_2 g_\tau)$. The last term in (2.2.17), after an integration by parts and use of (2.1.10) and (2.1.11), is seen to equal

$$\begin{aligned} \kappa \int_{\mathbf{x}} \epsilon^{ij} \operatorname{tr} (g_\tau^{-1} \partial_i g_\tau \partial_j \lambda) &= -\kappa \int_{\mathbf{x}} \operatorname{tr} (g_\tau^{-1} \partial_0 g_\tau g_\tau^{-1} \partial_i g_\tau g_\tau^{-1} \partial_j g_\tau) \\ &= -\frac{\kappa}{3} \int_{\mathbf{x}} \epsilon^{\alpha\beta\gamma} \operatorname{tr} (g_\tau^{-1} \partial_\alpha g_\tau g_\tau^{-1} \partial_\beta g_\tau g_\tau^{-1} \partial_\gamma g_\tau) \\ &= -8\pi^2 \kappa \int_{\mathbf{x}} w(g_\tau) \\ &= -8\pi^2 \kappa \frac{d}{d\tau} \int_{\mathbf{x}} w^0(g_\tau) \ . \end{aligned} \qquad (2.2.18)$$

Thus $e^{iG}e^{-iG_\varphi} = e^{-2\pi i \alpha_1(\varphi, g)}$ where

$$\alpha_1(\varphi, g) = -\frac{\kappa}{2\pi}\int_{\mathbf{x}} \{\text{tr}\ (2\varphi\partial_2 g\, g^{-1} + g^{-1}\partial_1 g\, g^{-1}\partial_2 g)\} + 4\pi\kappa\int_{\mathbf{x}} w^0(g) \tag{2.2.19}$$

and (2.2.15) becomes

$$U(g)\Psi(\varphi) = e^{-2\pi i\alpha_1(\varphi;g)}\Psi(\varphi^g)\ . \tag{2.2.20}$$

The last term in (2.2.19), which appears also in the 1-cocycle of topologically massive gauge theories [see (2.1.20b)], is multivalued for the same reason.

The Gauss law constraint (2.2.10) requires that physical states $\Psi(\varphi)$ be left unchanged by the action of $U(g)$, since the generator annihilates them.

$$U(g)\Psi(\varphi) = \Psi(\varphi) \tag{2.2.21}$$

Therefore in this theory, as in topologically massive gauge theories, functionals describing physical states are not gauge invariant; rather, according to (2.2.20), they satisfy

$$\Psi(\varphi^g) = e^{2\pi i\alpha_1(\varphi;g)}\Psi(\varphi)\ . \tag{2.2.22}$$

Only when $4\pi\kappa$ is an integer can this condition be met with single-valued functionals.

As indicated earlier, it is generally true that when a symmetric theory is described by a Lagrangian that changes by a total time derivative under a finite symmetry transformation, $L \to L + \frac{d}{dt}2\pi\alpha$, the 1-cocycle is just α, evaluated at fixed time. To verify this for the Chern–Simons theory, we must cast the Lagrangian in phase space form: the kinetic term should involve $p\dot{q}$, *i.e.* $\kappa A_2^a \dot{A}_1^a$ rather than $\frac{\kappa}{2}\epsilon^{ij}\dot{A}_i^a A_j^a$. This is achieved by subtracting $\kappa\frac{d}{dt}$ tr $(A_1 A_2)$ from (2.2.3).

$$\tilde{L}_{\text{CS}} = L_{\text{CS}} - \kappa\frac{d}{dt}\int_{\mathbf{x}} \text{tr}\ (A_1 A_2) \tag{2.2.23}$$

Then from (2.1.8), (2.1.9), (2.1.10) and (2.1.11) it follows that $\tilde{L}_{\text{CS}}$ transforms as

$$\begin{aligned}
\tilde{L}_{\text{CS}} &\to \tilde{L}_{\text{CS}} \\
&\quad + \frac{d}{dt}\kappa\int_{\mathbf{x}}\Big\{ \text{tr}\left(\epsilon^{ij}\partial_i g\, g^{-1}A_j - \partial_1 g\, g^{-1}A_2 - A_1\partial_2 g\, g^{-1} - g^{-1}\partial_1 g\, g^{-1}\partial_2 g\right) \\
&\quad + 8\pi^2\omega^0(g)\Big\} \\
&= \tilde{L}_{\text{CS}} + \frac{d}{dt}\kappa\int_{\mathbf{x}}\{-\text{tr}\ (2\varphi\partial_2 g\, g^{-1} + g^{-1}\partial_1 g\, g^{-1}\partial_2 g) + 8\pi^2 w^0(g)\}\ ,
\end{aligned} \tag{2.2.24}$$

in agreement with the general result and with (2.2.19).

We see that the 1-cocycle for the Chern–Simons theory is essentially the same as the one in(2.1.20) for the topologically massive gauge theory, once differences in polarization are taken into account: in the former there is a single field variable $\varphi^a = A_1^a$, while the latter is described by a pair A_i^a, $i = 1, 2$. In particular, the multivalued contribution to each is the same.

Next we construct explicitly states that obey (2.2.22), thus solving the Gauss law constraint. To this end we write

$$\Psi(\varphi) = e^{2\pi i\alpha_0(\varphi)}\psi(\varphi) \tag{2.2.25}$$

and seek a quantity $\alpha_0(\varphi)$, called a *cochain*, that satisfies

$$\alpha_0(\varphi^g) - \alpha_0(\varphi) = \alpha_1(\varphi; g) \tag{2.2.26}$$

Then (2.2.25) solves (2.2.22) with gauge invariant $\psi(\varphi)$.

If Eq. (2.2.26) holds the 1-cocycle α_1 is *trivial* — it is a *coboundary*. It is known that α_1 is non-trivial in *local* cohomology, but a *spatially non-local* functional that trivializes α_1 can be constructed. It is easy to verify that

$$\alpha_0(\varphi) = 4\pi\kappa \int_{\mathbf{x}} w^0(h) - \frac{\kappa}{2\pi} \int_{\mathbf{x}} \operatorname{tr}\,(\varphi h^{-1}\partial_2 h) \quad , \tag{2.2.27}$$

where h is defined by the non-local relation

$$\varphi \equiv h^{-1}\partial_1 h \quad , \tag{2.2.28}$$

solves (2.2.26) and therefore trivialized the cocycle (2.2.19).

It is to be remembered that the multivalued contribution to the trivializing functional (2.2.27) is related to the effective action of chiral fermions coupled to an external gauge field in two [Euclidean] dimensions. This connection arises because the fermion determinant is not gauge invariant; under a gauge transformation its change is related to the Chern–Simons 1-cocycle [9].

The gauge invariant functional $\psi(\varphi)$ in (2.2.25) is formed solely from $\varphi = A_1$. It must be constructed from path-ordered exponential integrals of φ along x^1 at fixed x^2; *e.g.* closed Wilson loops $\Phi(C_{x^1}; x^2) = \operatorname{tr} P \exp \int_{C_{x^1}} dx^1 \varphi(x^1, x^2)$ where P denotes path ordering. [In two-dimensional, sourceless electrodynamics analogous holonomies around closed loops in the single spatial direction are the only surviving degrees of freedom in the quantized theory; they give rise to the vacuum angle and probe a possible vacuum electric field [2h]] Whether such one-dimensional closed loops, or other gauge invariant constructions, exist depends on the topology of the two-dimensional space-like manifold on which the fixed-time canonical formalism is defined. Apart from this gauge-invariant functional, physical states of the non-Abelian Chern–Simons theory are given by

$$\Psi(\varphi) = N\, e^{2\pi i\alpha_0(\varphi)} \tag{2.2.29}$$

where N provides normalization and the other factor is related to the two-dimensional chiral fermion determinant — a non-local functional, as is normal for a physical wave functional. The necessary quantization of κ, (2.1.14), is again evident: from (2.2.25) and (2.2.29) we see that $\alpha_0(\varphi)$ contains the multi-valued term $4\pi\kappa \int_{\mathbf{x}} \omega^0(h)$, which with unquantixed κ would render $\Psi(\varphi)$ in (2.2.25) or (2.2.29) multi-valued.

Since the Hamiltonian vanishes, the physical states (2.2.25) or (2.2.29) solve the Chern–Simons theory for all times. It is instructive to demonstrate explicitly that the vector potential A_i acting on the state (2.2.29) is a pure gauge. To exhibit the action of $A_2^a = \delta/i\kappa\delta\varphi^a$ on $\Psi(\varphi)$ we need $\delta\alpha_0(\varphi)/i\kappa\delta\varphi^a$. This can be found from the definition (2.2.28), which implies $\partial_1(\delta h\, h^{-1}) = h\delta\varphi h^{-1}$, and from the formula (2.2.27) for $2\pi\alpha_0$, which has the consequence that $\delta(2\pi\alpha_0) = -2\kappa\int_{\mathbf{x}} \operatorname{tr}\left\{\partial_2 h\, h^{-1}\partial_1(\delta h\, h^{-1})\right\}$. Thus

$$\begin{aligned} A_2(\mathbf{x})\Psi(\varphi) &= \frac{2\pi}{\kappa} T^a \frac{\delta\alpha_0(\varphi)}{\delta\varphi^a(\mathbf{x})}\Psi(\varphi) \\ &= h^{-1}(\mathbf{x})\partial_2 h(\mathbf{x})\Psi(\varphi) \ . \end{aligned} \tag{2.2.30a}$$

Together with (2.2.28)

$$A_1(\mathbf{x})\Psi(\varphi) = h^{-1}(\mathbf{x})\partial_1 h(\mathbf{x})\Psi(\varphi) \tag{2.2.30b}$$

the desired result is obtained.

$$A_i\Psi(\varphi) = h^{-1}\partial_i h\Psi(\varphi) \tag{2.2.30c}$$

The wave-functional (2.2.29) has φ-independent norm $|N|^2$, hence it cannot be normlized by [funcitonal] integration over φ. This is to be expected of states that satisfy Gauss' law, because the Gauss law operator has a continuous spectrum. The resolution of course is that the integration measure $\mathcal{D}\varphi$ must be gauge fixed — $\delta(\varphi)$ is a natural choice leading to trivial integrals.

Note that it is possible to formulate the theory in terms of a gauge invariant, spatially non-local Lagrangian. From (2.2.24) it follows that the gauge transform of $\tilde{L}_{\mathrm{CS}}(A)$ is

$$\tilde{L}_{\mathrm{CS}}(A^g) = \tilde{L}_{\mathrm{CS}}(A) + \frac{d}{dt}2\pi\alpha_1(\varphi; g) \tag{2.2.31a}$$

With (2.2.26), this can be presented as

$$\tilde{L}_{\mathrm{CS}}(A^g) - \frac{d}{dt}2\pi\alpha_0(\varphi^g) = \tilde{L}_{\mathrm{CS}}(A) - \frac{d}{dt}2\pi\alpha_0(\varphi) \tag{2.2.31b}$$

This means that the equivalent Lagrangian

$$\begin{aligned} L_{\mathrm{CS}}^{\mathrm{invariant}}(A) &\equiv \tilde{L}_{\mathrm{CS}}(A) - \frac{d}{dt}2\pi\alpha_0(\varphi) \\ &= \tilde{L}_{\mathrm{CS}}(A) - 2\pi\int_{\mathbf{x}} \dot{\varphi}^a \frac{\delta\alpha_0(\varphi)}{\delta\varphi^a} \\ &= -2\kappa\int_{\mathbf{x}} \operatorname{tr}\left(A_2 - h^{-1}\partial_2 h\right)\dot{A}_1 \end{aligned} \tag{2.2.31c}$$

is invariant against time-independent gauge transformations ($h^g \equiv hg$). In (2.2.31c), $h^{-1}\partial_2 h$ is the non-local functional of $\varphi = A_1$, defined by (2.2.28); $A_2 - h^{-1}\partial_2 h$ does *not* vanish here, only when acting on physical states.

The same results may be presented in the holomorphic representation, wherein states are functionals of a complex function $\mathcal{A}^a$ and $A^{a*} \equiv \frac{1}{\sqrt{2}}(A_1^a$

$-iA_2^a)$ is realized by multiplication by $\mathcal{A}^a$ while $A^a \equiv \frac{1}{\sqrt{2}}(A_1^a + iA_2^a)$ acts by functional differentiation.

$$|\Psi\rangle \longrightarrow \Psi(\mathcal{A}) \tag{2.2.32a}$$

$$A^{a*}(\mathbf{x})|\Psi\rangle \longrightarrow \mathcal{A}^a(\mathbf{x})\Psi(\mathcal{A}) \tag{2.2.32b}$$

$$A^a(\mathbf{x})|\Psi\rangle \longrightarrow \frac{\delta}{\delta \mathcal{A}^a(\mathbf{z})}\Psi(\mathcal{A}) \tag{2.2.32c}$$

This action reproduces the commutator between A and A^*.

$$\left[A^a(\mathbf{x}), A^{*b}(\mathbf{y})\right] = \frac{1}{\kappa}\delta^{ab}\delta(\mathbf{x}-\mathbf{y}) \tag{2.2.33}$$

The adjoint relationship between the two operators is maintained, provided inner products involve a non-trivial measure in the functional integral

$$\langle \Psi_1|\Psi_2\rangle = \int \mathcal{D}\mathcal{A}\mathcal{D}\mathcal{A}^*\, e^{-\kappa \int_{\mathbf{x}} \mathcal{A}^{c*}(\mathbf{x})\mathcal{A}^c(\mathbf{x})}\Psi_1^*(\mathcal{A})\Psi_2(\mathcal{A}) \tag{2.2.34a}$$

where

$$\mathcal{D}\mathcal{A}\mathcal{D}\mathcal{A}^* = \mathcal{D}A_1^a\,\mathcal{D}A_2^a/\det(2\pi i) \tag{2.2.34b}$$

A development paralleling the previous discussion in the Cartesian polarization results in a wave functional in the holomorphic polarization analogous to (2.2.27) and (2.2.28).

$$\Psi(\mathcal{A}) = N\, e^{2\pi i \alpha_0(\mathcal{A})} \tag{2.2.35a}$$

$$\alpha_0(\mathcal{A}) = \frac{i\kappa}{2\pi}\int_{\mathbf{x}} \operatorname{tr}\,(\mathcal{A}h^{-1}\partial_+ h) + 4\pi\kappa\int_{\mathbf{x}}\omega^0(h) \tag{2.2.35b}$$

Here h is defined through

$$\mathcal{A} \equiv h^{-1}\partial_- h \tag{2.2.36}$$

and $\partial_\pm \equiv \frac{1}{\sqrt{2}}(\partial_1 \pm i\partial_2)$. The multivalued phase is of course encountered once again, while the integration measure $\exp -\kappa \int_{\mathbf{x}} \mathcal{A}^{c*}\mathcal{A}^c$ insures convergence of the functional integrals — no further gauge fixing is needed.

We conclude this presentation of the quantum Chern–Simons theory by noting that the Gauss law constraint is here solved after quantization. One may alternatively solve it classically, and quantize the remaining degrees of freedom. In general the two procedures do not commute, as is seen from the following example [2d]

Consider the quantum mechanics for planar motion of a point particle described by the two-vector $\mathbf{q} = (q^1, q^2)$. The Lagrangian

$$L = \frac{1}{2}\dot{\mathbf{q}}^2 - V(q) \tag{2.2.37}$$

is invariant under rotations through the angle θ,

$$\delta q^i = -\theta\epsilon^{ij}q^j \tag{2.2.38}$$

when $V(q)$ depends only on the magnitude of $\mathbf{q}$. Here q^i is a function of time t, and the overdot denotes differentiation with respect to t. However, θ is time-independent — the rotation in (2.2.38) is "global." In the usual way, one knows that invariance under (2.2.38) implies conservation of angular momentum $J = \mathbf{q} \times \mathbf{p}$, where $\mathbf{p} \equiv \frac{\partial L}{\partial \dot{\mathbf{q}}} = \dot{\mathbf{q}}$, and the familiar Hamiltonian operator for fixed angular momentum ℓ is

$$H_\ell = -\frac{\hbar^2}{2}\frac{\partial^2}{\partial r^2} + \frac{\hbar^2\left(\ell^2 - \frac{1}{4}\right)}{2r^2} + V(r) \ , \qquad r \geq 0 \ . \tag{2.2.39}$$

[We temporarily restore Planck's constant.] One may promote the global symmetry under rotations (2.2.38) to a "local" gauge symmetry by introducing into (2.2.37) a "gauge potential" $a(t)$,

$$L_a = \frac{1}{2}\left(\dot{q}^i + a\epsilon^{ij}q^j\right)\left(\dot{q}^i + a\epsilon^{ik}q^k\right) - V(q) \tag{2.2.40}$$

that transforms under time-dependent rotations.

$$\delta a = \dot{\theta} \tag{2.2.41}$$

In the Weyl gauge, $a = 0$, L_a coincides with (2.2.37), but now there is an additional constraint that captures the Lagrangian equation of motion obtained from (2.2.40) by varying a: the rotation generator J must annihilate physical states, which therefore are only s-states. If the constraint is solved classically, the classical Hamiltonian for rotationally invariant motion with vanishing angular momentum is

$$H_{\ell=0}^{\text{classical}} = \frac{p_r^2}{2} + V(r) \ , \tag{2.2.42a}$$

whose quantized form, obtained with the naive replacement $p_r^2 \to -\hbar^2\partial_r^2$,

$$H_{\ell=0}^{\text{classical}} \longrightarrow -\frac{\hbar^2}{2}\frac{\partial^2}{\partial r^2} + V(r) \tag{2.2.42b}$$

does not reproduce the quantum s-wave Hamiltonian that survives from (2.2.39) if the constraint is imposed *after* quantization.

$$H_{\ell=0} = -\frac{\hbar^2}{2}\frac{\partial^2}{\partial r^2} - \frac{\hbar^2}{8r^2} + V(r) \tag{2.2.43}$$

The $\mathcal{O}(\hbar^2)$ difference between (2.2.42b) and (2.2.43) can be viewed as an ordering ambiguity *i.e.* (2.2.43) follows from (2.2.42a) if p_r^2 is taken to be $-\frac{\hbar^2}{\sqrt{r}}\partial_r r \partial_r \frac{1}{\sqrt{r}}$, but without further information there is no way to justify the "correct" choice. No such ambiguity arises if one imposes the constraint after quantization. [Henceforth we return $\hbar$ to unity.]

This is not to say that there will always be a discrepancy when phase space is reduced before or after quantization. For example, in many-body quantum mechanics, the passage to the center-of-mass rest frame [which may be formulated as a gauge principle for translations [2g]] produces the same quantum theory whether it is carried out before or after quantization. Similarly, enforcing the Gauss law in quantum electrodynamics does not involve ordering ambiguities. It is therefore surprising that we find non-commutativity of quantization and phase-space reduction already for the Abelian Chern–Simons theory in flat space, which I shall now describe.

2.3 Abelian Chern–Simons Gauge Theory with Sources

We begin with the Lagrange density

$$\mathcal{L}_{\text{CS}} = \frac{\kappa}{2}\epsilon^{\alpha\beta\gamma}\partial_\alpha A_\beta A_\gamma - A_\mu j^\mu \tag{2.3.1}$$

where $j^\mu = (\rho, \mathbf{j})$ is the conserved matter current with time-independent charge.

$$Q = \int_{\mathbf{x}} \rho(t, \mathbf{x}) \tag{2.3.2}$$

Here the fields are real functions, and the coupling constant is absorbed in the definition of the current j^μ. We leave the matter Lagrangian unspecified; indeed we take the current to be an external, conserved, c-number source. The previous canonical development holds in this theory, except that the Hamiltonian does not vanish,

$$H = \int_{\mathbf{x}} A_i j^i \tag{2.3.3}$$

and the Gauss law constraint acquires an inhomogeneous term.

$$\epsilon^{ij}\partial_i A_j = -B = \frac{1}{\kappa}\rho \tag{2.3.4}$$

This constraint on the magnetic field B implies that particles with charge Q are also flux-tubes for A-flux, $\Phi = -Q/\kappa$, and leads to exotic statistics and angular momentum of the charge- and flux-carrying particles [10].

In contrast to the non-Abelian theory, here the gauge field contribution to the constraint (2.3.4) is linear, and may be chosen to be the momentum conjugate to a coordinate θ. This is achieved by decomposing A_i into its longitudinal and transverse components,

$$\begin{aligned} A_i &= \partial_i\theta + \epsilon^{ij}\partial_j^{-1}B \\ \partial_j^{-1} &\equiv \partial_j/\nabla^2 \end{aligned} \tag{2.3.5}$$

The decomposition (2.3.5) is unique and well-defined provided there are no zero modes of the two-dimensional Laplacian ∇^2. This we assume here; indeed, we consider space-time to be Minkowskian.

The commutation relation (2.2.8) implies that B and θ form a canonical pair.

$$[\theta(\mathbf{x}), B(\mathbf{y})] = \frac{i}{\kappa}\delta^2(\mathbf{x} - \mathbf{y}) \tag{2.3.6}$$

In the Schrödinger representation we realize B as a functional derivative with respect to the coordinate, θ, $B = \delta/i\kappa\delta\theta$. This is the ***rotationally invariant polarization***.

The constraint (2.3.4) reduces to

$$\left[\frac{1}{i}\frac{\delta}{\delta\theta(\mathbf{x})}+\rho(\mathbf{x})\right]\Psi=0\ , \tag{2.3.7}$$

and is solved by

$$\Psi(\theta;t)=N(t)\exp\left[-i\int_{\mathbf{x}}\rho\theta\right]\ , \tag{2.3.8}$$

where $N(t)$ is a θ-independent, but possibly time-dependent normalization factor. $N(t)$ is then determined by requiring that $\Psi(\theta;t)$ satisfies the time-dependent Schrödinger equation.

$$i\partial_t\Psi(\theta;t)=\left[\int_{\mathbf{x}}j^iA_i\right]\Psi(\theta;t) \tag{2.3.9}$$

Inserting the decomposition (2.3.5) and using the continuity equation for the matter current, we find

$$N(t)=\exp-\frac{i}{\kappa}\int_0^t dt'\int_{\mathbf{x}}\rho(t',\mathbf{x})\jmath(t',\mathbf{x})\ , \tag{2.3.10}$$

where we have written j^i in terms of its longitudinal and transverse parts.

$$j^i=-\partial_i^{-1}\dot{\rho}+\epsilon^{ij}\partial_j\jmath \tag{2.3.11}$$

$N(t)$ is normalized to 1 at $t=0$. Note that for static matter the state Ψ is an eigenstate of the Hamiltonian with energy eigenvalue

$$E=\frac{1}{\kappa}\int_{\mathbf{x}}\rho\jmath\ . \tag{2.3.12}$$

We see that the theory admits a unique physical state, which in the absence of sources is described by the wave functional $\Psi=1$.

Of course, the above development can alternatively be presented in the Cartesian polarization $A_1=\varphi$, $A_2=\delta/i\kappa\delta\varphi$, which we used for the non-Abelian theory. In the absence of external sources, the unique physical state that satisfies Gauss's law is

$$\Psi(\varphi)=N\exp\left[\frac{i\kappa}{2}\int\varphi\frac{\partial_2}{\partial_1}\varphi\right]\ , \tag{2.3.13}$$

in agreement with (2.2.25), (2.2.27) and (2.2.29). Ψ responds to a gauge transformation by

$$\begin{aligned}\Psi\left(\varphi^g\right)&=\Psi\left(\varphi+\partial_1\lambda\right)=e^{i2\pi\alpha_1(\varphi;\lambda)}\Psi(\varphi)\ ,\\ \alpha_1\left(\varphi;\lambda\right)&=\frac{\kappa}{2\pi}\int_{\mathbf{x}}\left[\varphi\partial_2\lambda+\frac{1}{2}\partial_1\lambda\partial_2\lambda\right]\ ,\end{aligned} \tag{2.3.14}$$

in agreement with (2.2.22).

The unitary transformation functional that connects the Cartesian polarization to the rotationally invariant one is

$$\langle\theta|\varphi\rangle = \det{}^{1/2}\left(\frac{\kappa}{2\pi}\frac{\Delta}{\partial_2}\right)$$
$$\times \exp\left[\frac{i\kappa}{2}\int\left\{\left(\partial_i\theta - \frac{\partial_i}{\partial_1}\varphi\right)\frac{\partial_1}{\partial_2}\left(\partial_i\theta - \frac{\partial_i}{\partial_1}\varphi\right) - \varphi\frac{\partial_2}{\partial_1}\varphi\right\}\right] \tag{2.3.15}$$
$$\Psi(\theta) = \langle\theta|\Psi\rangle = 1 \ , \qquad \Psi(\varphi) = \langle\varphi|\Psi\rangle = N\exp\left[\frac{i\kappa}{2}\int\varphi\frac{\partial_2}{\partial_1}\varphi\right]$$

In the rotationally invariant polarization the physical state, in the absence of external sources, $\Psi = 1$, is obviously gauge invariant. This fact realizes the gauge invariant formulation described previously: in the rotationally invariant polarization the Lagrangian, apart from a total time derivative, is invariant against time-independent gauge transformations.

$$\tilde{L}_{\mathrm{CS}} = \kappa\int_{\mathbf{x}} B\dot{\theta} - \frac{\kappa}{2}\frac{d}{dt}\int_{\mathbf{x}} B\theta \tag{2.3.16a}$$

$$L_{\mathrm{CS}}^{\mathrm{invariant}} = \kappa\int_{\mathbf{x}} B\dot{\theta} \tag{2.3.16b}$$

Finally in the holomorphic polarization (2.2.32) the state of the Abelian theory is

$$\Psi(\mathcal{A}) = N\exp\frac{\kappa}{2}\int\mathcal{A}\frac{\partial_+}{\partial_-}\mathcal{A} \tag{2.3.17}$$

We have already remarked that the Chern–Simons theory may be viewed as the $\kappa \to \infty$ limit of the topologically massive model. It is interesting to examine in detail how one theory passes to the other, and this we can do explicitly for the wave functionals of the Abelian theory. For $\mathcal{L}$ of (2.1.6a) to pass into $\mathcal{L}_{\mathrm{CS}}$ of (2.1.6b) it suffices [in the Abelian case] to rescale the potential by $\sqrt{\frac{\kappa'}{\kappa}}$ and set κ to infinity, with κ' remaining as the coefficient of the Chern–Simons Lagrange density. Performing this limit on the ground state wave functional (2.1.22a) leaves

$$\Psi_0(\mathbf{A}) \underset{\kappa\to\infty}{\longrightarrow} \left(\exp i\frac{\kappa'}{2}\int BA_L\right)\left(\exp -\frac{\kappa'}{2}\int \mathbf{A}_T\cdot\mathbf{A}_T\right) \tag{2.3.18a}$$

In terms of complex variables $\mathcal{A} \equiv \frac{1}{\sqrt{2}}(A_1 - iA_2)$ and $\mathcal{A}^* = \frac{1}{\sqrt{2}}(A_1 + iA_2)$, (2.3.18a) reads

$$\Psi_0(\mathbf{A}) \underset{\kappa\to\infty}{\longrightarrow} \exp\left(\frac{\kappa'}{2}\int\mathcal{A}\frac{\partial_+}{\partial_-}\mathcal{A}\right)\exp -\frac{\kappa'}{2}\int\mathcal{A}^*\mathcal{A} \tag{2.3.18b}$$

Comparison with (2.3.17) shows that in the limit, the ground state wave functional of the topologically massive theory tends to the unique state of the Chern–Simons theory, times the square root of the measure factor, so that the probability measures correctly pass into each other [2e]. Other details on this limit can be found in the literature [2e].

2.4 Quantum Holonomy

In the Abelian Chern–Simons theory defined on the topologically trivial plane there is very little structure. Indeed when the constraints are solved before quantization, there is no structure at all if there are no sources. However, by quantizing first, and computing the quantum holonomy around closed loops at fixed time, we encounter non-trivial results that show an explicit difference between solving the constraints before and after quantization [2d].

The holonomy operator $\Phi(C)$ is defined by the parallel transport equation around a closed planar loop C parametrized by $\mathbf{x}(\tau)$ for $\tau \in [0,1]$, with $\mathbf{x}(0) = \mathbf{x}(1) \equiv \mathbf{x}_0$, which serves as a marked point on the loop, where we also specify the initial and final unit tangent vectors, $\hat{\mathbf{v}}_0 = \dot{\mathbf{x}}(0)/|\dot{\mathbf{x}}(0)|$ and $\hat{\mathbf{v}}_1 = \dot{\mathbf{x}}(1)/|\dot{\mathbf{x}}(1)|$, respectively. The marked point is on a smooth segment of the loop if $\hat{\mathbf{v}}_0 = \hat{\mathbf{v}}_1$; otherwise it is at a cusp with opening angle $\pi \mp \cos^{-1} \hat{\mathbf{v}}_0 \cdot \hat{\mathbf{v}}_1$, where $\mp$ refer to opening angles $< \pi$ and $\geq \pi$, respectively, and $0 \leq \cos^{-1} \hat{\mathbf{v}}_0 \cdot \hat{\mathbf{v}}_1 < 2\pi$.

The equation for the holonomy

$$[i\partial_\tau + V(\tau)]\Phi(C) = 0 \ , \quad V(\tau) \equiv \dot{x}^i(\tau) A_i(\mathbf{x}(\tau)) \tag{2.4.1}$$

is solved at the classical level by the Aharonov–Bohm phase.

$$\Phi^{\text{classical}} = \exp\left[i \int_0^1 d\tau\, V(\tau)\right] = \exp\left[i \int_C dx^i\, A_i\right] \tag{2.4.2}$$

The constraint (2.3.4) in the theory without sources forces A_i to be a pure gauge, and therefore $\Phi^{\text{classical}} = 1$, independent of the loop C.

To solve Eq. (2.4.1) at the quantum level we must recall that $[A_1(\mathbf{x})], A_2(\mathbf{x})] \neq 0$, therefore $[V(\tau), V(\tau')] \neq 0$, and the quantum holonomy operator is given by a path-ordered expression.

$$\Phi(C) = P \exp\left[i \int_0^1 d\tau\, V(\tau)\right] \tag{2.4.3}$$

To determine the action of $\Phi(C)$ on states we first need to undo the path ordering. Since the commutator $[V(\tau), V(\tau')]$ is a c-number, this yields

$$\begin{aligned}\Phi(C) &= \exp\left[-\frac{1}{2}\int_0^1 d\tau \int_0^\tau d\tau' \,[V(\tau), V(\tau')]\right] \exp\left[i\int_0^1 d\tau\, V(\tau)\right] \\ &= \exp\left[-\frac{i}{2\kappa}\int_0^1 d\tau \int_0^\tau d\tau'\, \dot{x}^i(\tau)\epsilon^{ij}\dot{x}^j(\tau')\delta^2\left(\mathbf{x}(\tau) - \mathbf{x}(\tau')\right)\right] \\ &\times \exp\left[i\int_0^1 d\tau\, \dot{x}^i(\tau) A_i\left(\mathbf{x}(\tau)\right)\right] \ .\end{aligned} \tag{2.4.4}$$

In our chosen polarization, $\Phi(C)$ acts on functionals of the "coordinate" θ, and it is therefore convenient to reorganize (2.4.4) so that the "momentum" B stands on the right. To this end, we split the operator $V(\tau)$ into a self-commuting pair

$$
\begin{aligned}
V(\tau) &= V_1(\tau) + V_2(\tau) \\
[V_1(\tau), V_1(\tau')] &= [V_2(\tau), V_2(\tau')] = 0 \\
[V_1(\tau), V_2(\tau')] &= c\text{-number}
\end{aligned}
\tag{2.4.5}
$$

Splitting the operator-valued exponential in (2.4.4) gives an additional phase, which combines with the first to yield

$$
\begin{aligned}
\Phi(C) &= \exp[i\gamma] \exp\left[i \int_0^1 d\tau\, V_1(\tau)\right] \exp\left[i \int_0^1 d\tau\, V_2(\tau)\right] \\
\gamma &= i \int_0^1 d\tau \int_0^\tau d\tau' \, [V_1(\tau), V_2(\tau')]
\end{aligned}
\tag{2.4.6}
$$

With our polarization

$$
\begin{aligned}
V_1(\tau) &= \dot{x}^i(\tau) \partial_i \theta\,(\mathbf{x}(\tau)) = \frac{d}{d\tau} \theta\,(\mathbf{x}(\tau)) \quad , \\
V_2(\tau) &= \dot{x}^i(\tau) \epsilon^{ij} \partial_j^{-1} B\,(\mathbf{x}(\tau)) \quad ,
\end{aligned}
\tag{2.4.7}
$$

γ becomes

$$
\gamma = \frac{1}{\kappa} \int_0^1 d\tau \int_0^\tau d\tau' \frac{\partial}{\partial \tau} \dot{x}^i(\tau') \epsilon^{ij} \partial_j^{-1} \delta^2\,(\mathbf{x}(\tau) - \mathbf{x}(\tau')) \quad , \tag{2.4.8a}
$$

where $\partial_j^{-1}\delta$ is the derivative of the two-dimensional Green's function.

$$
\partial_j^{-1} \delta^2(\mathbf{r}) = \frac{1}{2\pi} \frac{r^j}{|\mathbf{r}|} = \frac{\epsilon^{jk}}{2\pi} \partial_k \tan^{-1} \frac{r^2}{r^1}
$$

The formula relating $\partial_j^{-1}\delta$ to a derivative of $\tan^{-1}$ gives an alternative expression for γ.

$$
\gamma = \frac{1}{2\pi\kappa} \int_0^1 d\tau \int_0^\tau d\tau' \frac{\partial}{\partial \tau} \frac{\partial}{\partial \tau'} \tan^{-1} \frac{x^2(\tau) - x^2(\tau')}{x^1(\tau) - x^1(\tau')} \tag{2.4.8b}
$$

The derivation may be organized differently. We recognize that the phase arising from splitting $\exp\left[i \int_0^1 d\tau\, V(\tau)\right]$ into $\exp\left[i \int_0^1 d\tau\, V_1(\tau)\right] \exp\left[i \int_0^1 d\tau\, V_2(\tau)\right]$ vanishes because the loop is closed, leaving for the entire contribution to γ just the last quantity in (2.4.4), which arises from undoing the path-ordering, and which is determined by the polarization-independent $[A_i, A_j]$ commutator.

$$
\gamma = -\frac{1}{2\kappa} \int_0^1 d\tau \int_0^\tau d\tau' \dot{x}^i(\tau) \epsilon^{ij} \dot{x}^j(\tau') \delta^2\,(\mathbf{x}(\tau) - \mathbf{x}(\tau')) \tag{2.4.8c}
$$

Using the identity $2\pi\delta^2(\mathbf{r}) = \epsilon^{ij}\partial_i\partial_j \tan^{-1} r^2/r^1$ we can rewrite (2.4.8c).

$$
\gamma = \frac{1}{4\pi\kappa} \int_0^1 d\tau \int_0^\tau d\tau' \left(\frac{\partial}{\partial \tau} \frac{\partial}{\partial \tau'} - \frac{\partial}{\partial \tau'} \frac{\partial}{\partial \tau} \right) \tan^{-1} \frac{x^2(\tau) - x^2(\tau')}{x^1(\tau) - x^1(\tau')} \tag{2.4.8d}
$$

Equation (2.4.8c) exhibits the elusive nature of γ. Owing to the δ-function enforcing $\mathbf{x}(\tau) = \mathbf{x}(\tau')$, one might conclude that $\dot{x}^i(\tau)\epsilon^{ij}\dot{x}^j(\tau')$, and hence γ, vanishes. However, upon closer examination we recognize that the two-dimensional δ-function is a product of two one-dimensional δ-functions, each enforcing the same constraint on the one-dimensional variables τ, τ'. Thus the integrand in (2.4.8c) involves the ambiguous quantity $\dot{x}^i(\tau)\epsilon^{ij}\dot{x}^j(\tau)/\left|\dot{x}^1(\tau)\dot{x}^2(\tau)\right| \times\delta(\tau-\tau')\delta(\tau-\tau')$, and in the following a careful analysis is performed to obtain an unambiguous result. But it is already clear that γ is non-trivial owing to the continuum properties of space, which give rise to δ-functions. In a discretized world with Kronecker delta's, (2.4.8) does indeed vanish.

Our evaluation of γ is based on the following observation. In spite of the singular nature of (2.4.8c), one can begin with any of the other formulas (2.4.8a), (2.4.8b) and (2.4.8d), manipulate finite expressions and arrive at an unambiguous answer for γ. By this procedure, we shall derive below the result

$$\gamma = \frac{1}{2\pi\kappa}\Delta\Theta_{\mathbf{r}} - \frac{1}{4\pi\kappa}\Delta\Theta_{\mathbf{v}} \quad . \tag{2.4.9}$$

Here $\Delta\Theta_{\mathbf{r}}$ is the total angle accumulated at the marked point $\mathbf{x}_0$ when the loop is traversed by a vector based at $\mathbf{x}_0$,

$$\begin{aligned} \Theta_{\mathbf{r}}(\tau) &= \tan^{-1}\frac{r^2(\tau)}{r^1(\tau)} \quad , \\ \mathbf{r} &= \mathbf{x} - \mathbf{x}_0 \quad , \\ \Delta\Theta_{\mathbf{r}} &\equiv \Theta_{\mathbf{r}}(1) - \Theta_{\mathbf{r}}(0) \quad , \end{aligned} \tag{2.4.10a}$$

while the quantity $\Delta\Theta_{\mathbf{v}}$ is the accumulated angular change in the tangent to the curve.

$$\begin{aligned} \Theta_{\mathbf{v}}(\tau) &= \tan^{-1}\frac{\dot{r}^2(\tau)}{\dot{r}^1(\tau)} = \tan^{-1}\frac{\hat{v}^2(\tau)}{\hat{v}^1(\tau)} \\ \Delta\Theta_{\mathbf{v}} &\equiv \Theta_{\mathbf{v}}(1) - \Theta_{\mathbf{v}}(0) \end{aligned} \tag{2.4.10b}$$

The evaluation of (2.4.8) that gives (2.4.9) is performed without using regulators. However, to illustrate the subtlety of (2.4.8) we remark here that if regulators *are* introduced, for example by regulating the δ-function, and if the regularization preserves the fact that the Green's function derivative is odd under interchange of argument, $\partial_j^{-1}\delta^2(\mathbf{x}-\mathbf{y}) = -\partial_j^{-1}\delta^2(\mathbf{y}-\mathbf{x})$, so that $\partial_j^{-1}\delta^2(0)$ vanishes, then we again obtain a unique answer that does not depend on the details of the regularization. However, this answer differs from (2.4.9): $\gamma^{\text{reg}} = \frac{1}{2\pi\kappa}\Delta\Theta_{\mathbf{r}}$, the contribution from the change in the tangent is missing.

We now present the derivation of (2.4.9). In (2.4.8a) or (2.4.8b) the τ derivative is interchanged with the τ' integral. Starting from (2.4.8a) we have

$$\begin{aligned} \gamma &= \frac{1}{\kappa}\int_0^1 d\tau\frac{\partial}{\partial\tau}\int_0^\tau d\tau'\,\dot{x}^i(\tau')\epsilon^{ij}\delta^2\left(\mathbf{x}(\tau)-\mathbf{x}(\tau')\right) \\ &\quad -\frac{1}{\kappa}\int_0^1 d\tau\left(\dot{x}^i(\tau')\epsilon^{ij}\partial_j^{-1}\delta^2\left(\mathbf{x}(\tau)-\mathbf{x}(\tau')\right)\right)\Big|_{\tau'=\tau} \\ &= -\frac{1}{\kappa}\int_0^1 d\tau'\,\dot{x}^i(\tau')\epsilon^{ij}\partial_j^{-1}\delta^2\left(\mathbf{x}(\tau')-\mathbf{x}_0\right) \\ &\quad +\frac{1}{\kappa}\int_0^1 d\tau\left(\dot{x}^i(\tau')\epsilon^{ij}\partial_j^{-1}\delta^2\left(\mathbf{x}(\tau')-\mathbf{x}(\tau)\right)\right)\Big|_{\tau'=\tau} \quad . \end{aligned} \tag{2.4.11a}$$

By expressing the derivative of the Green's functions in terms of the inverse tangent, or alternatively by beginning with (2.4.8b), we find

$$\begin{aligned}\gamma = & \frac{1}{2\pi\kappa}\int_0^1 d\tau' \frac{\partial}{\partial\tau'}\tan^{-1}\frac{x^2(\tau')-x_0^2}{x^1(\tau')-x_0^1} \\ & -\frac{1}{2\pi\kappa}\int_0^1 d\tau \left(\frac{\partial}{\partial\tau'}\tan^{-1}\frac{x^2(\tau')-x^2(\tau)}{x^1(\tau')-x^1(\tau)}\right)\bigg|_{\tau'=\tau}\end{aligned} \tag{2.4.11b}$$

[It is easy to see that the same formula emerges if one begins with (2.4.8d), treats the $\partial_\tau\partial_{\tau'}$ contribution as in (2.4.11b) and evaluates the $\partial_{\tau'}\partial_\tau$ contribution by first performing the τ' integral.] The integrand of the last term in (2.4.11b) is

$$\left(\frac{\partial}{\partial\tau'}\tan^{-1}\frac{x^2(\tau')-x^2(\tau)}{x^1(\tau')-x^1(\tau)}\right)\bigg|_{\tau'=\tau} = \frac{1}{2}\frac{\partial}{\partial\tau}\tan^{-1}\frac{\dot{x}^2(\tau)}{\dot{x}^1(\tau)}$$

and so γ becomes

$$\gamma = \frac{1}{2\pi\kappa}\int_0^1 d\tau\frac{\partial}{\partial\tau}\tan^{-1}\frac{x^2(\tau)-x_0^2}{x^1(\tau)-x_0^1} - \frac{1}{4\pi\kappa}\int_0^1 d\tau\frac{\partial}{\partial\tau}\tan^{-1}\frac{\dot{x}^2(\tau)}{\dot{x}^1(\tau)} \tag{2.4.11c}$$

This establishes (2.4.9). [Note that if a regulator had been used for the δ function in (2.4.11a), then the last term of each equation in (2.4.11) would vanish, with natural regulators such that $\partial_j^{-1}\delta^2(0) = 0$, leaving, as commented above, for γ^{reg} just the first term on the right-hand side of (2.4.11c). A possible regularization would be to retain the kinetic Maxwell term, and then decouple it by passing with κ to infinity. It should be no surprise, in view of earlier discussions on this limit [see (2.3.18)], that different answers can be obtained. For more discussion, see the literature [2e].

The value of γ depends on the curve. Consider first simple curves, traversed in the counterclockwise direction, without self-intersections. For smooth, simple curves, the angle swept out as the marked point is π, and

$$\frac{1}{2\pi\kappa}\Delta\Theta_{\mathbf{r}} = \frac{1}{2\kappa} \ . \tag{2.4.12a}$$

The angular change in the tangent is 2π, so

$$\frac{1}{4\pi\kappa}\Delta\Theta_{\mathbf{v}} = \frac{1}{2\kappa} \tag{2.4.12b}$$

and therefore

$$\gamma = 0 \tag{2.4.12c}$$

The same results hold if the curve has cusps, provided the marked point is not a cusp. However, when $\mathbf{x}_0$ does lie at a cusp where $\hat{\mathbf{v}}_0 \neq \hat{\mathbf{v}}_1$, there are shortfalls in the angular traversals and

$$\frac{1}{2\pi\kappa}\Delta\Theta_{\mathbf{r}} = \frac{1}{2\pi\kappa}\left(\pi \mp \cos^{-1}\hat{\mathbf{v}}_0\cdot\hat{\mathbf{v}}_1\right) \tag{2.4.13a}$$

$$\frac{1}{3\pi\kappa}\Delta\Theta_{\mathbf{v}} = \frac{1}{4\pi\kappa}\left(2\pi \mp \cos^{-1}\hat{\mathbf{v}}_0\cdot\hat{\mathbf{v}}_1\right) \tag{2.4.13b}$$

$$\gamma = \mp\frac{1}{4\pi\kappa}\cos^{-1}\hat{\mathbf{v}}_0\cdot\hat{\mathbf{v}}_1 \tag{2.4.13c}$$

where $\mp$ refer to opening angles of the cusp $< \pi$ and $\leq \pi$, respectively.

We now discuss loops with intersections. We set the following conventions: the marked point lies on an outermost smooth segment of the loop, insuring that $|\Delta\Theta_{\mathbf{r}}| = \pi$, regardless of the number of intersections and cusps, since $\mathbf{x}_0$ does not lie at any of these exceptional points; the parametrization takes the contour through $\mathbf{x}_0$ in a counterclockwise direction thus fixing $\Delta\Theta_{\mathbf{r}} = \pi$; an intersection is defined by an actual crossing — touching contours are not intersections; the total number of intersections is ν and we do not consider loops with multiple intersections at the same point.

Only $\Delta\Theta_{\mathbf{v}}$ varies with the intersection number ν. For $\nu = 1$, there are two elementary intersections: the "figure eight" where the two sub-loops are traversed in opposite directions, so that $\Delta\Theta_{\mathbf{v}} = 0$, and the "nested loop" where the two sub-loops are traversed in the same direction, with $\Delta\Theta_{\mathbf{v}} = 4\pi$. Loops with higher ν can be constructed by superposing in various ways these two elementary blocks. For a given ν, the highest possible value for $\Delta\Theta_{\mathbf{v}}$, $\Delta\Theta_{\mathbf{v}}^{\max} = (\nu+1)2\pi$, is achieved by superposing like-oriented nested intersections. The lowest possible value, $\Delta\Theta_{\mathbf{v}}^{\min} = -(\nu-1)2\pi$, is obtained by building out of ν like-oriented intersections a loop whose direction has been reversed by a single figure eight intersection. The possible values of $\Delta\Theta_{\mathbf{v}}$, for fixed ν, interpolate between $\Delta\Theta_{\mathbf{v}}^{\min}$ and $\Delta\Theta_{\mathbf{v}}^{\max}$ in steps of 4π. The different allowed values for $\Delta\Theta_{\mathbf{v}}$, combined with the unique $\Delta\Theta_{\mathbf{r}} = \pi$, give a table of values for γ at fixed ν.

$$\gamma = \frac{m}{\kappa} \quad , \quad m \in \left[-\frac{\nu}{2}, -\frac{\nu}{2}+1, \ldots, \frac{\nu}{2}-1, \frac{\nu}{2}\right] \tag{2.4.14}$$

Finally, we return to the full holonomy operator, and determine its action on the state (2.3.8). In our polarization, we have

$$\Phi(C)\Psi(\theta) = \exp\left[i\gamma(C)\right]\exp\left[i\int_C dx^i\partial_i\theta\right]\exp\left[-\frac{i}{\kappa}\int_C dx^i\epsilon^{ij}\partial_j^{-1}\rho\right]\Psi(\theta) \ . \tag{2.4.15}$$

Since the holonomy of the pure gauge $\partial_i\theta$ is trivial, $\exp\left[i\int_C dx^i\partial_i\theta\right] = 1$. The integral in the exponent of the last factor is evaluated by Stoke's law and gives

$$-\int_C dx^i\epsilon^{ij}\partial_j^{-1}\rho = \int_{S_C} d\mathbf{x}\,\rho = Q(C) \tag{2.4.16}$$

where $Q(C)$ is the total charge contained within the closed curve, with contributions appropriately signed if C is self-intersecting. [For a single point source of charge Q surrounded by a counterclockwise loop, $Q(C) = Q$.] Thus the physical state (2.3.8) is an eigenstate of the holonomy operator with eigenvalue $\phi(C)$.

$$\phi(C) = \exp\left(i\left[\gamma(C) + \frac{1}{\kappa}Q(C)\right]\right) \tag{2.4.17}$$

[Note that the holonomy operator commutes with the Gauss law operator because it is gauge invariant, and with the Hamiltonian (2.3.3) if $Q(C)$ is time-independent.]

Since holonomies around closed loops are the only gauge invariant and generally covariant observables, this shows that the $U(1)$ Chern–Simons theory in Minkowski space is characterized by the strengths of external charges, and by the vacuum holonomy $e^{i\gamma(C)}$, which is missed when the constraints are imposed before quantization. The vacuum holonomy, which can be evaluated without regularization, vanishes for simple loops provided the marked point is not at a cusp; otherwise, it is determined by the opening angle of the cusp. This rich structure is already present for the $U(1)$ Chern–Simons theory in Minkowski space.

The gauge structure in the Chern–Simons theory follows closely that of its topologically massive antecedent: both involve essentially the same cocycle in the action of the gauge group. However, the behavior of the quantum holonomy Φ is quite different. In the latter theory, the vacuum ground state — for the non-interacting Abelian model, the Gaussian (2.1.22) in $\mathbf{A}$ — is not an eigenstate of Φ, and the vacuum expectation value of the holonomy operator is infinite. In the Abelian Chern–Simons theory, the vacuum state — the only state for the source-free model in Minkowski space — is a Φ eigenstate with finite non-trivial eigenvalue, which is not seen when constraints are solved before quantization.

If expectation values of the holonomy operator in a topologically massive theory are to possess physical significance, Φ must be renormalized and its vacuum value is undefined. Neither regularization nor renormalization is needed in the Chern–Simons model; indeed since γ carries a rich loop dependence, a universal renormalization cannot remove the effect.

2.5 Anomalous Statistics and the Spin of Charged Particles

We saw earlier that charged particles interacting through a Chern–Simons [Abelian] gauge field carry flux. This has the consequence that their spin and statistics is modified by the gauge-field interaction [10] — a result which can be established without reference to the detailed nature of the particle dynamics [2d]. Here we first show how the holonomy modifies statistics, and that spin adjusts so that the spin-statistics theorem is preserved. Later, we shall take a point-particle model for the matter and regain these results in an explicit manner.

Consider two identical particles, each with charge Q, and imagine a fixed-time test of statistics by carrying one particle around the other, corresponding to a double interchange of the particles. The wave function of the test particle will acquire, in addition to the conventional statistical factor, the phase

$$P\exp\left[iQ\int_C dx^i\,A_i\right] \quad ,$$

where C is a loop without self-intersections surrounding the particle. The state (2.3.8) is an eigenstate of this operator with eigenvalue $\exp\left[iQ^2/\kappa\right]$, apart from the vacuum contribution, which is absent provided the marked point is not at a cusp. The phase acquired by the wave function under a single interchange of the two particles is half the above, *i.e.* $Q^2/2\kappa$. Thus the statistics phase is an observable, whose value satisfies the spin-statistics theorem because particles in this theory carry anomalous spin S,

$$2\pi S = \frac{Q^2}{2\kappa} \tag{2.5.1}$$

as we now demonstrate.

Spin fractionization for the charged matter particles is due to the angular momentum of the gauge field associated with the magnetic flux created by the charged particle. The variation of the gauge field under an infinitesimal spatial rotation $\delta x^i = -\epsilon^{ij}x^j$,

$$\delta A_i = -x^j \epsilon^{jk} \partial_k A_i - \epsilon^{ij} A_j \quad , \tag{2.5.2}$$

leaves the Lagrangian corresponding to (2.3.1) invariant, provided the external current $\mathbf{j}$ is rotationally covariant.

$$x^j \epsilon^{jk} \partial_k j^i + \epsilon^{ij} j^j = 0 \tag{2.5.3}$$

For the conserved angular momentum operator we take

$$J = -\frac{\kappa}{2} \int_{\mathbf{x}} x^i \epsilon^{ij} \{A_j, B\} \quad . \tag{2.5.4}$$

J generates the transformation (2.5.2),

$$\delta A_i = i[J, A_i] \tag{2.5.5}$$

and commutes with the Hamiltonian (2.3.3) when the current is rotationally covariant, *i.e.*, (2.5.3) is satisfied. Also J is gauge invariant, as is seen by replacing A_j in (2.5.4) by $\partial_j \lambda$ and B by $-\rho/\kappa$ according to (2.3.4). When the charge density is spherically symmetric, an integration by parts yields a vanishing response to the gauge transformation. J is not obtained from Noether's energy-momentum tensor, rather from the symmetric tensor, which has no pure gauge field contribution owing to the topological nature of the Chern–Simons term; only the interaction contributes.

$$T^{\mu\nu} = g^{\mu\nu} j^\alpha A_\alpha - j^\mu A^\nu - j^\nu A^\mu \tag{2.5.6}$$

Thus we have

$$\begin{aligned} J = \int_{\mathbf{x}} \epsilon^{ij} x^i T^{0j} &= \int_{\mathbf{x}} \epsilon^{ij} x^i A_j \rho \\ &= -\kappa \int_{\mathbf{x}} \epsilon^{ij} x^i A_j B \quad , \end{aligned} \tag{2.5.7}$$

where (2.3.4) was used. In (2.5.4) the expression is symmetrized to insure Hermiticity; however, it differs from (2.5.7) by a commutator $i \int_{\mathbf{x}} \epsilon^{ij} \partial_j \delta^2(\mathbf{x}-\mathbf{x})$, which although involving $\delta^2(0)$ can be set to zero by an integration by parts. [Alternatively, $\partial_j \delta^2(\mathbf{x})$ is odd in its argument.]

Consider now the action of J on the state (2.3.8)

$$J\Psi(\theta) = \left[\int_{\mathbf{x}} \epsilon^{ij} x^i A_j \rho\right] \Psi(\theta) = \int \rho \epsilon^{ij} x^i \left[\partial_j \theta - \frac{1}{\kappa} \epsilon^{jk} \partial_k^{-1} \rho\right] \Psi(\theta) \tag{2.5.8}$$

The contribution proportional to $\partial_j \theta$ vanishes upon an integration by parts for spherically symmetric ρ. This shows that the physical state (2.3.8) is an angular momentum eigenstate with eigenvalue S.

$$\begin{aligned} S &= \frac{1}{\kappa} \int \rho x^i \partial_i^{-1} \rho \\ &= \frac{1}{2\pi\kappa} \int_{\mathbf{x}} \int_{\mathbf{y}} \rho(t, \mathbf{x}) \frac{\mathbf{x} \cdot (\mathbf{x} - \mathbf{y})}{|\mathbf{x} - \mathbf{y}|^2} \rho(t, \mathbf{y}) \\ &= \frac{1}{4\pi\kappa} \int_{\mathbf{x}} \int_{\mathbf{y}} \rho(t, \mathbf{x}) \left[1 + \frac{\mathbf{x}^2 - \mathbf{y}^2}{|\mathbf{x} - \mathbf{y}|^2}\right] \rho(t, \mathbf{y}) \\ &= \frac{Q^2}{4\pi\kappa} \ , \end{aligned} \tag{2.5.9}$$

Equation (2.5.9) gives the fractional spin carried by the charged matter particles, and agrees with previous results. S is a sharp observable, which satisfies the spin statistics relation (2.5.1).

2.6 Point-Particles with Abelian Chern–Simons Gauge Fields

A possible model for charged matter consists of point particles [11]. The total Lagrangian is

$$L = L_{\text{matter}} + L_{\text{interaction}} + L_{\text{CS}} \tag{2.6.1}$$

where

$$L_{\text{matter}} = \frac{1}{2} \sum_{p=1}^{N} m_p v_p^2(t) \tag{2.6.2a}$$

$$\begin{aligned} L_{\text{interaction}} &= \sum_{p=1}^{N} e_p \Big(\mathbf{v}_p(t) \cdot \mathbf{A}\left(t, \mathbf{r}_p(t)\right) - A_0\left(t, \mathbf{r}_p(t)\right) \Big) \\ &= - \int_{\mathbf{x}} A_\mu(x) j^\mu(x) \\ j^\mu &= \sum_{p=1}^{N} e_p v_p^\mu(t) \delta^2\left(\mathbf{x} - \mathbf{r}_p(t)\right) = \left(\rho(x), \mathbf{j}(x)\right) \\ v_p^\mu &= (1, \mathbf{v}_p) \end{aligned} \tag{2.6.2b}$$

$$L_{\text{CS}} = \frac{\kappa}{2} \int_{\mathbf{x}} \epsilon^{ij} \dot{A}_i(x) A_j(x) - \kappa \int_{\mathbf{x}} A_0(x) B(x) \tag{2.6.2c}$$

We are considering N point-particles with coordinates $\mathbf{r}_p(t)$, $p = 1, \ldots, N$, which are the particle dynamical variables, and $\mathbf{v}_p(t) = \dot{\mathbf{r}}_p(t)$ are the velocities.

The masses and charges are m_p and e_p, respectively. The second expression for the interaction Lagrangian makes use of the point-particle current, which is a δ-function. Thus the integral over all space evaluates $x = (t, \mathbf{x})$, the field point argument of $A_\mu(x)$, at $x = (t, \mathbf{r}_p(t))$. The time component A_0 has not been set to zero.

The [unordered] Euler–Lagrange equations of motion consist of the Lorentz force equation for the matter variables

$$m_p \dot{v}_p^i = e_p \left(E^i(\mathbf{r}_p) + \epsilon^{ij} v_p^j B(\mathbf{r}_p) \right) \tag{2.6.3a}$$

and a field-current identity that relates the electromagnetic fields to the matter currents.

$$E^i(x) = \frac{1}{\kappa} \epsilon^{ij} j^j(x) \tag{2.6.3b}$$

$$B(x) = -\frac{1}{\kappa} \rho(x) \tag{2.6.3c}$$

Point-particle electrodynamics suffers from well-known self-energy problems. Let us observe that these are absent here. Consider the equation of motion for a single particle, $N = 1$ and subscript p suppressed. The force in Eq. (2.6.3a) arises from the electromagnetic fields at the particle position $\mathbf{r}$; by (2.6.3b) and (2.6.3c) they are given by the charge and current densities evaluated at $\mathbf{x} = \mathbf{r}$. However from (2.6.2b) we see that at $\mathbf{x} = \mathbf{r}$ there appears the undefined quantity $\delta^2(\mathbf{r} - \mathbf{r})$ — the density function of a point-particle at its position. Fortunately this singular object is multiplied by a factor that vanishes, since according to the field-current identity, the quantity

$$E^i(\mathbf{x}) + \epsilon^{ij} v^j B(\mathbf{x}) = \frac{1}{\kappa} \epsilon^{ij} \left(j^j(x) - v^j \rho(x) \right) = \frac{1}{\kappa} \epsilon^{ij} \left(v^j - v^j \right) \delta^2 (\mathbf{x} - \mathbf{r})$$

vanishes unambiguously; therefore we shall take it to be zero also at $\mathbf{x} = \mathbf{r}$. In other words the charge and current densities are regulated by non-singular expressions for the evaluation of the self-interaction, which is then shown to vanish.

With this prescription, particles interacting through Chern–Simons gauge fields do not experience self-interactions. Equations (2.6.3) combine to give for the particle coordinates a closed equation of motion, free from undefined quantities.

$$m_p \dot{v}_p^i = \epsilon^{ij} \frac{e_p}{\kappa} \sum_{q \neq p} e_q \left(v_q^j - v_p^j \right) \delta \left(\mathbf{r}_p - \mathbf{r}_q \right) \tag{2.6.4}$$

The Hamiltonian arising from the Lagrangian (2.6.1) and (2.6.2) is

$$H = \frac{1}{2} \sum_{p=1}^{N} m_p v_p^2 + \int_{\mathbf{x}} A_0(x) \left(\kappa B(x) + \rho(x) \right) \tag{2.6.5}$$

It is recognized that the Lagrange multiplier A_0 may be set to zero [by choosing the Weyl gauge] provided Eq. (2.6.3c) is imposed as a constraint. Thus the Hamiltonian is just the free particle one.

$$H = \frac{1}{2} \sum_{p=1}^{N} m_p v_p^2 \tag{2.6.6}$$

Although the gauge field is invisible in (2.6.6), its presence is felt in the commutator algebra.

The commutation relations between canonical variables follow in the usual way, except that vector potentials satisfy the Abelian version of (2.2.8). However, it is useful to present the algebra in terms of the gauge invariant velocity operator, which occurs in the Hamiltonian, rather than in terms of the canonical momentum.

$$\mathbf{p}_p = \frac{\partial L}{\partial \mathbf{v}_p} = m_p \mathbf{v}_p + e_p \mathbf{A}(\mathbf{r}_p) \tag{2.6.7}$$

Thus we have from (2.2.8),

$$\left[A^i(\mathbf{x}), A^j(\mathbf{y})\right] = \frac{i}{\kappa} \epsilon^{ij} \delta^2(\mathbf{x} - \mathbf{y}) \tag{2.6.8}$$

The particle variables satisfy

$$\left[r_p^i, r_q^j\right] = 0 \tag{2.6.9a}$$

$$\left[r_p^i, m_q v_q^j\right] = i\delta^{ij}\delta_{pq} \tag{2.6.9b}$$

$$\left[m_p v_p^i, m_q v_q^j\right] = i\epsilon^{ij} \left(\delta_{pq} e_p B(\mathbf{r}_q) + \frac{1}{\kappa} e_p e_q \delta^2(\mathbf{r}_p - \mathbf{r}_q)\right) \tag{2.6.9c}$$

The velocity commutator does not vanish; rather it contains terms that arise from: (i) the fact that $\mathbf{v}$ differs from $\mathbf{p}$ by $\mathbf{A}(\mathbf{r})$, (2.6.7), and $\mathbf{p}$ does not commute with $\mathbf{A}(\mathbf{r})$ but produces the first term in the parenthesis of (2.6.9c); also (ii) the vector potentials do not commute, (2.6.8), giving rise to the second term in parenthesis of (2.6.9c). There is also the non-vanishing commutator between velocity and field.

$$\left[m_p v_p^i, A^j(\mathbf{x})\right] = -\frac{i}{\kappa} \epsilon^{ij} e_p \delta^2(\mathbf{x} - \mathbf{r}_p) \tag{2.6.10}$$

Finally we remind that (2.6.3c) is imposed as a constraint.

Before proceeding with an analysis of the dynamical problem, it is interesting to record the symmetries of our theory.

First there are the spatial translation and rotation symmetries, under which the coordinates and fields change as

$$\begin{aligned} \delta r_p^i &= a^i \\ \delta A_\mu(t, \mathbf{x}) &= -a^i \partial_i A_\mu(t, \mathbf{x}) \qquad \text{(translations)} \end{aligned} \tag{2.6.11}$$

$$\begin{aligned} \delta r_p^i &= -\epsilon^{ij} r_p^j \\ \delta A_0(t, \mathbf{x}) &= -\epsilon^{jk} x^j \partial_\kappa A_0(t, \mathbf{x}) \\ \delta A_i(t, \mathbf{x}) &= -\epsilon^{jk} x^j \partial_\kappa A_i(t, \mathbf{x}) - \epsilon^{ij} A_j(t, \mathbf{x}) \quad \text{(rotations)} \end{aligned} \tag{2.6.12}$$

The Lagrangian (2.6.1), (2.6.2) is invariant, and the conserved constants of motion are the momentum $\mathbf{P}$ and angular momentum J, respectively.

$$\mathbf{P} = \sum_{p=1}^{N} (m_p \mathbf{v}_p + e_p \mathbf{A}(\mathbf{r}_p)) + \kappa \int_{\mathbf{x}} \mathbf{A} B$$
$$= \sum_{p=1}^{N} \mathbf{p}_p + \kappa \int_{\mathbf{x}} \mathbf{A} B \tag{2.6.13}$$

$$J = \sum_{p=1}^{N} (\mathbf{r}_p \times (m_p \mathbf{v}_p + e_p \mathbf{A}(\mathbf{r}_p))) + \kappa \int_{\mathbf{x}} (\mathbf{x} \times \mathbf{A}) B$$
$$= \sum_{p=1}^{N} \mathbf{r}_p \times \mathbf{p}_p + \kappa \int_{\mathbf{x}} (\mathbf{x} \times \mathbf{A}) B \ . \tag{2.6.14}$$

The second formula in both expressions makes use of the canonical momentum, (2.6.7). The vector potential, evaluated at $\mathbf{x} = \mathbf{r}_p$, which is combined with $m_p \mathbf{v}_p$ to form the canonical momentum, arises from the integration Lagrangian (2.6.2b); the last term in (2.6.13) and (2.6.14), proportional to κ, arises from the Chern–Simons kinetic term.

Let us observe that $\sum_{p=1}^{N} \mathbf{r}_p \times \mathbf{p}_p$ possesses integer eigenvalues when acting on single valued wave functions. Thus point-particles, interacting with gauge fields that are governed by Chern–Simons dynamics, possess in their angular momentum a contribution additional to the usual integer. The extra term is not quantized but is determined by the gauge field and the Chern–Simons coupling strength κ. This is consistent with the results of the external source analysis, presented earlier, and it will be shown later that in fact there is complete agreement.

Note that for point-particles moving in ***prescribed external*** gauge fields [rather than ***dynamical*** ones], the last term in (2.6.13) and (2.6.14) is missing, and $\mathbf{A}(\mathbf{r}_p)$ is a given function of $\mathbf{r}_p$. Therefore, in the external field problem the angular momentum spectrum comprises the conventional integers, even though it differs from the kinematical momentum [12].

$$\sum_{p=1}^{N} \mathbf{r}_p \times \mathbf{p}_p = \sum_{p=1}^{N} \mathbf{r}_p \times m_p \mathbf{v}_p + \sum_{p=1}^{N} e_p \mathbf{r}_p \times \mathbf{A}(\mathbf{r}_p)$$

[This point is occasionally confused in the literature.]

In addition to the symmetries under the above spatial transformations, the theory is also invariant against transformations of time: obviously time translation $[t \to t + t_0]$ is a symmetry leading to energy [= Hamiltonian] conservation; but there are two further, unexpected time transformations that leave the action invariant: time dilation $[t \to \lambda t]$, and time special conformal transformation $[1/t \to 1/t + 1/t_0]$. Together, the three form a dynamical $SO(2,1)$ symmetry group of conformal transformations [2f].

Infinitesimally we have for these time reparametrizations

$$\delta t = -f(t) \tag{2.6.15a}$$

$$f(t) = \begin{cases} 1 & \text{translation} \\ t & \text{dilation} \\ t^2 & \text{special conformal transformation} \end{cases} \tag{2.6.15b}$$

The dynamical variables transform as

$$\begin{aligned}
\delta_f \mathbf{r}_p(t) &= f(t)\mathbf{v}_p(t) - \frac{1}{2}\dot{f}(t)\mathbf{r}_p(t) \\
\delta_f A_0(t,\mathbf{x}) &= \partial_t\big(f(t)A_0(t,\mathbf{x})\big) + \frac{1}{2}\dot{f}(t)\mathbf{x}\cdot\nabla A_0(t,\mathbf{x}) - \frac{1}{2}\ddot{f}(t)\mathbf{x}\cdot\mathbf{A}(t,\mathbf{x}) \\
\delta_f \mathbf{A}(t,\mathbf{x}) &= \partial_t\big(f(t)\mathbf{A}(t,\mathbf{x})\big) - \frac{1}{2}\dot{f}(t)\mathbf{A}(t,\mathbf{x}) + \dot{f}(t)\mathbf{x}\cdot\nabla\mathbf{A}(t,\mathbf{x})
\end{aligned} \tag{2.6.16}$$

and the Lagrangian changes by a total time derivative. The constants of motion arising from the three transformations are, respectively

$$H = \frac{1}{2}\sum_{p=1}^{N} m_p v_p^2 + \int_{\mathbf{x}} A_0(\kappa B + \rho) \tag{2.6.17a}$$

$$D = tH - \frac{1}{4}\sum_{p=1}^{N} m_p\left(\mathbf{r}_p\cdot\mathbf{v}_p + \mathbf{v}_p\cdot\mathbf{r}_p\right) - \frac{1}{2}\int_{\mathbf{x}} \mathbf{x}\cdot\mathbf{A}(\kappa B + \rho) \tag{2.6.17b}$$

$$K = -t^2 H + 2tD + \frac{1}{2}\sum_{p=1}^{N} m_p r_p^2 \tag{2.6.17c}$$

Of course (2.6.17a) coincides with the Hamiltonian of (2.6.5).

Returning now to the dynamical problem, we observe that when the constraint (2.6.3c) is imposed, all reference to gauge fields disappears from the equation of motion (2.6.4). Hence, we can set $\kappa B + \rho$ to zero throughout, and suppress the gauge degrees of freedom. Therefore, only the dynamical algebra (2.6.9) is relevant, and now it takes the form

$$\left[r_p^i, r_q^j\right] = 0 \tag{2.6.18a}$$

$$\left[r_p^i, m_q v_q^j\right] = i\delta^{ij}\delta_{pq} \tag{2.6.18b}$$

$$\begin{aligned}
\left[m_p v_p^i, m_q v_q^j\right] &= i\frac{\epsilon^{ij}}{\kappa}\Big((1-\delta_{pq})e_p e_q \delta^2(\mathbf{r}_q - \mathbf{r}_q) \\
&\qquad - \delta_{pq}\sum_{n\neq p} e_p e_n \delta^2(\mathbf{r}_p - \mathbf{r}_n)\Big)
\end{aligned} \tag{2.6.18c}$$

The consistency of these commutation relations is established by realizing the operators through their action on functions of $\mathbf{r}_p$, with $\mathbf{r}_p$ acting by multiplication, while $m_p\mathbf{v}_p$ is redefined as

$$m_p v_p^i = p_p^i - \frac{e_p}{2\pi\kappa}\epsilon^{ij}\sum_{q\neq p} e_q \frac{\left(r_p^j - r_q^j\right)}{|\mathbf{r}_p - \mathbf{r}_q|^2} \tag{2.6.19}$$

with $\mathbf{p}_p$ acting as $-i\nabla_{\mathbf{r}_p}$.

The symmetry generators (2.6.13), (2.6.14) and (2.6.17) become

$$\mathbf{P} = \sum_{p=1}^{N} m_p \mathbf{v}_p \tag{2.6.20}$$

$$J = \sum_{p=1}^{N} \mathbf{r}_p \times m_p \mathbf{v}_p \tag{2.6.21}$$

$$H = \frac{1}{2} \sum_{p=1}^{N} m_p v_p^2 \tag{2.6.22a}$$

$$D = tH - \frac{1}{4} \sum_{p=1}^{N} m_p (\mathbf{r}_p \cdot \mathbf{v}_p + \mathbf{v}_p \cdot \mathbf{r}_p) \tag{2.6.22b}$$

$$K = -t^2 H + 2tD + \frac{1}{2} \sum_{p=1}^{N} m_p r_p^2 \tag{2.6.22c}$$

The presence of the interaction is hidden, but it is in evidence in the commutators (2.6.18) and in the relation (2.6.19) between canonical momentum and velocity, which implies that the effective particle Lagrangian is [11]

$$L_{\text{effective}} = \sum_{p=1}^{N} \left(\frac{1}{2} m_p v_p^2 + e_p \mathbf{v}_p \cdot \mathbf{a}_p \right) \tag{2.6.23a}$$

$$\begin{aligned} a_p^i(\mathbf{r}_1, \ldots, \mathbf{r}_N) &= \frac{1}{2\pi\kappa} \epsilon^{ij} \sum_{q \neq p} e_q \frac{(r_p^j - r_q^j)}{|\mathbf{r}_p - \mathbf{r}_q|^2} \\ &= \frac{1}{2\pi\kappa} \epsilon^{ij} \frac{\partial}{\partial r_p^j} \sum_{q \neq p} e_q \ln |\mathbf{r}_p - \mathbf{r}_q| \\ &= -\frac{1}{2\pi\kappa} \frac{\partial}{\partial r_p^i} \sum_{q \neq p} e_q \theta_{pq} \\ \tan\theta_{pq} &= \frac{y_p - y_q}{x_p - x_q} \end{aligned} \tag{2.6.23b}$$

Explicitly, (2.6.23a) reads

$$L_{\text{effective}} = \frac{1}{2} \sum_{p=1}^{N} m_p v_p^2 + \frac{1}{2\pi\kappa} \sum_{\substack{p,q=1 \\ p<q}}^{N} e_p e_q \frac{(\mathbf{v}_p - \mathbf{v}_q) \times (\mathbf{r}_p - \mathbf{r}_q)}{|\mathbf{r}_p - \mathbf{r}_q|^2} \tag{2.6.24}$$

Note that the angular momentum (2.6.21) is not constructed from the canonical momentum, hence as already remarked, its eigenvalues are non-integral. All the constants of motion $C = \mathbf{P}$, J, H, D and K generate the appropriate transformations (2.6.11), (2.6.12) and (2.6.16) upon commutation.

$$\delta \mathbf{r}_p = i\,[C, .\mathbf{r}_p] \tag{2.6.25}$$

In particular, the equation of motion (2.6.4), properly symmetrized, emerges upon commuting $\mathbf{v}_p$ with H. Also the $SO(2,1)$ generators satisfy the conformal Lie algebra,

$$[D,H] = -iH \ , \quad [D,K] = iK \ , \quad [H,K] = 2iD \tag{2.6.26}$$

which may be presented in the more familiar Cartan basis by forming linear combinations with the help of a fixed, positive parameter a of time dimensionality

$$\mathcal{R} = \frac{1}{2}\left(\frac{1}{a}K + aH\right) \tag{2.6.27a}$$

$$\mathcal{S} = \frac{1}{2}\left(\frac{1}{a}K - aH\right) \tag{2.6.27b}$$

$$L_\pm = (\mathcal{S} \pm iD) \tag{2.6.27c}$$

$\mathcal{S}$ and D act as non-compact two-dimensional boost generators, while $\mathcal{R}$ generates the rotations that form the compact $SO(2)$ subgroup of $SO(2,1)$.

$$[\mathcal{R}, L_\pm] = \pm L_\pm \tag{2.6.28a}$$

$$[L_+, L_-] = -2\mathcal{R} \tag{2.6.28b}$$

J commutes with the conformal generators; it rotates $\mathbf{P}$ in the proper manner.

$$\left[J, P^i\right] = i\epsilon^{ij} P^j \tag{2.6.29}$$

The momentum commutes with H; the commutators with the remaining conformal generators are

$$[D, \mathbf{P}] = -\frac{i}{2}\mathbf{P} \tag{2.6.30}$$

$$[K, \mathbf{P}] = -i\left(t\mathbf{P} - \sum_{p=1}^{N} m_p \mathbf{r}_p\right) \tag{2.6.31}$$

Equation (2.6.30) shows that the scale dimension of the momentum is 1/2, opposite to that of the coordinate $\mathbf{r}_p$. Since the commutator of two constants of motion is again a constant of motion, the right-hand side of (2.6.31) shows that center-of-mass motion is free. This is a consequence of the evident invariance of our theory against Galileo boosts, which are generated by $t\mathbf{P} - \sum_{p=1}^{N} m_p \mathbf{r}_p$.

That the generators are indeed constants of motion may be established by use of the formula

$$\frac{dC}{dt} = \frac{i}{\hbar}[H, C] + \frac{\partial C}{\partial t} \tag{2.6.32}$$

It follows from (2.6.32) that all three generators are time-dependent. Note however, D and K do not commute with the Hamiltonian; their total time derivative vanishes owing to the explicit time-dependence seen in (2.6.22b) and (2.6.22c).

We conclude this discussion of point-particle/Chern–Simons dynamics by recording the Casimir operator $\mathcal{J}^2$ of the SO(2,1) group.

$$\mathcal{J}^2 = \mathcal{R}^2 - \mathcal{S}^2 - D^2 = \frac{1}{2}(KH + HK) - D^2 \tag{2.6.33}$$

2.7 Quantum Dynamics

The Schrödinger equation governing dynamics of particles interacting with Chern–Simons gauge fields is inferred from (2.6.19), (2.6.22a) and (2.6.23).

$$i\frac{\partial}{\partial t}\Psi(t;\mathbf{r}_1,\ldots,\mathbf{r}_N) = H\Psi(t;\mathbf{r}_1,\ldots,\mathbf{r}_N) \tag{2.7.1a}$$

$$H = \sum_{p=1}^{N}\frac{1}{2m_p}\left(\frac{1}{i}\nabla_{\mathbf{r}_p} - e_p\mathbf{a}_p\right)^2 \tag{2.7.1b}$$

The wave function Ψ is single-valued. We may however make use of the formulas in (2.6.23b) to express $\mathbf{a}_p$ as a gradient, and remove the interaction in (2.7.1) by redefining the phase of the wave function [13]

$$\Psi(\mathbf{r}_1,\ldots,\mathbf{r}_N) = e^{i\Theta}\Psi^0(\mathbf{r}_1,\ldots,\mathbf{r}_N) \tag{2.7.2}$$

$$\Theta = \sum_{\substack{p,q=1\\ p<q}}^{N}\nu_{pq}\theta_{pq} \tag{2.7.3a}$$

$$\nu_{pq} = -\frac{e_p e_q}{2\pi\kappa} \tag{2.7.3b}$$

Then Ψ^0 satisfies the *free* Schrödinger equation.

$$i\frac{\partial}{\partial t}\Psi^0(t;\mathbf{r}_1,\ldots,\mathbf{r}_N) = H^0\Psi^0(t;\mathbf{r}_1,\ldots,\mathbf{r}_n) \tag{2.7.4a}$$

$$H^0 = e^{-i\Theta}H^0e^{i\Theta} = \sum_{p=1}^{N}\left(-\frac{1}{2m_p}\nabla^2_{\mathbf{r}_p}\right) \tag{2.7.4b}$$

Even though H^0 is a sum of one-body Hamiltonians, Ψ^0 cannot be chosen as a simple product of one-body eigenstates [plane waves] because Ψ^0 satisfies complicated aperiodicity conditions, which must hold so that Ψ be single-valued. Of course Ψ^0 is a ***superposition*** of plane waves, however, determining the precise superposition, which when multiplied by $e^{i\Theta}$ gives a single-valued wave function, is a challenging, non-trivial problem that has been solved only for the two-body case.

The two-body problem is tractable because of the center-of-mass reduction, wherein only the relative coordinate, $\mathbf{r} = \mathbf{r}_1 - \mathbf{r}_2$, experiences the interaction, while the center-of-mass coordinate $\mathbf{R} = \frac{m_1\mathbf{r}_1+m_2\mathbf{r}_2}{m_1+m_2}$ moves freely. By setting total momentum to zero, the two-body wave function depends only on $\mathbf{r}$ and satisfies

$$i\frac{\partial}{\partial t}\psi(t;\mathbf{r}) = h\psi(t;\mathbf{r}) \tag{2.7.5}$$

where the Hamiltonian for relative motion is

$$h = \frac{1}{2M}\left(\mathbf{p} - \mathbf{a}(\mathbf{r})\right)^2 \tag{2.7.6}$$

M is the reduced mass, $M^{-1} = m_1^{-1} + m_2^{-1}$, and the vector potential **a** gives rise to a vortex with flux $\Phi = -\frac{e_1 e_2}{\kappa}$

$$
\begin{aligned}
a^i(\mathbf{r}) &= -\frac{\Phi}{2\pi}\epsilon^{ij}\frac{\hat{r}^j}{r} = -\frac{\Phi}{2\pi}\epsilon^{ij}\partial_j \ln r \\
&= \frac{\Phi}{2\pi}\partial_i\theta = \nu\partial_i\theta
\end{aligned}
\tag{2.7.7}
$$

$$\mathbf{r} = (r\cos\theta, r\sin\theta)$$

Hence (2.7.5) requires, after the usual separation of time,

$$\psi(t;\mathbf{r}) = e^{-iEt}\psi_E(\mathbf{r}) \tag{2.7.8}$$

solving the following eigenvalue problem.

$$-\frac{1}{2M}\left(\nabla - i\nu\nabla\theta\right)^2 \psi_E(\mathbf{r}) = E\psi_E(\mathbf{r}) \tag{2.7.9}$$

The eigenfunctions necessarily have theform

$$\psi_E(\mathbf{r}) = e^{i\nu\theta}\psi_k^0(\mathbf{r}) \tag{2.7.10}$$

where $\psi_k^0(\mathbf{r})$, though governed by the free Hamiltonian

$$h^0 = e^{-i\nu\theta}h\,e^{i\nu\theta} \tag{2.7.11}$$

$$h^0\psi_k^0 = \frac{k^2}{2M}\psi_k^0 \ , \qquad E = \frac{k^2}{2M} \tag{2.7.12}$$

is not a conventional plane wave, owing to a non-trivial boundary condition,

$$\psi_k^0(r, \theta = 0) = e^{-i2\pi\nu}\psi_k^0(r, \theta = 2\pi) \tag{2.7.13}$$

which must be met so that $\psi_E(\mathbf{r})$ is single-valued.

Since rotation by 2π corresponds to double exchange of particles, we see that ψ_k^0 acquires a statistics factor $-\pi\nu = e^2/2\kappa$, in agreement with (2.5.1) for $e_1 = e_2$. Moreover, the [relative] angular momentum

$$
\begin{aligned}
J &= \mathbf{r}\times M\mathbf{v} = \mathbf{r}\times\mathbf{p} - \mathbf{r}\times\mathbf{a} \\
&= \mathbf{r}\times\mathbf{p} + \frac{e^2}{2\pi\kappa}
\end{aligned}
\tag{2.7.14}
$$

indicates that each particle possesses additional spin $e^2/4\pi\kappa$ again in agreement with (2.5.1).

[The angular momentum operator acting on the multi-valued wavefunction ψ_k^0 is

$$
\begin{aligned}
J^0 &= e^{-i\nu\theta}J\,e^{i\nu\theta} = e^{-i\nu\theta}\frac{1}{i}\frac{\partial}{\partial\theta}e^{i\nu\theta} + \frac{e^2}{2\pi\kappa} \\
&= \frac{1}{i}\frac{\partial}{\partial\theta} + \nu + \frac{e^2}{2\pi\kappa} = \frac{1}{i}\frac{\partial}{\partial\theta} \ ,
\end{aligned}
$$

i.e. it is just the angular derivative. Nevertheless, its eigenvalues are non-integral, just as those of J in (2.7.14), since $J^0 = \frac{1}{i}\frac{\partial}{\partial\theta}$ acts on multi-valued functions which satisfy (2.7.13). It should further be emphasized that, as we have already stated, the reason that J is just $r \times M\mathbf{v}$ and not $\mathbf{r} \times \mathbf{p}$ is because our effective particle theory arises from a *dynamical* model for the gauge potential, where the dynamics is governed by the Chern–Simons term. If the problem (2.7.5) and (2.7.6) is viewed as describing single particle motion in an *externally* prescribed gauge potential $\mathbf{a}(\mathbf{r})$, then the correct angular momentum *is* $\mathbf{r} \times \mathbf{p}$, with integral eigenvalues [12]].

The Schrödinger equation (2.7.5), (2.7.6) and (2.7.7) leads only to scattering, and the scattering amplitude has been obtained long ago by Aharonov and Bohm, and later by Ruijsenaars [14]. More recently the problem has re-emerged in the context of planar gravity [3e]. As we shall see, there too one seeks free solutions with unconventional boundary conditions.

I shall now present the solution, using the gravitational techniques, which have the advantage of giving an explicit wave function, in the form of a contour integral [2f]

$$\psi_k^0(\mathbf{r}) = \oint \frac{dz}{2\pi} e^{i\mathbf{k}(z)\cdot\mathbf{r}} \rho(z) \tag{2.7.15}$$

Here $\mathbf{k}(z) = (k\cos z, k\sin z)$ and it is obvious that $\psi_k^0(\mathbf{r})$ satisfies the free equation. That it also satisfies (2.7.13) requires a special contour C and weight function ρ, which we now derive, by considering the scattering problem, the radial equation and the phase shifts.

The *Ansatz* $u_E^j(r)\frac{e^{ij\theta}}{\sqrt{2\pi}}$ is made for $\psi_E(\mathbf{r})$ in (2.7.9); j is an arbitrary integer to insure single valuedness, $j = 0, \pm1, \pm2, \ldots$; $u_E^j(r)$ satisfies the radial equation,

$$\left(-\frac{1}{r}\frac{d}{dr}r\frac{d}{dr} + \frac{(j-\nu)^2}{r^2}\right) u_E^j(r) = k^2 u_E^j(r) \tag{2.7.16}$$

and is given by a Bessel function.

$$u_E^j(r) = \sqrt{M} J_{|j-\nu|}(kr) \tag{2.7.17}$$

The angular momentum of this partial wave is $j-\nu$; see (2.7.14). The normalization is fixed by

$$\int_0^\infty r dr\, u_E^j(r) u_E^j(r) = \delta(E - E') \tag{2.7.18a}$$

and insures

$$\int_0^\infty dE\, u_E^j(r) u^j E(r') = \frac{1}{r}\delta(r - r') \tag{2.7.18b}$$

When the plane wave identity

$$e^{ikr\cos\theta} = \sum_{j=-\infty}^{\infty} e^{ij(\theta+\frac{\pi}{2})} J_j(kr) \tag{2.7.19}$$

is recalled, we are led to form the scattering solution by

$$\psi_E(\mathbf{r}) = \sum_{j=-\infty}^{\infty} e^{i(\delta_j + \frac{\pi}{2}j)} u_E^j(r) \frac{e^{ij\theta}}{\sqrt{2\pi}} \tag{2.7.20}$$

where the phase shift δ_j of $u_E^j(r)$ relative to $J_j(kr)$ is identified from their large r asymptotes.

$$\delta_j = \begin{cases} \nu\frac{\pi}{2} & j > [\nu] \\ -\nu\frac{\pi}{2} - j\pi & j \leq [\nu] \end{cases} \tag{2.7.21}$$

Here the brackets [ ] indicate integer part. The energy independence of the phase shifts is a consequence of scale invariance [15].

From (2.7.20) and (2.7.21), the wave function is constructed as

$$\begin{aligned} \psi_E(\mathbf{r}) = &\sqrt{\frac{M}{2\pi}} \sum_{j=[\nu]+1}^{\infty} e^{i\frac{\pi}{2}(\nu+j)} J_{j-\nu}(kr) e^{ij\theta} \\ &+ \sqrt{\frac{M}{2\pi}} \sum_{j=-\infty}^{[\nu]} e^{-i\frac{\pi}{2}(\nu+j)} J_{\nu-j}(kr) e^{ij\theta} \end{aligned} \tag{2.7.22}$$

The sums are performed by using the Schläfli contour representation for the Bessel function.

$$J_\alpha(kr) = e^{i\frac{\pi}{2}\alpha} \oint_s \frac{dz}{2\pi} e^{-ikr\cos z} e^{iz\alpha} \tag{2.7.23}$$

The Schläfli contour C_s begins at $z = -3\pi/2 + i\infty$, descends to slightly above the real axis, passes from $z = -3\pi/2 + i(0^+)$ to $z = \pi/2 + i(0^+)$ and ascends to $z = \pi/2 + i\infty$. The sums are now geometric, and give for the two terms in (2.7.22), respectively,

$$\begin{aligned} \psi_E(\mathbf{r}) = &\left(\frac{M}{2\pi}\right)^{1/2} \oint_s \frac{dz}{2\pi} e^{-ikr\cos s} e^{i([\nu]\pi - \{\nu\}z + [\nu]\theta)} \frac{-1}{1+e^{-i(z+\theta)}} \\ &+ \left(\frac{M}{2\pi}\right)^{1/2} \oint_s \frac{dz}{2\pi} e^{-ikr\cos s} e^{i([\nu]\pi - \{\nu\}z - [\nu]\theta)} \frac{1}{1+e^{i(z-\theta)}} \\ = &\left(\frac{M}{2\pi}\right)^{1/2} \oint_{-s} \frac{dz}{2\pi} e^{-ikr\cos s} e^{i([\nu]\pi + \{\nu\}z + [\nu]\theta)} \frac{1}{1+e^{i(z-\theta)}} \\ &+ \left(\frac{M}{2\pi}\right)^{1/2} \oint_s \frac{dz}{2\pi} e^{ikr\cos s} e^{i([\nu]\pi + \{\nu\}z + [\nu]\theta)} \frac{1}{1+e^{i(z-\theta)}} \end{aligned} \tag{2.7.24a}$$

$$\{\nu\} \equiv \nu - [\nu]$$

In passing from the first to the second equality, the change of variables $z \to -z$ is performed in the first integral. As a consequence, the integration contour for that integral, now called C_{-s}, becomes the mirror image of the Schläfli contour C_s. [C_{-s} starts from $3\pi/2 - i\infty$, ascends to $3\pi/2$ below the real axis, passes to $-\pi/2$ and descends to $-\pi/2 - i\infty$.] As a further consequence, the integrands of the two contour integrals become identical [we use $e^{i[\nu]\pi} = e^{-i[\nu]\pi}$]. To proceed, contours are shifted: C_{-s} is shifted by $\pi/2$ to the left, and C_s by $\pi/2$ to right, so that the vertical portions of both contours are at $z = \pm\pi$. The last step is to redefine the integration variable by $z = -z' + \theta - \pi$. The z' integral now runs over the contour C depicted in Fig. 1a, and $\psi_E(\mathbf{r})$ is represented by [z' is renamed z]

$$\begin{aligned} \psi_E(\mathbf{r}) &= \left(\frac{M}{2\pi}\right)^{1/2} e^{i\nu\theta} \oint \frac{dz}{2\pi} e^{ikr\cos(z-\theta)} \frac{e^{-i\{\nu\}z}}{1-e^{-iz}} \\ &= \left(\frac{M}{2\pi}\right)^{1/2} e^{i\nu\theta} \oint \frac{dz}{2\pi} e^{i\mathbf{k}(z)\cdot\mathbf{r}} \frac{e^{-i\{\nu\}z}}{1-e^{-iz}} \end{aligned} \tag{2.7.24b}$$

[A constant phase factor has been suppressed.] Thus we have derived the representation (2.7.15), with contour C as in Fig. 1a and $\rho(z)$ given by

$$\rho(z) = \left(\frac{M}{2\pi}\right)^{1/2} \frac{e^{-i\{\nu\}z}}{1-e^{-iz}} \tag{2.7.25}$$

That $\psi_E(\mathbf{r})$ as given by (2.7.24b) satisfies the free Schrödinger equation is obvious; that it is single-valued — periodic in θ with 2π period — is more easily seen in (2.7.24a).

The contour C avoids the pole in $\rho(z)$ at $z = 0$. However, we may alternatively enclose the pole and replace the contour C by the three-segmented contour $\tilde{C}$, depicted in Fig. 1b. The portion encircling the pole is evaluated by Cauchy's residue theorem, contributing $(M/2\pi)^{1/2} e^{i(kr\cos\theta + \nu\theta)}$; the portions arising from the vertical axes are presented in terms of real integrals by setting $z = \theta - 2\pi + iy$ and $z = \theta + iy$. Evidently, this separation decomposes the total scattering wave function $\psi_E(\mathbf{r})$ into an incoming wave [pole contribution] and the scattered wave [vertical contour contributions].

$$\psi_E(\mathbf{r}) = \left(\frac{M}{2\pi}\right)^{1/2} (\psi^{\text{in}}(\mathbf{r}) + \psi_E^{\text{sc}}(\mathbf{r}) \tag{2.7.26a}$$

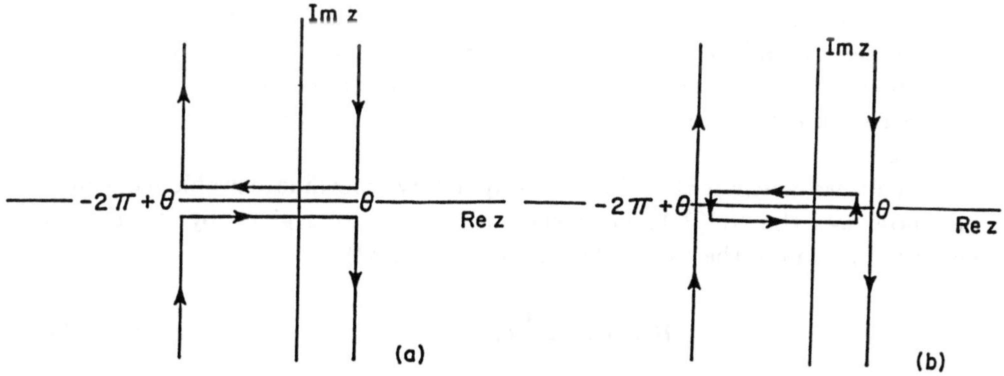

Fig. 1. (a) Integration contour C for the wave function (2.7.24b). The pole at the origin is avoided. (b) Contour $\tilde{C}$ equivalent to contour C. The pole at the origin is enclosed, giving rise to the incoming wave. The vertical contours produce the scattered wave.

$$\psi_E^{\rm in}(\mathbf{r}) = e^{i(kr\cos\theta+\nu\theta)} \tag{2.7.26b}$$

$$\psi_E^{\rm sc}(\mathbf{r}) = e^{i[\nu]\theta} e^{i\nu\pi} \sin\nu\pi \int_{-\infty}^{\infty} \frac{dy}{\pi} e^{ikr\cosh y} \frac{e^{\{\nu\}y}}{e^{y-i\theta}-1} \tag{2.7.26c}$$

The large r asymptote of $\psi_E^{\rm sc}(\mathbf{r})$ defines the scattering amplitude $f(\theta)$ through the formula

$$\psi_E^{\rm sc}(\mathbf{r}) \underset{r\to\infty}{\longrightarrow} \sqrt{\frac{i}{r}} f(\theta) e^{ikr} \tag{2.7.27}$$

Although the integral (2.7.26c) for $\psi_E^{\rm sc}(\mathbf{r})$ cannot be performed, its limit at large r can be evaluated. The circular wave formula (2.7.27) is found, with scattering amplitude

$$f(\theta) = \frac{1}{\sqrt{2\pi k}} e^{-i\{\nu\}\theta} e^{i(\nu+1/2)(\theta+\pi)} \frac{\sin\nu\pi}{\sin\theta/2} \tag{2.7.28}$$

Equations (2.7.26b) and (2.7.28) are essentially the results of Aharonov and Bohm and Ruijsenaars [14]; note especially that the incoming wave is not a plane wave, but is modulated by the additional phase $e^{i\nu\theta}$.

Since $\mathbf{r} = \mathbf{r}_1 - \mathbf{r}_2$, $e^{i\mathbf{k}(z)\cdot\mathbf{r}}$ is a product of two plane waves

$$\begin{aligned} e^{i\mathbf{k}(z)\cdot(\mathbf{r}_1-\mathbf{r}_2)} &= \psi_1^z(\mathbf{r}_1)\psi_2^z(\mathbf{r}_2) \\ \psi_1^z(\mathbf{r}) &= e^{i\mathbf{k}(z)\cdot\mathbf{r}} \\ \psi_2^z(\mathbf{r}) &= e^{-i\mathbf{k}(z)\cdot\mathbf{r}} \end{aligned} \tag{2.7.29}$$

Hence the representation (2.7.24b) shows explicitly how products of one-body plane waves are superposed to form our solution [2f].

$$\psi_E^0(\mathbf{r}) = \left(\frac{M}{2\pi}\right)^{1/2} \oint \frac{dz}{2\pi} \psi_1^z(\mathbf{r}_1)\psi_2^z(\mathbf{r}_2) \frac{e^{-i\{\nu\}z}}{1-e^{-iz}} \tag{2.7.30}$$

As yet we do not have a similar closed form for the N-body wave function. The problem is reminiscent of the δ-function interaction on a line. There too the many-body wave function is obtained by superposing one-body wave functions in a fashion prescribed by the Bethe *Ansatz*. Perhaps similar ideas will prove useful here.

Finally, we conclude that the two-body relative coordinate problem of course also possesses the $SO(2,1)$ symmetry, with generators given by the relative coordinate parts of the two-body generators (2.6.22)

$$H = (h) = \frac{1}{2} M v^2 \tag{2.7.31a}$$

$$D = tH - \frac{1}{4} M(\mathbf{r}\cdot\mathbf{v} + \mathbf{v}\cdot\mathbf{r}) \tag{2.7.31b}$$

$$K = -t^2 H + 2tD + \frac{1}{2} M r^2 \tag{2.7.31c}$$

The algebraic properties of these quantities hold as before; now they are based on the dynamical algebra

$$[r^i, r^j] = 0 \ , \quad [r^i, Mv^j] = i\delta^{ij} \ , \quad [Mv^i, Mv^j] = i\epsilon^{ij} 2\pi\nu\, \delta(\mathbf{r}) \tag{2.7.32}$$

The Casimir in (2.6.33) can be expressed in terms of the angular momentum (2.7.14)

$$\mathcal{J}^2 = \frac{1}{4}(J^2 - 1) \tag{2.7.33}$$

Since the eigenvalues of J are $j - \nu$, those of $\mathcal{J}^2$ are $\frac{1}{4}\left((j-\nu)^2 - 1\right)$, and the entire motion at fixed angular momentum is described by a single, irreducible, unitary and infinite-dimensional representation of $SO(2,1)$. We have already remarked that the energy independence of the phase shifts is a consequence of the symmetry. Because of the higher symmetry, the time-dependent Schrödinger equation (2.7.5) can be separated in coordinates other than the usual time and space. Indeed group theory may be used to give a complete, alternative analysis of the problem [2f].

3. Planar Gravity

3.1 Introduction

The equations for Einstein's theory of gravity — general relativity — can be presented in any space-time with dimension d equal to or greater than three: The Einstein tensor

$$G_{\mu\nu} \equiv R_{\mu\nu} - \frac{1}{2} g_{\mu\nu} R \tag{3.1.1}$$

vanishes in the absence of matter sources,

$$G_{\mu\nu} = 0 \tag{3.1.2a}$$

while in their presence it is proportional to the energy-momentum tensor of matter, $T_{\mu\nu}$.

$$G_{\mu\nu} = 2\pi G\, T_{\mu\nu} \tag{3.1.2b}$$

Here $R_{\mu\nu}$ and R are traces of the four-index Riemann tensor $R_{\alpha\mu\beta\nu}$ in which all local geometrical information about the space-time is encoded. G is the gravitational coupling constant — the generalization to other dimensions of Newton's constant; in (3.1.2b) G enters with an unconventional normalization that is convenient for the subsequent analysis. The reason that (3.1.2) cannot be posited in two space-time dimensions is because there $G_{\mu\nu}$ vanishes identically. [However other geometrical equations have been proposed at $d = 2$ [16]].

It is obvious from (3.1.2b) that when space-time is flat, *i.e.* when the Riemann tensor vanishes, so also does the Einstein tensor and $T_{\mu\nu}$ must be zero. In general, the converse does not hold: absence of matter implies vanishing Einstein tensor, but the Riemann tensor need not be zero so that empty space-time need not be flat. However, in three dimensions the Riemann tensor is linearly related to the Einstein tensor,

$$R^{\alpha\mu}_{\beta\nu} = \epsilon^{\alpha\mu\gamma}\epsilon_{\beta\nu\delta}G^{\delta}_{\gamma} \tag{3.1.3}$$

so that the vanishing of the latter implies the vanishing of the former: empty space-time is necessarily locally flat [17].

Several consequences follow immediately: since the vacuum state [empty space-time] is locally flat, there are no gravitational waves in the classical theory, and upon quantization there are no quantum gravitons. Sources produce curvature, but only locally at the location in space-time of the sources. Forces between sources are not mediated by graviton exchange, since there are no gravitons. Rather interactions arise because the locally flat space-time possesses in the large non-trivial geometrical and topological structure that gives rise to non-trivial motions. It also follows that the non-relativistic limit of Einstein's general relativity in three-dimensions is not three-dimensional Newtonian gravity, which involves a conventional force law that decreases with the inverse power of the distance.

It is the purpose of our research program to study in three-dimensional space-time the classical and quantum motions of matter that interacts gravitationally [3]. Since there are no propagating gravitational degrees of freedom, the problem is tractable, and we can learn much about the puzzles that are encountered when a geometrical theory is confronted by quantum mechanics. In four dimensions these puzzles exist as well, and it is my opinion that understanding them is important for understanding quantum gravity; a task quite independent of and perhaps more fundamental than the task of overcoming the unrenormalizable infinities that pollute four-dimensional gravity, but are absent in three dimensions since non-renormalizable graviton exchange does not occur. To conclude these introductory remarks, I note the following points.

(a) The theory can be elaborated by adding a cosmological constant to the field equation. The vacuum is then a space of constant curvature, whose sign depends on the sign of the cosmological constant. While some investigations of such models have been performed, I shall not further discussthem here [3c].

(b) Another elaboration of the conventional theory involves adding a topological term, analogous to the gauge theoretic modification [3a]. This Chern–Simons addition will be discussed below.

3.2 Classical Space-Times

We record several interesting space-times that arise from classical sources [3b]. We begin with a single massive but spinless point-particle. Without loss of generality the particle is taken to be at rest at the origin of the coordinate system, *i.e.* it is described by an energy-momentum tensor all whose components, except the energy density, vanish,

$$\begin{aligned}\sqrt{\det g_{\mu\nu}}T^{00} &= M\delta(X)\delta(Y) \\ T^{0i} &= T^{ij} = 0\end{aligned} \tag{3.2.1}$$

Here M is the particle mass.

The task is to find the metric or equivalently to give a formula for the line element. Clearly it is non-trivial only in its spatial components,

$$(ds)^2 = (dt)^2 - (d\ell)^2 \tag{3.2.2}$$

and we need to find expressions for $(d\ell)^2$.

We recognize that we seek a space which is everywhere flat [$T^{\mu\nu}$ vanishes] except at the origin where a δ-function singularity concentrates the curvature. It is clear that the desired space is a cone, with the source particle positioned at the apex [3b,4d,17]. It remains to give an analytic description of this obvious geometrical fact.

To solve the Einstein equation (3.1.2b) with sources given by (3.2.1), it is necessary to choose a coordinate system, and the conical solution looks different in different coordinates. Of course only the two-dimensional spatial section needs to be considered.

Particularly useful coordinates, which lend themselves to a many-body generalization, are the conformal ones where the metric tensor is a multiple of the flat metric tensor; this can always be locally achieved in two dimensions. The conformally flat spatial metric that solves Einstein's equation then leads to the following spatial interval.

$$(d\ell)^2 = \frac{1}{R^{2GM}}\Big((dR)^2 + R^2(d\Theta)^2\Big) \tag{3.2.3}$$

Here the variables range over the conventional circles.

$$\begin{aligned}0 \le\ &R \le \infty \\ -\pi \le\ &\Theta \le \pi\end{aligned} \tag{3.2.4}$$

While (3.2.3) certainly provides the desired solution, it does not seem to produce the cone described earlier. Nor is it manifest that the space is flat except at the origin.

All this can be seen by passing to another coordinate system, attained from (3.2.3) and (3.2.4) by a change of variables.

$$\begin{aligned}r &= \frac{R^{1-GM}}{1-GM} \\ \theta &= (1-GM)\Theta\end{aligned} \tag{3.2.5}$$

In terms of r and θ the spatial metric is flat, and the line-element is trivial.

$$(d\ell)^2 = (dr)^2 + r^2(d\theta)^2 \tag{3.2.6}$$

However, the range of the new variables is unconventional — an angular region is excised, since according to (3.2.4) the range of (r,θ) is

$$\begin{gathered} 0 \le r \le \infty \\ -\pi(1-GM) \le \theta \le \pi(1-GM) \end{gathered} \tag{3.2.7}$$

This describes a cone, with apex determined by GM. [Henceforth we take $GM \le 1$. For $GM > 1$, the space changes character and the description becomes more complicated [3b]. At $GM = 1$, it is seen from (3.2.3) that space becomes a cylinder in the variable $r = \ln R$.]

In summary, we say that a point particle of mass M at the origin gives rise to a locally flat space-time, but the global identification of coordinate variables is unconventional and reveals the presence of a massive point-particle: the point (t, r, θ) is identified with

$$(t, r, \theta) \approx (t, r, \theta + 2\pi(1 - GM) \tag{3.2.8a}$$

In terms of a complex variable description of the space, $z = x + iy$, we identify z with

$$z \approx e^{-2\pi i GM} z \tag{3.2.8b}$$

This is the analog in planar gravity of the Schwarzschild solution.

To find the planar analog of the Kerr solution, we endow our point-particle at the origin with spin S, *i.e.* now the energy-momentum tensor possess non-trivial energy density and momentum density, the latter giving rise to no momentum but to angular momentum S.

$$\begin{aligned} \sqrt{\det g_{\mu\nu}} T^{00} &= M\delta(X)\delta(Y) \\ \sqrt{\det g_{\mu\nu}} T^{0i} &= \sqrt{\det g_{\mu\nu}} T^{i0} = S\epsilon^{ij}\partial_j \delta(X)\delta(Y) \\ T^{ij} &= 0 \end{aligned} \tag{3.2.9}$$

In the spatially conformal coordinate system, the metric that solves the field equation leads to a space-time interval, which is non-trivial in time as well as space.

$$(ds)^2 = (dt + GSd\Theta)^2 - \frac{1}{R^{2GM}}\Big((dR)^2 + R^2(d\Theta)^2\Big) \tag{3.2.10}$$

Once again, by a change of variables one may pass to a locally flat space-time, where the presence of a massive, spinning source is encoded in a non-trivial identification of coordinate variables. Defining new spatial variables as in (3.2.5) and also a new time variable τ by

$$\tau = t + GS\Theta = t + \frac{GS}{1 - GM}\theta \tag{3.2.11}$$

we see that (3.2.10) becomes flat,

$$(ds)^2 = (d\tau)^2 - (dr)^2 - r^2(d\theta)^2 \tag{3.2.12}$$

but the required identification is

$$(\tau, r, \theta) \approx (\tau + 2\pi GS, r, \theta + 2\pi(1 - GM)) \tag{3.2.13}$$

Time is helical, space is conical and there are closed time-like contours.

Note that specifying a solution is equivalent to specifying an element of the $2+1$-dimensional Poincaré group that effects the identification (3.2.13).

The static one-body solution can be generalized to describe N particles located at $\mathbf{R}_i$, with masses M_i and spins S_i, $i = 1, \ldots, N$ [18]. One finds in spatially conformal coordinates

$$(ds)^2 = \left(dt + G \sum_{i=1}^{N} S_i \frac{(\mathbf{R}-\mathbf{R}_i)}{|\mathbf{R}-\mathbf{R}_i|^2} \times d\mathbf{R} \right)^2 - \prod_{i=1}^{N} \frac{1}{|\mathbf{R}-\mathbf{R}_i|^{2GM_i}} (d\mathbf{R})^2 \quad (3.2.14)$$

The passage to locally flat coordinates is effected by first defining a new time τ.

$$d\tau = dt + G \sum_{i=1}^{N} S_i \frac{(\mathbf{R}-\mathbf{R}_i)}{|\mathbf{R}-\mathbf{R}_i|^2} \times d\mathbf{R} \quad (3.2.15)$$

This hides the spins in complicated identifications on τ. To flatten the spatial interval, it is useful to express it in complex variables $Z = X + iY$, *etc.*

$$(d\ell)^2 = \left(\prod_{i=1}^{N} \frac{1}{(Z - Z_i)^{GM_i}} \right) dZ \left(\prod_{i=1}^{N} \frac{1}{(\bar{Z} - \bar{Z}_i)^{GM_I}} \right) d\bar{Z} \quad (3.2.16)$$

Thus the definition

$$dz = \left(\prod_{i=1}^{N} \frac{1}{(Z - Z_i)^{GM_i}} \right) dZ \quad (3.2.17)$$

gives the flat spatial interval

$$(d\ell)^2 = dz\, d\bar{z} \quad (3.2.18)$$

but complicated identifications on the complex plane, which generalize (3.2.8b), reveal the presence of N particles with masses M_i. Unlike in the one-body problem, we cannot express z as a closed form function of Z, but for most purposes the integral expression suffices,

$$z = \int^{Z} dZ' \prod_{i=1}^{N} \frac{1}{(Z' - Z_i)^{GM_i}} \quad (3.2.19)$$

and it can be explicitly evaluated in special cases.

It is easy to show that the above solution also satisfies self-consistently the geodesic equation [3b]. Thus a static N-body configuration exists and is stable in three-dimensional space-time, in contrast to higher dimensions where gravitational attraction would prevent this. This demonstrates vividly the absence of Newtonian attraction in our theory.

With point-particle sources, the two-dimensional space is flat, but curvature is concentrated on a lower-dimensional sub-space: the zero-dimensional collection of points where the particles are located. One may next consider flat space with curvature concentrated on one-dimensional lines; *i.e.* string sources in the plane, which presumably correspond to domain walls in four-dimensional space-time, just as points on the plane correspond to strings in four-dimensional space-time.

When considering strings, it is natural to allow for tension along the string; otherwise the source is an uninteresting pulvarization of the point-particle — a "dust" string.

In the spinless case the results are simple and startling [3f,19]. There are no open strings, only closed ones. A circular source at $r = a$ is described by an energy-momentum tensor whose non-vanishing components are

$$\sqrt{\det g_{\mu\nu}} T_0^0 = \mu \delta(r-a) \tag{3.2.20a}$$

$$\sqrt{\det g_{\mu\nu}} T_\theta^\theta = \tau \delta(r-a) \tag{3.2.20b}$$

Here μ and τ are mass and stress density/per unit length; the total mass is $M = 2\pi a\mu$; for a relativistic string $\tau = \mu$. The momentum density and the other stress components vanish. With this source the space-time interval in conformally flat spatial coordinates is

$$(ds)^2 = \begin{cases} \left(1 - 2\pi G a\tau \ln \frac{r}{a}\right)(dt)^2 - \left(\frac{a}{r}\right)^2 \left((dr)^2 + r^2(d\theta)^2\right) & r \geq a \\ (dt)^2 - (dr)^2 - r^2(d\theta)^2 & r \leq a \end{cases} \tag{3.2.21}$$

The exterior spatial interval also reads $\left(d\, a \ln \frac{r}{a}\right)^2 + (ad\theta)^2$, which is a half-cylinder of radius a extending from infinity to $r = a$, where it is capped by the flat disk of the $r \leq a$ region. Moreover, the total mass $M = 2\pi a\mu$ is given by G^{-1}, so that

$$GM = 1 \tag{3.2.22}$$

We have seen earlier that for point-particles obeying (3.2.22) the space is a cylinder; here (3.2.22) is always obeyed for spinless strings under tension and the space is a capped cylinder.

Although for $\tau > 0$, g_{00} vanishes at a finite distance, this is not a conventional horizon because g_{00} does not change sign, but time does "stand still" there. Clearly there exist solutions with either sign of τ and unrelated to μ. However, for a relativistic string $\tau = \mu > 0$.

For more discussion on these extended objects and inclusion of spin, please consult the research papers [3f,19].

3.3 Quantum Dynamics

The simplest non-trivial dynamics arises when we consider the interaction of two point-particles with each other. As in other contexts, it is possible to pass to the center-of-mass frame where the relative coordinate moves in an effective potential that describes the interaction [3d]. The same problem arises without the center-of-mass reduction, but in the limit when one particle's mass becomes much larger than the other [3e].

In view of this, it suffices to consider the problem of a test particle [mass m] moving in the field produced by the source particle [mass M] located at the origin.

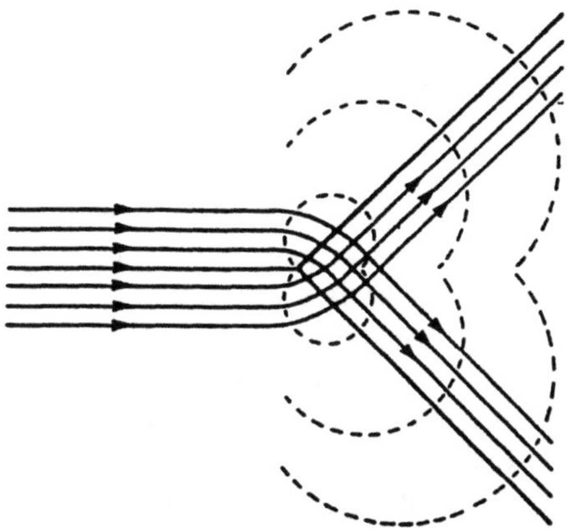

Fig. 2. Qualitatative pictorialization for scattering of waves on an obstacle at the origin. The two sharp lines are classical trajectories with scattering angle $\pm\pi GM$ in (3.3.1), the sign depending on which side the trajectory passes the source. The envelope to the right of the source, formed by heavy diagonal lines, is the sharp geometrical shadow. Broken lines represent diffraction on two sharp edges, even though no edge is actually present — the source [conical defect] produced the "edges."

The classical motion of a spinless test particle is easy to describe: in flat coordinates there is no deviation from straight-line motion. However, when the unconventional identification (3.2.13) is performed, we find a classical scattering angle,

$$\Delta\theta_{\text{classical}} = \pm\pi GM \tag{3.3.1}$$

and a classical time delay,

$$\Delta t_{\text{classical}} = \mp\frac{GS\pi}{1-GM} \tag{3.3.2}$$

where S is the source particle's spin, and the sign depends on which side the source is passed. The classical trajectories are depicted in Fig. 2 [ignore the dotted lines for the moment]. They depend only on the impact parameter, but not on the energy; the scattering angle does not vary with impact parameter, except in its sign.

Next we give a quantum mechanical description and to this end we solve a quantum mechanical equation appropriate to the test particle: Schrödinger or Klein–Gordon for spinless test particles; Dirac for spin 1/2 test particles, *etc.* [We do not second quantize the matter degrees of freedom.] The question that still must be considered is what interaction should we use to describe the influence of the source on the test particle.

The answer that we propose is that no interaction need be considered; rather we solve the free, non-interacting equation but impose on the solution a coordinate condition that reflects the identification (3.2.13).

For example, let us consider the simplest case first — a spinless test particle in a spinless source. The equation we propose to solve is the free [square-root] Klein–Gordon,

$$i\frac{\partial}{\partial t}\psi(t;r,\theta) = \sqrt{-\nabla^2 + m^2}\,\psi(t;r,\theta) \tag{3.3.3}$$

with the requirement that

$$\begin{aligned}\psi(t;r,\theta) &= \psi(t;r,\theta+2\pi\alpha)\\ \alpha &= 1-GM\end{aligned} \tag{3.3.4}$$

[If non-relativistic motion is of interest, the non-local "square root" operator is replaced by $m - \nabla^2/2m$, which leads to the free Schrödinger equation, with boundary conditions (3.3.4). The mathematical analysis is identical.]

The solution of (3.3.3), which satisfies (3.3.4) is constructed along the same lines as the Aharonov–Bohm scattering solution discussed in detail earlier. I shall not repeat that presentation, beyond remarking that time, radial and angular variables are separated in the usual way, with partial waves carrying angular momentum, ℓ, which is not integer quantized, rather $\alpha\ell$ is an integer. This of course is a consequence of the fact that the angular range is $2\pi\alpha$, not 2π.

The scattering solution is given by a contour integral in which plane waves are superposed, with definite weight [3d,e]

$$\begin{aligned}\psi(t;r,\theta) &= e^{-iEt}\oint\frac{dz}{2\pi}e^{i\mathbf{k}(z)\cdot\mathbf{r}}\frac{1}{1-e^{iz/\alpha}}\\ &\equiv e^{-iEt}\psi(r,\theta)\end{aligned} \tag{3.3.5}$$

Here $E=\sqrt{k^2+m^2}$ and $\mathbf{k}$ is the vector of magnitude k, rotated by the contour integration variable z: $\mathbf{k} = (k\cos z,\ k\sin z)$. That (3.3.5) satisfies (3.3.3) is obvious, that also the boundary condition (3.3.4) is obeyed depends on the specific weight function in (3.3.5) and also on the contour, which is depicted in Fig. 3a.

The weight function has poles on the real axis at $z = z_n = 2\pi n\alpha$ and the contour C avoids them. However, the contour may be deformed as in the discussion of the Aharonov–Bohm problem. We can consider the equivalent, three segment contour $\tilde{C}$, depicted in Fig. 3b, where the poles are encircled and also there are integrals along the vertical lines. The contribution from the encircled poles is evaluated by Cauchy's theorem; it gives the incoming wave. The remaining integrals along the vertical lines give the scattered wave, but the integrations cannot be evaluated, so no closed form is available. Nevertheless, the large r asymptote is accessible, and the scattering amplitude is determined explicitly [3d,e]

$$\psi(r,\theta) = \psi^{\text{in}}(r,\theta) + \psi^{\text{sc}}(r,\theta) \tag{3.3.6}$$

$$\psi^{\text{in}}(r,\theta) = \alpha\sum_n{}' e^{i\mathbf{k}(z_n)\cdot\mathbf{r}} \tag{3.3.7a}$$

$$\begin{aligned}\psi^{\text{sc}}(r,\theta) &= i\int_{-\infty}^{\infty}\frac{dy}{2\pi}e^{ikr\cosh y}\left[\frac{1}{1-e^{i\frac{\pi}{\alpha}}e^{-\frac{1}{\alpha}(y+i\theta)}} - \frac{1}{1-e^{-i\frac{\pi}{\alpha}}e^{-\frac{1}{\alpha}(y+i\theta)}}\right]\\ &\underset{r\to\infty}{\longrightarrow}\sqrt{\frac{i}{r}}f(\theta)e^{ikr}\end{aligned} \tag{3.3.7b}$$

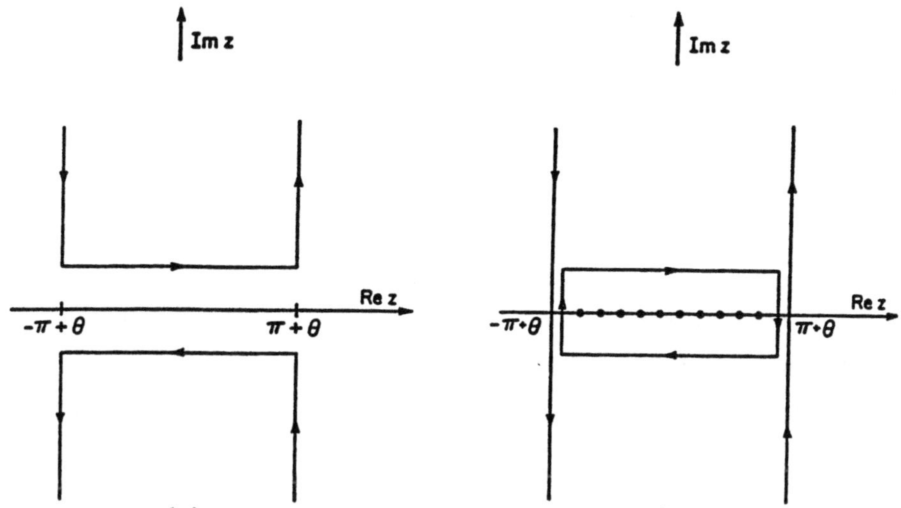

Fig. 3. (a) Integration contour C for the representation of $\psi(r,\theta)$ in (3.3.5). (b) Integration contour $\tilde{C}$ for the representation of $\psi(r,\theta)$ equivalent to that in (a) but giving rise to the decomposition $\psi = \psi^{\text{in}} + \psi^{\text{sc}}$. The incoming wave ψ^{in} is given by the [negative] Cauchy contour around the poles at $z = 2\pi n\alpha$, indicated by heavy dots. The integrals along the left and right verticle contours determine the scattered wave ψ^{sc}, whose large distance asymptote defines the scattering amplitude f.

$$f(\theta) = \frac{1}{2\sqrt{2\pi k}} \left[\left(\text{ctn}\frac{\theta - \pi}{2\alpha} - i \right) - \left(\text{ctn}\frac{\theta + \pi}{2\alpha} - i \right) \right] \tag{3.3.8}$$

The prime on the sum in (3.3.7a) indicates that z_n must lie in the interval $[-\pi + \theta, \pi + \theta]$. Note that the incoming wave is not a plane wave, rather it is a superposition of variously rotated plane waves. This is analogous to the modulated plane wave found in the Aharonov–Bohm analysis.

We observe that the scattering amplitude $f(\theta)$ in (3.3.8) is real and vanishes when $1/\alpha$ is an integer. Also there are singularities at finite values of θ, where either of the two cotangents blows up. Finally, the optical theorem, which in two dimensions and with our normalization reads

$$\text{Im}\, f(0) = \sqrt{\frac{k}{4\pi}} \int d\theta\, |f(\theta)|^2 \tag{3.3.9}$$

fails because the left-hand side vanishes and the right-hand side diverges. Nevertheless, there is no loss of unitarity: one can verify from the exact solution (3.3.6) – (3.3.7) that the probability current is conserved. The peculiarities of the scattering amplitude are presumably related to the long-range nature of the "interaction": no matter how far the scattered particle is from the source, it remains on a cone. An interesting problem that here remains is the study of how a wave packet evolves in time.

Going beyond the simplest case, we consider the situation that arises when both the source and the test particle are spinning. The source spin S is arbitrary; for the test particle we consider spins 0 and 1/2, solving the Klein–Gordon and Dirac equations, respectively, but now with the more elaborate identification (3.2.13). One may again give a contour integral representation for the wave function, obtain the incoming wave by performing a Cauchy contour integral, and deduce an explicit formula for the scattering amplitude. The result is an elegant generalization of (3.3.8), which can be presented in universal form, provided the following definitions are made.

S^s = spin of source [can be arbitrary, previously called S].

S^t = spin of test particle [actual calculations done only for $S^t = 0, 1/2$].

E^s = energy of source [taken to be M].

$$E^t = \text{energy of test particle } \left(E^t = \sqrt{k^2 + m^2}\right) \ . \tag{3.3.10}$$

The scattering amplitude is [3d,e]

$$f(\theta) = \frac{e^{-i[\omega]\theta/\alpha}}{2\sqrt{2\pi k}} \left[e^{-i\{\omega\}\pi/\alpha} \left(\text{ctn}\frac{\theta - \pi}{2\alpha} - i \right) - e^{i\{\omega\}\pi/\alpha} \left(\text{ctn}\frac{\theta + \pi}{2\alpha} - i \right) \right] \tag{3.3.11}$$

Here ω is the symmetric cross product

$$\omega = E^s S^t + E^t S^w \tag{3.3.12}$$

while the square and curly brackets denote integer and fractional part, respectively.

$$\omega = [\omega] + \{\omega\} \tag{3.3.13}$$

For the spinless test particle, $S^t = 0$, one can determine from the phase shift $\delta(E)$ the time-delay by Wigner's formula. Agreement with the classical result (3.3.2) is found.

$$\Delta T = 2\frac{\partial \delta(E)}{\partial E} \tag{3.3.14}$$

We may understand the scattering amplitude as arising from diffraction effects [like in physical optics] which supplement the classical trajectories [whose analogy is geometrical optics]. These diffraction patterns are indicated by the dotted arcs in Fig. 2 and the two terms in (3.3.8) correspond to the two branches. We observe that scattering consists of a rotation through the angle $\pm\pi GM$, and we recall that in the presence of spin a rotation is accompanied by a phase change in the wave function. This explains the emergence of the additional phases in (3.3.11) as compared to (3.3.8).

The analysis of the Dirac equation is especially interesting owing to the fact that the Dirac Hamiltonian ceases to be self-adjoint on a conical, time-helical space time [3e] [The same malady afflicts the Dirac equation in the presence of a vortex — the spinning Aharonov–Bohm effect [20].] Of course the derivatives are formally Hermitian, but consideration of the boundary conditions indicates that a self-adjoint extension, depending on parameters, must be made and different physical results emerge with different values for the parameters. [In deriving Eq. (3.3.11) a definite choice is made to insure universality — but other choices are possible.]

In physical terms what is seen here is the failure of the point-particle description. Extended, smooth objects — described *e.g.* by fields — would lead to a self-adjoint Hamiltonian and in the point-particle limit various parameters, characterizing the extended object, survive as boundary terms on the particle surface and provide the missing information. The situation is similar to what is found for the Dirac equation with a [Dirac] point magnetic monopole. The Hamiltonian needs a one-parameter self-adjoint extension [21]. When a smooth 't Hooft–Polyakov monopole is considered, the parameter is identified as the QCD vacuum angle [22]. For the gravitational [and vortex] problems it remains an open question what model for the extended particle gives a physical origin to the mathematically necessary self-adjoint extension parameters.

The loss of self-adjointness appears to be related to the closed time-like curves that are present in a background metric arising from a spinning source.

We conclude this discussion of quantum motion by remarking that the true two-body problem — in contrast to its test particle source-particle equivalent description — is solved on a space with deficit angle given by the eigenvalues of the two-body Hamiltonian [3d]. This truly "Machian" behavior raises conceptual puzzles — for example it is impossible to superpose or compare energy eigensolutions. Moreover, the three- or more-body problem has thus far not been resolved [apart from a very easy special case [3e]] owing in part to difficulties in describing the multi-conical space on which the physical motion takes place.

3.4 Topological Elaborations

Up to now the discussion has been based on the three-dimensional version of the Einstein equation (3.1.2). However, in complete analogy to three-dimensional gauge theories, it is possible to modify (3.1.2) by an additional term, because in three dimensions there exists another second rank tensor that is symmetric and covariantly conserved. Sometimes called the *Cotton tensor*, its form is

$$C^{\mu\nu} = \frac{1}{2\sqrt{\det g_{\mu\nu}}} \epsilon^{\mu\alpha\beta} D_\alpha R^\nu_\beta + \mu \leftrightarrow \nu \tag{3.4.1}$$

Symmetry is manifest, covariant conservation follows from the Bianchi identities. $C^{\mu\nu}$ is traceless as follows from (3.4.1). also with the help of Bianchi identities.

$$C^\mu_\mu = 0 \tag{3.4.2}$$

Moreover, $C^{\mu\nu}$ may be viewed as the three-dimensional conformal tensor — an odd-parity analog of the Weyl tensor, the latter vanishing identically at $d = 3$. [That is why the Riemann tensor is determined by the Einstein tensor.] $C^{\mu\nu}$ is invariant against conformal redefinition of the metric tensor $g^{\mu\nu}(x) \to \lambda(x) g^{\mu\nu}(x)$ and vanishes if and only if space-time is conformally flat, $g_{\mu\nu}(x) = \lambda(x)\eta_{\mu\nu}$. We may supplement/replace the left-hand of (3.1.2) by the addition of a multiple of $C^{\mu\nu}$ [3a]

$$G^{\mu\nu} + \frac{1}{\kappa} C^{\mu\nu} = 0 \tag{3.4.3a}$$

$$G^{\mu\nu} + \frac{1}{\kappa} C^{\mu\nu} = 2\pi G T^{\mu\nu} \tag{3.4.3b}$$

[Also a cosmological constant can of course be added to the equation with or without sources, (3.4.3a) or (3.4.3b) respectively — we shall not do so.]

From its definition (3.4.1), we see that $C^{\mu\nu}$ is of one derivative order higher than $G^{\mu\nu}$, hence the dimension of κ is mass. Analysis of the linearized approximation yields dramatic results. While in the absence of the modification, there are no gravitational excitations, the addition "liberates" a previously "confined" graviton, which now becomes a single propagating mode; moreover, it is massive, while retaining general covariance. The spin is ± 2, the sign being correlated with the sign of κ. [The triple derivative nature of the differential equations (3.4.3) does not give rise to acausality; here, the conformal invariance comes into play, removing possibly dangerous terms from $C^{\mu\nu}$.]

$G^{\mu\nu}$ is obtained variationally from the Einstein–Hilbert action. Similarly, $C^{\mu\nu}$ may be obtained variationally from the Chern–Simons action, for the local Lorentz group in 2+1 dimensions — $SO(2,1)$. Constructing that quantity as in a gauge theory from the connection — either Christoffel or spin — but viewing the connection as a function of the fundamental dynamical variable — either the metric tensor or the *dreibein* respectively — and varying the dynamical variable gives $C^{\mu\nu}$ [3a].

Thus we see that the proposed modification is the complete analog of the situation in the gauge theory, and for that reason the model (3.4.3) is called *topologically massive gravity*.

However, no quantization condition need be imposed on κ [23]. Non-trivial homotopies in a non-compact group like $SO(2,1)$ coincide with those of its maximal compact subgroup, here $SO(2)$; but $SO(2)$ is trivial in this respect, so the gravitational Chern–Simons action is invariant, just as the field equations are covariant, and κ is unrestricted.

It is not known whether topologically massive gravity is renormalizable.

Of course a theory based solely on the Chern–Simons action/Cotton tensor field equation may also be considered [3a]. Here again, there are no propagating degrees of freedom, and due to the tracelessness of $C^{\mu\nu}$, only massless sources, with trace-free energy-momentum tensor can be coupled. However, owing to its triple derivative structure, the topological term is *not* natural for a low energy description, in contrast to the gauge theoretic Chern–Simons term. On the contrary. The Einstein/Hilbert theory is dominant at low energies, while the Chern–Simons/Cotton term dominates at high energy.

I conclude this discussion of topological elaborations on planar gravity by the following observations.

(a) Just like the gauge theoretic Chern–Simons term, the gravitational $SO(2,1)$ Chern–Simons term is induced by virtual fermions [24]. This raises a puzzle about our treatment of quantum scattering, when the matter degrees of freedom are *second* quantized fermions and the "bare" gravitational action is just the conventional Einstein–Hilbert action. On the one hand the bare gravitational action suggests that there are no propagating gravitational degrees of freedom. On the other, fermion loops induce a Chern–Simons action which when considered together with the bare action indicates the presence of massive, propagating gravitons. So which viewpoint is correct? Is the emergent "graviton" a fermion/anti-fermion bound state? How should perturbative calculations be organized?

(b) The fact that in planar Einstein gravity, the gravitational field is locally determined by matter sources is analogous to the situation in gauge theoretic Chern–Simons theory. Indeed the analogy exposes an identity: the Einstein–Hilbert action is also the Chern–Simons term for $ISO(2,1)$, the inhomogeneous $(2+1)$-dimensional Lorentz group, *i.e.* the Poincaré group [25]. There are six generators: J^μ rotations and P^μ translations. With these we associate respectively the "gauge" connections ω^μ and e^μ — the spin connection and *dreibein* — and use an off-diagonal "trace," $\langle P^\mu P^\nu \rangle = 0$, $\langle J^\mu J^\nu \rangle = 0$, $\langle J^\mu P^\mu \rangle = \delta^{\mu\nu}$, to construct the Chern–Simons term. The result is the Einstein–Hilbert action in first-order form.

(c) The Lagrangian for topologically massive gravity consists of $\mathcal{L}_{\mathrm{EH}} + \frac{1}{\kappa}\mathcal{L}_{\mathrm{CS}}$, the Einstein–Hilbert Lagrangian summed with κ^{-1} times the Chern–Simons term. Equivalently we may write it as $\mathcal{L}_{\mathrm{CS}} + \kappa\mathcal{L}_{\mathrm{EH}}$, and view the higher derivative $\mathcal{L}_{\mathrm{CS}}$ as the "kinetic" term and $\kappa\mathcal{L}_{\mathrm{EH}}$ as the "mass" term. The former possesses more symmetry than the latter — it is conformally invariant. In some sense that is "too much" symmetry, and no propagation is possible with just the kinetic term. Inclusion of the less symmetric mass [Einstein–Hilbert] term lowers the symmetry and "liberates" the previously confined graviton. One may even promote κ to a scalar field φ with its own [unspecified] dynamics. The combination $\mathcal{L}_{\mathrm{CS}} + \varphi\mathcal{L}_{\mathrm{EH}} + \mathcal{L}_\varphi$ can be conformally invariant for suitably chosen $\mathcal{L}_\varphi$. Then an expansion about $\langle\varphi\rangle = 0$ contains no propagating gravitons, while the symmetry breaking starting point $\langle\varphi\rangle = \kappa$ liberates the graviton [26].

(d) Some classical solutions to topologically massive gravity have been found. They are planar analogs of Gödel universes [27].

References

1 . G. 't Hooft, in *Proceedings of XIX Schladming School, Acta. Phys. Austr. Suppl. XXII*, **531** (1980)

2 Papers from which the lectures on planar gauge theories are drawn:

a. R. Jackiw and S. Templeton: Phys. Rev. D **23**, 2291 (1981); J. Schonfeld: Nucl. Phys. **B185**, 157 (1981)

b. S. Deser, R. Jackiw and S. Templeton: Phys. Rev. Lett. **48**, 975 (1982); Ann. Phys. (NY) **140**, 372 (1982); (E) **185**, 406 (1988)

c. S. Deser and R. Jackiw: Phys. Lett. **B139**, 371 (1984); L. Faddeev and R. Jackiw: Phys. Rev. Lett. **60**, 1692 (1988)

d. G. Dunne, R. Jackiw and C. Trugenberger: Ann. Phys. (NY) **194**, 197 (1989)

e. G. Dunne, R. Jackiw and C. Trugenberger: Phys. Rev. D **41**, xxx (1990)
f. R. Jackiw: Ann. Phys. (NY) (in press)
Much of this material is summarized in:
g. R. Jackiw, in S.Treiman, R. Jackiw, B. Zumimo and E. Witten, *Current Algebra and Anomalies* (Princeton University Press/World Scientific, Princeton, NJ/Singapore, 1985)
h. R. Jackiw in Lectures presented at the *Fifth Jorge Swieca School* (Campos de Jordão, São Paulo, Brazil, 1989), to be published in the Proceedings.
The above papers and reviews should be consulted for reference to other literature on ths subject.

3 Papers from which the lectures on planar gravity are drawn:
a. S. Deser, R. Jackiw and S. Templeton: Phys. Rev. Lett. **48**, 975 (1982); Ann. Phys. (NY) **140**, 372 (1982); (E) **185**, 406 (1988)
b. S. Deser, R. Jackiw and G. 't Hooft: Ann. Phys. (NY) **152**, 220 (1984)
c. S. Deser and R. Jackiw: Ann. Phys. (NY) **153**, 405 (1984)
d. G. 't Hooft: Comm. Math. Phys. **117**, 685 (1988)
e. S. Deser and R. Jackiw: Comm. Math. Phys. **118**, 495 (1988);
P. Gerbert and R. Jackiw: Comm. Math. Phys. **124**, 229 (1989)
f. S. Deser and R. Jackiw: Ann. Phys. (NY) **192**, 352 (1989)
Much of this material is summarized in:
g. R. Jackiw: Nucl. Phys. **B252**, 343 (1985)
h. R. Jackiw, in *Proceedings of the XVII International Colloquium on Group Theoretic Methods in Physics* (in press)
The above papers and reviews should be consulted for reference to other literature on the subject.

4
a. L. Alvarez-Gaumé and E. Witten: Nucl. Phys. **B234**, 269 (1983)
b. E. Witten: Comm. Math. Phys. **121**, 351 (1989);
Y.-H. Chen, F. Wilczek, E. Witten and B. Halperin, Intl. Jnl. Mod. Phys. **B3**, 1001 (1989)
c. A. Polyakov: Mod. Phys. Lett. **A3**, 325 (1988)
d. J. Gott and M. Alpert: Gen. Rel. Grav. **16**, 243 (1984);
J. Gott, Ap. J. **288**, 42 (1985)
e. E. Witten: Nucl. Phys. **B311**, 46 (1988/89);
J. Horne and E. Witten: Phys. Rev. Lett. **62**, 501 (1989);
S. Carlip: Nucl. Phys. **B324**, 106 (1989)

5 B. Rosenstein, B. Warr and S.-H. Park: Phys. Rev. Lett. **62**, 1433 (1989), Phys. Rev. D **39**, 3088 (1989);
B. Rosenstein and B. Warr: Phys. Lett. **B218**, 465 (1989); **B219**, 469 (1989); Texas preprint UTTG-18-89 (1989)

6 N. Redlich: Phys. Rev. Lett. **52**, 18 (1984); Phys. Rev. D **29**, 2366 (1984)

7 W. Siegel: Nucl. Phys. **B156**, 135 (1979)

8 C. Hagen: Ann. Phys. (NY) **157**, 342 (1984); Phys. Rev. D **31**, 2135 (1985)

9 D. Gonzales and N. Redlich: Ann. Phys. (NY) **169**, 104 (1984)

10 Polyakov: Ref. [4c];
G. Semenoff: Phys. Rev. Lett. **61**, 517 (1988);
Dunne et al.: Ref. [2d];
T. Matsuyama: Phys. Lett. **B228**, 99 (1989)

11 S. Zhang, T. Hanson and S. Kivelson: Phys. Rev. Lett. **62**, 8 (1989);
Chen et al.: Ref. [4b];
Jackiw, Ref. [2h]

12 R. Jackiw and N. Redlich: Phys. Rev. Lett. **50**, 555 (1983)

13 D. Arovas, J. Schrieffer, F. Wilczek and A. Zee: Nucl. Phys. **B251**, 117 (1985)

14 Y. Aharonov and D. Bohm: Phys. Rev. **115**, 485 (1959);
S. Ruijsenaars: Ann. Phys. (NY) **146**, 1 (1983)

15 R. Jackiw: Phys. Today **25** No. 1, 23 (1972)

16 R. Jackiw and C. Teitelboim in *Quantum Theory of Gravity*, S. Christensen, ed. (Adam Hilgar, Bristol, UK, 1984);
A. Polyakov: Mod. Phys. Lett. **A2**, 893 (1987);
K. Isler and C. Trugenberger: Phys. Rev. Lett. **63**, 834 (1989);
A. Chamseddine and D. Wyler: Phys. Lett. **B228**, 75 (1989); Zürich University preprint (1989).

17 A. Staruszkiewicz: Act. Phys. Pol. **24**, 735 (1963)
18 G. Clément: Int. Jnl. Theor. Phys. **24**, 267 (1985)
19 G. Grignani and C.-K. Lee: Ann. Phys. (NY) (in press);
G. Clément: Ann. Phys. (NY) (in press)
20 P. Gerbert: Phys. Rev. D **40**, 1346 (1989)
21 A. Goldhaber: Phys. Rev. D **16**, 1815 (1977);
C. Callias: Phys. Rev. D **16**, 3068 (1977)
22 B. Grossman: Phys. Rev. Lett. **50**, 464 (1983);
H. Yamagishi: Phys. Rev. D **27**, 2383 (1983);
E. D'Hoker and E. Farhi: Phys. Lett. **127B**, 360 (1983)
23 R. Percacci: Ann. Phys. (NY) **177**, 27 (1987)
24 L. Alvarez-Gaumé, S. Della Pietra and G. Moore: Ann. Phys. (NY) **163**, 288 (1985);
M. Goni and M. Valle: Phys. Rev. D **34**, 648 (1986);
I. Vuorio: Phys. Lett. **B175**, 176 (1986);
J. van der Bij, R. Pisarski and S. Rao: Phys. Lett. **B179**, 87 (1986)
25 A. Achúcarro and P. Townsend: Phys. Lett. **B180**, 89 (1986);
Witten: Ref. [4e]
26 S. Deser and Z. Yang: Brandeis University preprint BRZ TH-273 (1989)
27 I. Vuorio: Phys. Lett. **B163**, 91 (1985);
R. Percacci, P. Sodano and I. Vuorio: Ann. Phys. (NY) **176**, 344 (1987);
M. Ortiz, Cambridge University preprints, DAMTP R-89/13, 17 (1989)

Boundary Terms, Long Range Effects, and Chiral Symmetry Breaking[1]

G. Morchio [1], *F. Strocchi* [2]

[1] Dipartimento di Fisica dell'Universita, Pisa
[2] Accademia dei Lincei, Roma
ISAS and INFN, Trieste

1. Introduction

The aim of these lectures is to discuss the rôle of *boundary terms* in the presence of long range interactions (typically Coulomb-like). Significant examples are provided by gauge theories and by many-body theories with Coulomb interactions (Coulomb systems). The main point is that, in the presence of long range interactions, boundary terms can give rise to volume effects and therefore be equivalent to (non negligible) external fields. As we will explain, this phenomenon has no counterpart in the case of short range interactions and this may explain why such terms have been generally regarded as irrelevant for the definition of the dynamics. One of the most interesting cases in which it has been realized that a boundary term (actually a four divergence term in the Lagrangean) has relevant implications and may give rise to physical effects is the θ term in QCD [1] [2].

Other examples in which the boundary terms give rise to volume effects are provided by the two-dimensional quantum electrodynamics (Schwinger model) [3] [4] and by the Stückelberg-Kibble model in four dimensions [5] [6] (usually regarded as a prototype model of the Higgs effect). Actually, this phenomenon is generic for Coulomb systems, as we will see more clearly below, and in fact it occurs in many-body theories like the jellium model [7] [9], the BCS model [11] [12], the Heisenberg-Weiss model [11] etc.

As we will see, the realization of the rôle played by the boundary terms will be crucial for the discussion of spontaneous symmetry breaking and in particular for the generalization of the Goldstone's theorem in order to include the case in which the symmetry breaking gives rise to a *mass* gap, namely the associated Goldstone boson spectrum has an energy gap for low momentum [11] [9] . Significant examples are the plasmon energy gap (the plasmons can be shown to be the generalized Goldstone bosons associated to the breaking of the Galilei boost [7]), the energy gap in superconductivity (BCS model [11]), the mass generation in the Higgs effect (the generalized Goldstone bosons associated to the breaking of the gauge symmetry in the Stückelberg-Kibble model have a non-zero mass corresponding to the vector boson mass [5] [6] [11]), the mass gap associated to chiral symmetry breaking in the Schwinger model [4] etc.

[1] Lectures given by the second author at the XXIX. INTERNATIONAL SCHLADMING WINTER SCHOOL, March 1-10,1990

Fields and Particles Editors: H. Mitter · W. Schweiger

In order to explain the general ideas and the specific mechanism, even at the risk of looking pedantic, we will start reviewing the problem already at the classical level, discuss the solution in the various approaches and work out explicitly the examples.

i) The Hamiltonian Approach: Coupling to the Boundary and Variables at Infinity

To clarify the mechanism one has to carefully handle the thermodynamical limit or more generally the removal of the infrared cut-off. No substantial problem occurs in the case of (sufficiently) short range interactions. In this case, if H_V denotes the finite volume (or the infrared cut-off) Hamiltonian, the corresponding time evolution induced on a variable A, well localized inside the volume V, is given by

$$i\frac{d}{dt}A = [H_V, A]$$

and the right-hand side involves operators localized in V, plus contributions from operators localized near the boundary of V, which vanish as $V \to \infty$, thanks to the short range of the interaction. For the discussion of the delocalization associated to the dynamics of non-relativistic systems see [8] [9]. This means that the boundary conditions (or the presence of boundary terms in H_V) do not affect the time evolution of localized variables in the limit $V \to \infty$. Since different "phases" of an infinitely extended system may be viewed as corresponding to preparations of the states with different boundary conditions, the above result implies that *different phases have the same dynamics* and their difference is solely due to the ground state correlations. The realization that the same Hamiltonian can be used to describe different phases is actually one of the deepest and beautiful results of the algebraic approach to phase transitions and spontaneous symmetry breaking [12] [13]. This picture is actually at the basis of the standard wisdom on spontaneous symmetry breaking, which is characterized by *symmetric equations of motion* and *non-symmetric ground state correlation functions* [14].

The situation changes substantially in the presence of long range interactions, since in this case the commutator $[H_V, A]$, contains, apart from terms which are essentially localized inside V, terms arising from the interaction with the boundary and such terms do not become irrelevant when $V \to \infty$. In general such additional terms contain, in a substantial way, operators localized on or near the boundary of V, which in the limit $V \to \infty$ converge to variables at infinity. This means that in the thermodynamical limit the coupling between A and the boundary does not vanish (the decrease of the coupling like e.g. $1/r^2$ can be compensated by the integration over the boundary surface, see below). The result is that the *boundary conditions not only affect the* (correlation functions of the) *ground state, but also the definition itself of the dynamics* . In particular, it is no longer true that different "phases" have the same dynamics and part of the standard wisdom on spontaneous symmetry breaking, phase transition etc., has to be revisited in this perspective.

This feature of long range interactions was first noticed by Haag [10] in his treatment of the BCS model of superconductivity and it underlies the mechanisms of ***seizing of the vacuum*** emphasized by Kogut and Susskind [3] in their discussion of the Schwinger model. For a general clarification of this phenomenon from first principles and its relation to quantum bifurcation, symmetry breaking, energy gap generation, generalized Goldstone boson spectrum, see [11] [9] [15] .

We will discuss examples of this phenomenon in quantum field theory (QFT), typically gauge theories, and in many body theories.

ii) The Lagrangean Approach: Boundary Ward Identities

For relativistic systems and in particular for QFT the most suitable (and fruitful) approach is that based on the Lagrangean formulation and its direct quantum implementation, namely the functional integral. It is therefore useful to reexamine the possible relevance of boundary terms, in particular of total divergence terms, in such approach. The main point is already visible at the classical level, which will be discussed first. The discussion of the quantum case is very similar.

It is often stated, even in standard textbooks, that total divergence terms are ineffective, but this is not completely correct. The point is that in deriving the classical equations of motion the stationarity of the action under local variations of the fields, i.e. variations with compact support in space-time, gives rise to equations of motion which do not depend on the presence of total divergence terms in the Lagrangean (for simplicity we will call these equations ***local Ward identities***). However, the Hamilton variational (or least action) principle requires more, namely the stationarity of the action under variation of the fields with compact support in time, but possibly non compact support in space. In this way, for each finite volume V one gets additional equations, which we will briefly call ***boundary Ward identities***. Boundary terms and in particular four divergence terms contribute in general to such equations. In the infinite volume limit such boundary Ward identities (BWI) are not relevant for the definition of the dynamics in the case of short range interactions, but they are not irrelevant in general in the presence of long range interactions. It is in this way that the boundary terms come into play in the Lagrangean approach.

iii) The Schwinger Model and the θ Angle Problem

The above ideas will be applied to the analysis of the Schwinger model (in the bosonized form). As we shall see a careful bookkeeping of the boundary terms and of the boundary conditions leads to new structures of the model. In particular, the boundary terms completely screen off the topological term (θF) in the Lagrangean, so that the theory is independent of the Lagrangean parameter θ; its place is taken by the variable at infinity ϕ_∞

$$\theta \int E d^2 x \rightarrow \phi_\infty \int E d^2 x$$

and the pure phases are labelled by $\theta_{vac} = \langle \phi_\infty \rangle = \langle \phi \rangle$. The relevant point is that when an external field is added, like a fermion mass term, $M = m(\cos\theta_m + i\gamma^5 \sin\theta_m)$, the variable at infinity ϕ_∞ gets "aligned" to θ_m, $\langle \phi_\infty \rangle = \theta_m$, (the phase is unique), exactly as in spin models like the Heisenberg-Weiss model, the BCS model, the Heisenberg model with long range interaction with the boundary, etc. In this way one gets a theory invariant under the CP symmetry defined by the mass term, CP_m , and a natural solution of the strong CP problem.

Such striking difference with respect to the standard approach can be explained by the fact that, as a consequence of a long range coupling to the boundary, the boundary terms are turned into volume effects and boundary conditions have the same effect of a non-vanishing external field. Different boundary conditions (leading to different bookkeeping of the boundary terms) may give rise to different theories, i.e. to different dynamics, with different symmetry properties. This phenomenon is much more drastic than the usual influence of the boundary conditions on the choice of the ground state (and the corresponding pure phase) *without* affecting the dynamics (the same in each phase).

The above features of the Schwinger model are then compared with the standard picture of QCD and various speculations are made on the occurring of analogous structures.

iv) Boundary Terms and Spontaneous Symmetry Breaking

It need not to be emphasized here the rôle played by the mechanism of spontaneous symmetry breaking in the recent developments of theoretical physics, both at the level of many-body physics and for the unification of elementary particle interactions. Since long time, spontaneous symmetry breaking has been recognized to be the basis of many collective phenomena and in particular of phase transitions in statistical mechanics. The theoretical clarification of the mechanism has been achieved to a high degree of rigour in the last years and formalized in the so-called Goldstone's theorem [16]. The result is that the conditions for the applicability of the conclusions of the theorem, a subject of discussions in the early theoretical developments, are now out of question. Furthermore, the realization of the Goldstone mechanism in many branches of physics has provided a powerful non-perturbative tool to get exact information on the excitation spectrum and in particular to predict the low momentum behaviour of the energy of the elementary excitations associated to a symmetry breaking, $\omega(k) \rightarrow 0$ as $k \rightarrow 0$, (*Goldstone bosons*). We only mention here the examples of the spin waves in the theory of ferromagnetism, the Landau phonons in the theory of superfluidity, the phonon excitations in crystals, the pions as Goldstone particles of chiral symmetry breaking etc.

In the meantime, it has become more and more evident that a large class of physical systems (in particular many-body systems and elementary particles) exhibit spontaneous symmetry breaking without the zero energy gap and the Goldstone bosons predicted by the theorem. Clearly, some of the assumptions of the theorems must fail, but the long discussion on the possible mechanisms for evading the Goldstone's theorem seems to have led to a series of folklore catchwords or perturbative prescriptions rather than to a sharp and clear identification of the crucial points. For non-relativistic systems the folklore explanation for the absence of Goldstone bosons with zero energy gap is that the Coulomb potential leads to a shift of energy (at $k \to 0$) by a mechanism advocated more on the ground of ad hoc techniques and approximations in the treatment of several physical examples, rather than through a general non-perturbative control. Actually, such explanation does not appear completely convincing because long range correlations of the Coulomb type do not invalidate the applicability of the theorem in the case of relativistic local quantum field theory.

For the Higgs phenomenon, most of the heuristic analysis is based on perturbative expansions and the disappearance of the would-be Goldstone bosons to give mass to the gauge bosons has become the cheap explanation. Again no control of the mechanism at the same level of rigour and clarity as in the Goldstone's theorem has been made available in the past literature.

As emphasized by Swieca [17], the common feature underlying the "evasions" of the standard Goldstone's theorem can be traced back to the insufficiently fast decay in space of the commutator between the current, which generates the symmetry, and the variable A which gives rise to the symmetry breaking order parameter:

$$\lim_{|x|\to\infty} |x|^2 [j(x,t), A] \neq 0. \tag{1.1}$$

Swieca further argued on the basis of semiheuristic considerations [17], that such slow decay may be related to the long range of the interaction potential

$$\lim_{|x|\to\infty} |x|^2 V(x) \neq 0 \tag{1.2}$$

but he did not provide a clear and rigorous control of the phenomenon. Actually, condition (1.2) cannot be regarded as more than an indication of the possible relation between eq. (1.1) and the range of the potential: for non relativistic (continuum) systems the current involves a space derivative of the fields and therefore the critical power for the fall-off of the potential appears to be $V(x) \sim |x|^{-1}$ (*Coulomb potential*), rather than $|x|^{-2}$ (the latter being possibly related to the condition that the commutator of *any* two-field variables $[A_{x,t}, B]$ falls off faster than $|x|^{-2}$). On the other hand, for spin systems interacting through a two-body potential the existence of the dynamics in the C^* algebraic sense is already called in question when $V(x)$ decreases slower than $|x|^{-3}$ and therefore condition (1.2) may lose its relevance for eq. (1.1), in comparison with the problem of the definition itself of the (algebraic) dynamics. Furthermore, in the case in which the current is conserved but $j(x,0)$ is a non-local operator (a condition frequently occurring in spin models), the slow fall-off (1.1) may already occur for a potential decreasing like $|x|^{-3}$.

The above discussion indicates that the long range of the potential may play a relevant rôle for the evasion of the Goldstone's theorem, but it does not clearly say how and why this happens. The problem is not of secondary importance since most of many body systems involve long range interactions, as the Coulomb force is at the basis of the structure of matter. Thus, for large branches of physics (like Coulomb systems, electron gaps and plasmons, superconductors, etc.) the relation between spontaneous symmetry breaking and spectrum is not covered by the standard Goldstone's theorem. For gauge theories, especially for the Higgs phenomenon, the same problem arises, e.g. in the Coulomb gauge, where spontaneous symmetry breaking has been shown to occur [5], and more generally when the symmetry breaking is characterized by a non-local order parameter.

From a theoretical point of view it appears that for the non-relativistic systems mentioned above and for the elementary particle interactions, the existence of an energy gap compatible with spontaneous symmetry breaking, has been discussed with ad hoc approximations and no general unifying picture seems to emerge. This does not appear to us as a mere question of principle, since the identification of common structural features shared by different physical systems may shed light also on the discussion of concrete problems.

As we shall see, a careful handling of the boundary terms and their being turned into volume effects in the thermodynamical limit will lead us to a general mechanism of spontaneous symmetry breaking with associated energy gap. In this way one can prove a generalized Goldstone's theorem which provides an exact (non-perturbative) prediction of a non-trivial excitation (quasi-particle) spectrum at low momentum (*generalized Goldstone bosons with energy gap*).

Such mechanisms of symmetry breaking account for the mass gap in the Schwinger model, in the Stückelberg-Kibble model, in the BCS model and for the plasmon energy gap in the jellium model.

2. The Hamiltonian Approach: Coupling to the Boundary and Variables at Infinity

As is well known in QFT and in general for quantum systems with infinite degrees of freedom, the formal Hamiltonian which is supposed to define the model is not well defined. The general strategy, similar to that adopted in Statistical Mechanics, is to regularize the Hamiltonian with an ultraviolet and an infrared cut-off, calculate the corresponding time evolution of the field variables and then remove the cut-offs. For simplicity, we will (first) consider models for which the ultraviolet regularization is not important. We will then define the model by considering a finite volume (or infrared cut-off) Hamiltonian H_V. As already stressed above, the removal of the infrared cut-off, $V \to \infty$, is very delicate and especially the effect of the boundary terms requires a special care in the presence of long range interactions. In general, it might appear that a certain arbitrariness is involved in spelling out the boundary terms; actually it is not so, since symmetry properties can be used to decide which boundary terms should be present in H_V.

i) The Schwinger Model

The Schwinger model describes two-dimensional quantum electrodynamics (QED2) and it is formally defined by the following Lagrangean [18]

$$\mathcal{L} = \int dx \left(\bar{\psi} i \not\partial \psi - e\bar{\psi} \not{A} \psi - \frac{1}{4} F_{\mu\nu}F^{\mu\nu} \right). \tag{2.1}$$

In the Coulomb gauge one has

$$\partial_1 A_1 = 0$$

and one chooses

$$A_1 = 0.$$

The Maxwell equations then give

$$-\partial_1^2 A_0 = e\psi^\dagger \psi$$

and

$$A_0(x) = -\frac{1}{2} e \int |x - x'| \psi^\dagger(x')\psi(x')dx'. \tag{2.2}$$

The formal Lagrangean then becomes $\left(U(x) \equiv \frac{1}{2}|x|\right)$

$$\mathcal{L} = \int dx (\bar{\psi} i \not\partial \psi) + \frac{e^2}{4} \int dx dx' \psi^\dagger(x)\psi(x)U(x - x')\psi^\dagger(x')\psi(x').$$

The Lagrangean further simplifies if one adopts fermion bosonization, namely the remarkable property that in one space dimension a fermion system can be described by a bosonic field. The basic correspondence is [19]

$$\begin{aligned} j_\mu &=: \bar{\psi}\gamma_\mu\psi := \left(1/\sqrt{\pi}\right) \epsilon_{\mu\nu}\partial^\nu \phi, \\ j_\mu^5 &=: \bar{\psi}\gamma_\mu\gamma_5\psi := \left(1/\sqrt{\pi}\right) \partial_\mu \phi, \end{aligned} \tag{2.3}$$

$$\begin{aligned} &: \bar{\psi}\psi : = K : \cos 2\sqrt{\pi}\phi : , \qquad &&: \bar{\psi}\gamma_5\psi : = iK : \sin 2\sqrt{\pi}\varphi : , \\ &i\bar{\psi} \not\partial \psi = \frac{1}{2} \partial_\mu \phi \partial^\mu \phi , &&K = \text{const} . \end{aligned} \tag{2.3'}$$

If the fermion is massless, so is also the boson, as follows clearly from the last equation, which relates the two Lagrangeans. (It also follows from the conservation of j_μ and j_μ^5 and the relation $j_\mu = \epsilon_{\mu\nu} j^{5\nu}$.) The boson field is also a canonical field since the current commutation relation

$$[j_0(x), j_1(x')] = -\frac{i}{\pi} \delta'(x - x')$$

implies the CCR for ϕ:

$$[\phi(x), \dot{\phi}(x')] = i\delta(x - x').$$

For a rigorous discussion of the boson-fermion correspondence see Ref. [20] sec. 6C, and references therein.

In the bosonized form the Lagrangean then takes the form

$$\mathcal{L} = \frac{1}{2} \int \left[\dot{\phi}^2 - (\partial_1 \phi)^2 \right] dx + \frac{e^2}{4\pi} \int \partial_1 \phi(x) |x - x'| \partial_1 \phi(x') dx dx',$$

which yields the following formal Hamiltonian:

$$H = \frac{1}{2} \int \left[\pi^2 + (\partial_1 \phi)^2 \right] dx - \frac{e^2}{4\pi} \int \partial_1 \phi(x) \partial_1 \phi(y) |x - y| dx dy \ , \qquad (2.4)$$

where $\pi = \dot{\phi}$ is the canonical momentum conjugate to ϕ.

In the standard treatment of the model [18] [3] [21] one then argues that in discussing the (representation of the) field ϕ one is naturally led to the introduction of an "angle" θ and its conjugate variable Π_θ, which should in some way specify the boundary conditions at infinity. Then Π_θ is recognized to be the generator of chiral transformations and the ground state degeneracy with respect to chiral transformations is then interpreted in terms of chiral breaking (θ vacua and chiral breaking) without associated Goldstone bosons. The chiral transformations of the fermion fields

$$\psi(x) \to e^{i\alpha\gamma_5} \psi(x), \qquad \bar{\psi}(x) \to \bar{\psi}(x) e^{i\alpha\gamma_5}, \qquad (2.5)$$

imply the following transformations for the bosonic field ϕ

$$\phi(x) \to \phi(x) + \frac{\alpha}{\sqrt{\pi}} \ , \qquad (2.6)$$

as it follows easily from the correspondence (2.3). The chiral invariance of the Lagrangean (2.1) is reflected in the bosonic language by the invariance of the Hamiltonian (2.4) under the transformations (2.6).

To correctly define the dynamics and find out the rôle of the boundary terms we will consider the infrared cut-off Hamiltonian

$$H_V = \frac{1}{2} \int dx \left[\pi^2 + (\nabla \phi)^2 \right] - \frac{e^2}{4\pi} \int_V \partial_1 \phi(x) \partial_1 \phi(y) U(x - y) dx dy \ , \qquad (2.7)$$

(V being the interval $[-L, L]$). More correctly one should introduce a smooth infrared cut-off: $U(x) \to U_L(x) \equiv U(x) f_L(x)$, $f_L(x) = f(|x|/L)$, $f(x) = 1$ for $|x| < 1$, $f(x) = 0$ for $|x| > 1 + a$, $a << 1$. For the mathematical delicate points involved in defining the finite volume dynamics α_V^t, its strong limit with respect to a family of infrared regular states, etc. see Refs. [4] [15] .

The corresponding equations of motion are

$$\begin{aligned} \dot{\phi} &= \pi, \\ \dot{\pi} &= \Delta\phi - \frac{e^2}{\pi}\phi + \frac{e^2}{2\pi} \int_V dy \partial_y \left(\partial_y U(x - y) \phi(y) \right) \ . \end{aligned} \qquad (2.8)$$

The last term in the r.h.s. of the second equation converges (strongly) to $(e^2/\pi)\phi_\infty$

$$\frac{1}{2}\int_V dy\partial_y\left(\partial_y|x-y|\phi(y)\right)$$
$$=\frac{1}{2}\left(\phi(L)+\phi(-L)\right)\equiv\phi_B\stackrel{V\to\infty}{\longrightarrow}\phi_\infty\equiv\lim_{V\to\infty}\frac{1}{V}\int_V dy\phi(y). \tag{2.9}$$

Thus one gets the equations of motion

$$\Box\phi=-\frac{e^2}{\pi}\phi+\frac{e^2}{\pi}\phi_\infty\,. \tag{2.10}$$

From a rigorous point of view the convergence $\phi_B\to\phi_\infty$ is somewhat delicate: first of all one should use the smooth infrared regularization $U(x)\to U_L(x)$, by which the term in question would become the average of the field over a spherical shell of inner radius L and thickness La. Secondly, one should better discuss the convergence in terms of the exponentiated field (Weyl operators). Thirdly, and more substantially, the above convergence requires some regularity of the large distance behaviour of the states to which the operator is applied [4]. Such infrared regularity condition is satisfied by a large class of "phases" defined by a translationally invariant ground (or equilibrium) state. For the mathematical questions see Ref. [4] .

A few comments may be useful. First the variable ϕ_∞ is what is technically called a ***variable at infinity*** , i.e. a variable "localized" outside any bounded region [22]. One can easily show that such variable commutes with ϕ and π and therefore it must be a c-number in each irreducible representation of the field algebra generated by ϕ and π. Quite generally, variables at infinity commute with any local variable as a consequence of asymptotic abelianess in space

$$\lim_{|x|\to\infty}[A_x,B]=0,$$

(A,B local variables, $A_x\equiv$ the x-translated of A), a condition that can hardly be dispensed with for a reasonable quantum mechanical interpretation of the theory (see e.g. [14]). Typical examples of variables at infinity are the ***ergodic means***

$$A_\infty=\lim_{V\to\infty}\frac{1}{V}\int_V dxA_x\equiv\lim_{V\to\infty}A_V, \tag{2.11}$$

or the infinite volume limit of ***averages around the boundary*** , e.g.

$$\lim_{L\to\infty}\int dxf_L(x)A_x\equiv\lim_{V\to\infty}A_{\partial V}, \tag{2.12}$$

where $f_L(x)$ is a regular function with support in the region $L<|x|<L(1+\epsilon)$ and suitably normalized $\int f_L(x)dx=1$. One can show that the limit (2.12) is again A_∞.

If $\lim_{|x|\to\infty}A_x$ exists, independently of the direction, then it coincides with the ergodic mean since the values of A_x in any bounded region V_0 do not contribute to the limit in eq. (2.11). However, the ergodic mean exists also in more general situations.

As already mentioned, when A is an operator the existence of the limit (2.11) in the strong topology requires some infrared regularity of the states to which it is applied.

The second remark is that the non-trivial limit (2.9) crucially depends on the delicate interplay between the kinetic term and the long range of the Coulomb potential. For a short range potential the third term on the r.h.s. of the second equation (2.8) would vanish in the limit $V \to \infty$. This mechanism is in fact rather general; one can show that variables at infinity occur in the dynamics of field variables generically for Coulomb systems (see e.g. the jellium model [5] [8] [9]). The point is that in space dimension $d \geq 2$ the slow decrease of the derivative of the Coulomb potential is compensated by the integration over the surface of the volume V and one gets a non zero contribution in the limit $V \to \infty$ [7].

One would have got the same eq. (2.10) if instead of the original Hamiltonian (2.4) one had the Hamiltonian

$$\begin{aligned} H &= \frac{1}{2} \int \left[\pi^2 + (\nabla\phi)^2\right] dx - \frac{e^2}{2\pi} \int \phi^2 dx + \frac{e^2}{\pi} \phi_\infty \int \phi dx \\ &= H_{\text{free}} + \frac{e^2}{\pi} \phi_\infty \int \phi dx, \end{aligned} \tag{2.13}$$

which describes the interaction of a free massive field with the "external" field ϕ_∞. It is clear at this point that modifying the boundary conditions , and therefore ϕ_∞, leads to a volume effect. It is worthwhile to remark that the last term on the r.h.s. of eq. (2.13), yielding the equivalence between (2.4) and (2.13), would not be present if one would freely integrate by parts, neglecting boundary terms. The term would also be absent if, for finite volumes, one imposes the chiral breaking boundary condition $\phi_B = 0$, leading to a "phase" with $\langle\phi_\infty\rangle = \langle\phi\rangle = 0$, in the thermodynamical limit. In fact, since ϕ_∞ commutes with ϕ and π and it is a c-number in each pure phase, the value taken by ϕ_∞ can be used to label the QFT phases of the massless Schwinger model. As we shall see below, in the massless case it can be identified with the θ angle (in the bosonic language) and the last term in eq. (2.13) can be seen as the analogous of the θ-term. In the massive Schwinger model, the relation between ϕ_∞ and θ is more delicate and it plays a crucial rôle for understanding the mechanism of chiral symmetry breaking (the so-called U(1) problem) in QCD.

It is also worthwhile to remark that, as displayed by eq. (2.13), the Hamiltonian (2.4') is not equivalent to the free Hamiltonian H_{free} of a free massive field and that the additional term (with the $\phi_\infty\phi$ coupling) is crucial for keeping the chiral invariance, which is a property of H but not of H_{free}. (Note that $\phi \to \phi + c$ implies $\phi_\infty \to \phi_\infty + c$).

Since in each pure phase ϕ_∞ becomes a c-number, $\langle\phi\rangle$, correspondingly the Hamiltonian (2.13) takes the following form

$$H \to H_{\text{eff}} = H_{\text{free}} + \frac{e^2}{\pi} \langle\phi\rangle \int \phi(x) dx \tag{2.14}$$

and it clearly displays the dependence of the effective dynamics, in each pure phase, from the boundary conditions.

A natural question is whether quite generally one can make the boundary terms irrelevant by always adopting the strategy of imposing vanishing boundary conditions for the relevant variables (in this case $\phi_B = 0$). While this is certainly an admissible choice (corresponding in this case to the $\theta = 0$ vacuum, according to the standard language), one would in this way lose the access to other phases, those corresponding to $\theta \neq 0$. In the massless Schwinger model this restriction is not physically relevant, but, as we will see below, it does make a difference in the massive Schwinger model or more generally when the phase structure of the model is explored by means of external fields. This is also one of the general reasons why in Statistical Mechanics it is of interest to investigate the phase structure of the model even when the "phases" are related by a symmetry transformation and therefore are all isomorphic. Their physical distinction does in fact show up when an external field is introduced.

A possible alternative could be to bosonize the Schwinger model in terms of the variable $\widehat{\phi} = \phi - \phi_\infty$ (or $\widehat{\phi} = \phi - \langle\phi\rangle$) rather than ϕ. Clearly by construction $\langle\widehat{\phi}\rangle = 0$, there would be no boundary term contribution and one could freely integrate by parts. However, the description of chiral symmetry breaking would be more problematic. Strictly speaking $\widehat{\phi} = \phi - \phi_\infty$ is a chiral invariant variable and therefore to construct chirally charged operators (like $\bar{\psi}\psi$ or $\bar{\psi}\gamma_5\psi$) one has to introduce a further bosonic variable, which is not chirally invariant, for example by modifying the bosonization correspondence (2.3)

$$
\begin{aligned}
&: \bar{\psi}(1+\gamma_5)\psi := K : \exp 2i\sqrt{\pi}\widehat{\phi} : U \\
&: \bar{\psi}(1-\gamma_5)\psi := K : \exp -2i\sqrt{\pi}\widehat{\phi} : U^\dagger
\end{aligned}
\tag{2.15}
$$

with U a spurion field playing the rôle of $\exp 2i\sqrt{\pi}\phi_\infty$. In this way, however, since the chiral charge is carried by the spurion field U, which commutes with $\widehat{\phi}$, the chiral transformations of the fermion bilinears (2.15) cannot be generated by the local charge density $(1\sqrt{\pi})\,\partial_0\widehat{\phi}$ (not even at equal times). Similarly, the choice of the variable $\widehat{\phi} = \phi - \langle\phi\rangle$ would either be equivalent to restricting the attention to the "phase" $\langle\widehat{\phi}\,\rangle = 0$, or it would give rise to the same kind of problems for the generation of chiral transformations.

In our opinion, the advantage of the approach discussed in this section is that the bosonization correspondence is done at the level of local fields (at $t = 0$), with no spurion field involved, independently of the dynamics. The occurrence of variables at infinity is then strictly related to the dynamics and their rôle is therefore dependent on the form of the Hamiltonian, the existence of external fields etc. Furthermore, the so arising variables at infinity are defined in terms of the local fields, no further (external or spurion-like) degree of freedom being necessary. Since the spectrum of such variables at infinity can be used to parameterize the vacuum structure, it is also of some usefulness to have a systematic and constructive procedure to find out such variables at infinity, without having to use in advance the wisdom gained by knowing the solution.

In conclusion, the model described above shows that:

i) boundary terms are in general not irrelevant and should not be neglected (unless one adopts suitable boundary conditions),

ii) in the presence of long range interactions, boundary terms are in general turned into volume effects, i.e. they are equivalent to the introduction of a (non-negligible) external field,

iii) the boundary conditions (b.c.) do not only affect the ground state but also the dynamics; so that different "phases" have in general different effective dynamics. In contrast to the short range case, symmetry breaking induced by such boundary terms do not fulfil the general assumption of the Goldstone's theorem. As we shall see, this will in general give rise to "massive" Goldstone bosons; in the particular case of the massless Schwinger model, massive Goldstone bosons appear associated to the breaking of chiral symmetry [4],

iv) we also see an example of a general mechanism, (see also the jellium model), by which a long range interaction becomes effectively equivalent to a screened interaction plus a coupling to the boundary (or better to variables at infinity in the limit $V \to \infty$). In a certain sense the delocalization associated to the long range of the interaction disappears, leaving as a remnant of it the coupling to variables at infinity. For a general discussion of this mechanism from first principles and its relation to the seizing of the vacuum see refs. [11] [15].

ii) The Stückelberg-Kibble Model

The crucial rôle played by the boundary terms (b.t.) is also displayed by the four-dimensional Stückelberg-Kibble model considered as a prototype of the Higgs phenomenon [23] [24]. The model is defined by freezing the modulus of the Higgs field $\chi = |\chi| e^{i\phi}$, $|\chi| = \text{const}$ in the Abelian Higgs- Kibble Lagrangean; one then gets the S-K Lagrangean ($|\chi| = 1$):

$$\mathcal{L}_{SK} = -\frac{1}{4} F_{\mu\nu}^2 - \frac{1}{2} (\partial_\mu \phi + e A_\mu)^2$$

and in the Coulomb gauge the following infrared cut-off Hamiltonian:

$$H_L = \frac{1}{2} \int dx \left[(\nabla \phi)^2 + \pi^2 \right] + \frac{1}{2} \int \pi(x) U_L(x-y) \pi(y) dx dy, \tag{2.16}$$

where $U_L(x) \equiv (e^2/|x|) f_L(x)$, $f_L(x) = f(x/L)$ and $f(x) = 1$ for $|x| < 1$, $f(x) = 0$ for $|x| > 1 + \epsilon$. Again, a correct bookkeeping of the boundary terms, without neglecting those arising from integration by parts, yields the following equations of motion:

$$\dot{\phi} = \pi + U_L \star \pi \,, \qquad \dot{\pi} = \Delta \phi \,,$$

and

$$\begin{aligned} \ddot{\phi}(x) &= \Delta\phi(x) + \int U_L(x-y) \Delta\phi(y) dy \\ &= \Delta\phi(x) - 4\pi e^2 \phi(x) + 4\pi e^2 \int \sigma_L(x-y) \phi(y) dy, \end{aligned}$$

where

$$4\pi e^2 \sigma_L \equiv \Delta(f_L U) - f_L \Delta U.$$

Now, $\sigma_L \neq 0$ only in the spherical shell S_L around the boundary of inner and outer radii L and $L(1+\epsilon)$, respectively, and

$$4\pi e^2 \int \sigma_L(x)dx = \int \Delta(f_L U(x))dx - \int f_L(x)\Delta U(x)dx = 4\pi e^2.$$

Thus, $\sigma_L * \phi$ is the average of ϕ on the spherical shell S_L and therefore (see eq. (2.12))

$$\lim_{L\to\infty} \sigma_L * \phi = \phi_\infty. \tag{2.17}$$

(As in the case of the Schwinger model the existence of the limit in the strong topology requires some infrared regularity condition of the states to which it is applied [11] [15]. For this and other mathematical points see refs. [11] [15]).

Hence, one gets the following equations of motion [5] [11]

$$\Box \phi = -4\pi e^2 \phi + 4\pi e^2 \phi_\infty. \tag{2.18}$$

As in the case of the Schwinger model, the Hamiltonian is not exactly equivalent to that of a free field of mass $m^2 = 4\pi e^2$; the (non-negligible) boundary terms have given rise to a non trivial coupling between the field ϕ and the variable at infinity ϕ_∞. In each irreducible representation (pure phase) of the algebra generated by ϕ and π, such coupling leads to a volume effect which breaks the gauge symmetry $\phi \to \phi + c$.

iii) Boundary Terms and Volume Effects

The first important consequence of the mechanism outlined above is the following: in the case of short range interactions the boundary conditions do not affect the equations of motion, which are the same in each "phase". The effect of boundary conditions is only that of selecting the ground state. In presence of spontaneous symmetry breaking we therefore have symmetric equations of motion, but non-symmetric correlation functions. In the case of long range interactions the above basic structure of phase transitions and symmetry breaking requires revision: the boundary condition not only determines the ground state (i.e. the 'phase') but, if the infrared cut-off is properly removed, they crucially affect the dynamics, which therefore is not the same in each pure "phase". This is easily seen in the previous models where in each pure phase ϕ_∞ becomes a c-number, actually an order parameter depending on the phase, and it is determined by the boundary conditions which select the ground state: $\phi_\infty \to \langle\phi\rangle$. The net effect is that the boundary conditions play the same rôle of a non- negligible external field, whose value is given in each pure phase by the corresponding order parameter. This is very clear e.g. in the Schwinger model, where the term $\phi_\infty \phi$ becomes $\langle\phi\rangle\phi$, with $\langle\phi\rangle$ playing then the rôle of an external field. *The boundary conditions or boundary terms have been turned into volume effects* , as a consequence of the long range of the interaction.

The final picture is very close to what characterizes mean field models [11], like e.g. the Heisenberg-Weiss model of ferromagnetism, the BCS model of superconductivity [11] [8] etc.

One may at this point ask whether the realization of the above infrared structures, apart from qualifying non-trivial questions of principle, are also of some practical relevance or of some help when one treats QFT, like gauge theories, exhibiting long range interactions. One may think that, after all, in the models discussed above one is ending up with an effective Hamiltonian with an interaction term like $\langle\phi\rangle\phi$ and one might as well get *the same* result by introducing from the very beginning $\langle\phi\rangle$ as a free parameter, say θ, i.e. by formulating the model in such a way that the symmetry $\phi \rightarrow \phi+c$ (i.e. the chiral $U(1)$ symmetry in the Schwinger model) is broken from the very beginning in the definition of the Hamiltonian. (This would be the exact analogue of 't Hooft position with respect to the $U(1)$ problem in QCD [25] as we shall see more clearly below). However, the two strategies lead to substantially different results if the above models are perturbed by the introduction of an external field like the fermion mass term in the Schwinger model.

To illustrate the situation and further clarify the mechanism, in view of the discussion of the massive Schwinger model, we consider a Heisenberg-Weiss like model defined by the following finite volume Hamiltonian

$$H_V = -J\frac{1}{|\partial V|}\sum_{\substack{i\in V\\ j\in\partial V}} \boldsymbol{\sigma}_i \cdot \boldsymbol{\sigma}_j \equiv -J\sum_{j\in V} \boldsymbol{\sigma}_i \cdot \boldsymbol{\sigma}_{\partial V}. \tag{2.19}$$

where ∂V denotes the boundary of V.

Under very general conditions [11], the average around the boundary $\boldsymbol{\sigma}_{\partial V}$ strongly converges to the variable at infinity $\boldsymbol{\sigma}_\infty$

$$\boldsymbol{\sigma}_\infty \equiv \lim_{V\rightarrow\infty}\frac{1}{V}\sum_{i\in V}\boldsymbol{\sigma}_i \equiv \lim_{V\rightarrow\infty}\boldsymbol{\sigma}_V, \tag{2.20}$$

in the thermodynamical limit. Thus, the dynamics converges to that obtained by using the following Hamiltonian

$$H_V = -J\sum_{i\in V}\boldsymbol{\sigma}_i\cdot\boldsymbol{\sigma}_\infty\ . \tag{2.21}$$

Hence, in the thermodynamical limit the model is equivalent to the Heisenberg-Weiss model, obtained by replacing the short range spin-spin interaction

$$-\frac{1}{2}\sum_{i,j\in V} J_{ij}\boldsymbol{\sigma}_i\cdot\boldsymbol{\sigma}_j, \tag{2.22}$$

with J_{ij} a short range potential (with $J_{ii}=0$), by the long range interaction of the spin at the $i-th$ site with the operator average of the spin inside the volume V

$$\sum_{j\in V} J_{ij}\boldsymbol{\sigma}_j \rightarrow \frac{J_{av}}{V}\sum_{j\in V}\boldsymbol{\sigma}_j = J_{av}\boldsymbol{\sigma}_V. \tag{2.23}$$

It is not difficult to show that in the thermodynamical limit the dynamics of the Heisenberg-Weiss model is the same as that obtained by the Hamiltonian (2.21).

Since the variable at infinity $\boldsymbol{\sigma}_\infty$ commutes with $\boldsymbol{\sigma}_i$, in each pure phase it becomes a c-number; actually, for a translationally invariant ground state, $\boldsymbol{\sigma}_\infty$ coincides with the mean magnetization

$$\boldsymbol{\sigma}_\infty \to \langle \boldsymbol{\sigma}_\infty \rangle = \lim_{V\to\infty} \frac{1}{V} \sum_{i\in V} \langle \boldsymbol{\sigma}_i \rangle = \langle \boldsymbol{\sigma} \rangle. \tag{2.24}$$

Thus, in each pure phase the Hamiltonian (2.21) becomes

$$H_{\text{eff}} = -J\langle \boldsymbol{\sigma} \rangle \cdot \sum_{i\in V} \boldsymbol{\sigma}_i \tag{2.25}$$

with the mean field $\langle \boldsymbol{\sigma} \rangle$ playing the rôle of an external field. In conclusion, the *long range* interaction with the boundary (eq. (2.19)) has been shown to be equivalent to a volume interaction similar to that of an external field. More generally the same features are displayed by a Heisenberg model with a boundary term of the type (2.19), namely a model described by the following finite volume Hamiltonian

$$H_V = -\sum_{i,j\in V} J_{ij} \boldsymbol{\sigma}_i \cdot \boldsymbol{\sigma}_j - J \sum_{i\in V} \boldsymbol{\sigma}_i \cdot \boldsymbol{\sigma}_{\partial V} \tag{2.26}$$

(with J_{ij} a finite or short range interaction). The introduction of an external magnetic field interaction $\boldsymbol{h} \cdot \Sigma_{i\in V} \boldsymbol{\sigma}_i$ mimics very closely the structure of the massive Schwinger model and it can be used to illustrate the difference between the two-strategies (in the presence of an external field), as anticipated at the beginning of this paragraph. However, for the sake of simplicity, we will for the moment still use the oversimplified version (2.19) with an external field $\boldsymbol{h}$:

$$H_V = -J \sum_{i\in V} \boldsymbol{\sigma}_i \cdot \boldsymbol{\sigma}_{\partial V} + \boldsymbol{h} \cdot \sum_{i\in V} \boldsymbol{\sigma}_i\ . \tag{2.27}$$

According to the first strategy (the one advocated in these lectures), the effect of the boundary terms is obtained, according to the general principles of Statistical Mechanics, by removing the infrared cut-off in the dynamics defined by H_V. This leads to $\boldsymbol{\sigma}_{\partial V} \to \boldsymbol{\sigma}_\infty$ and for fixed $\boldsymbol{h}$ one has only one pure phase with the property that

$$\langle \boldsymbol{\sigma}_\infty \rangle = \langle \boldsymbol{\sigma} \rangle \equiv \boldsymbol{n},$$

$\boldsymbol{n}$ pointing in the direction of $\boldsymbol{h}$. In fact, the equations of motion give

$$\frac{d}{dt} \boldsymbol{\sigma}_i = 2(\boldsymbol{h} - J\boldsymbol{\sigma}_\infty) \wedge \boldsymbol{\sigma}_i. \tag{2.28}$$

The ergodic mean of eq. (2.28) (or the limit $i \to \infty$) yields

$$\frac{d}{dt} \boldsymbol{\sigma}_\infty = 2\boldsymbol{h} \wedge \boldsymbol{\sigma}_\infty$$

and the time translation invariance of the ground state gives

$$0 = \frac{d}{dt}\langle \boldsymbol{\sigma}_\infty \rangle = 2\, \boldsymbol{h} \wedge \langle \boldsymbol{\sigma}_\infty \rangle ,$$

i.e. $\langle \boldsymbol{\sigma}_\infty \rangle$ pointing in the direction of $\boldsymbol{h}$. In this way one gets a theory for which the rotations around $\boldsymbol{h}$ are not broken.

Quite different is the result obtained by the following other strategy: essentially one first treats the theory with $\boldsymbol{h} = 0$, performs the infinite volume limit, then decomposes into pure phases, then treats $\langle \boldsymbol{\sigma}_\infty \rangle$ as a free parameter $\boldsymbol{\theta}$, and finally introduces an external field. The result is the same as the one described by the following Hamiltonian

$$H_V = -J \sum_{i \in V} \boldsymbol{\sigma}_i \cdot \boldsymbol{\theta} + \boldsymbol{h} \cdot \sum_{i \in V} \boldsymbol{\sigma}_i \tag{2.29}$$

and the unique ground state is labelled by an order parameter $\langle \boldsymbol{\sigma} \rangle$ which points in the direction of $\boldsymbol{h} - J\boldsymbol{\theta}$. Now, the alinement of $\langle \boldsymbol{\sigma} \rangle$ with $\boldsymbol{h}$ is therefore "not natural" in the language of QFT and a "fine tuning" is needed to get this property and to have the rotations around $\boldsymbol{h}$ unbroken. The point is that the infinite volume limit and the introduction of a magnetic field do not commute (even if in the Hamiltonian (2.29) the external field can be treated as a small perturbation).

The two alternative strategies can also be viewed as the result of imposing different boundary conditions. The first case corresponds to using free (or periodic) boundary conditions, so that $\boldsymbol{\sigma}_{\partial V}$ is not fixed and, in the unique phase selected by the external field $\boldsymbol{h}$, takes the value $\boldsymbol{n} = \langle \boldsymbol{\sigma}_\infty \rangle$, pointing in the direction of $\boldsymbol{h}$. This strategy can be also characterized by the property of maintaining the rotational symmetry as much as possible, leaving its breaking only to the action of the external field $\boldsymbol{h}$.

The second case corresponds to using symmetry breaking boundary conditions $\langle \boldsymbol{\sigma}_{\partial V} \rangle = \boldsymbol{\theta}$, with $\boldsymbol{\theta}$ a fixed (free) parameter. The Hamiltonian (2.27) then takes the following form

$$H_V = -J \sum_{i \in V} \boldsymbol{\sigma}_i \cdot \boldsymbol{\theta} + \boldsymbol{h} \cdot \sum_{i \in V} \boldsymbol{\sigma}_i \quad . \tag{2.30}$$

Clearly such boundary conditions break rotational symmetry and open the possibility of a mismatch between the direction of $\boldsymbol{\theta}$ and that of $\boldsymbol{h}$.

One might be surprised or suspicious of such a dramatic consequence of the boundary conditions, much stronger than in the standard (short range) case, where their consequence is only that of choosing the ground state, if a degeneracy is present, and where they are ineffective in the presence of an external field.

Here, the point is that the long range interaction with the boundary turns boundary terms into volume effects which play exactly the *same rôle of an external field* . Its direction may be left to be determined by the external field perturbation $\boldsymbol{h}$ (case I), with symmetry breaking induced only by $\boldsymbol{h}$, or it may be fixed in advance (case II), breaking rotational invariance, independent of *h*. Actually, strategy I gives the correct infinite volume extrapolation of the behaviour of a spin system with long range coupling to the boundary, in the presence of a homogeneous external magnetic field which forces the spins on the boundary to get aligned to it.

3. The Lagrangean and Functional Integral Approach. Boundary Ward Identities

It is instructive (and useful) to recover the features discussed above, and in particular the rôle of the boundary terms, in the Lagrangean approach, both because it is more suitable for relativistic systems (for which typical boundary terms are the four divergences) and because this is the way to make natural the connection with the functional integral.

We will start with the classical case, where the main features are already present, but not sufficiently realized in the treatments given by standard textbooks.

i) The Classical Lagrangean Approach

The Hamilton variational principle in Classical Mechanics requires the stationarity of the action under arbitrary variations of the coordinates, provided such variations vanish at the end points of the time interval $[t_1, t_2]$

$$\delta q_i(t_1) = 0 = \delta q_i(t_2) \; ,$$
$$\delta \dot{q}_i(t_1) = 0 = \delta \dot{q}_i(t_2) \; .$$

In going from discrete to continuous systems one has to be careful about the space boundary.

In the standard version of the variational principle one considers an arbitrary space volume V and variations of the fields which vanish on the time boundary $t = t_1$, $t = t_2$, *as well as* on the space boundary. The stationarity of the action then yields the Euler-Lagrange equations and the presence of a boundary term, typically a four-divergence, does not affect such equations [26]. We will call such equations Local Ward Identities (LWI).

However, strictly speaking, the variational principle requires more, namely the stationarity of the action with respect to variations of the fields with compact support in time, but possibly non compact support in space. In this way, one gets further equations, which we will call Boundary Ward Identities (BWI); boundary terms in general affect such equations.

More explicitly, given the action in the space time region $V \times T, T = [t_1, t_2]$,

$$A_{VT} = \int_{VT} \mathcal{L}(\phi, \partial\phi) dx \; dt \tag{3.1}$$

its stationarity under variations of the fields

$$\phi \to \phi + \epsilon\chi,$$

with $\chi(x,t)$ a regular function satisfying

$$\chi(x,t_1) = 0 = \chi(x,t_2), \tag{3.2}$$

gives

$$\begin{aligned} 0 = \delta A_{VT} = \int_{VT} dx\ dt\ \epsilon\chi \left[\frac{\delta \mathcal{L}}{\delta\phi} - \partial_\mu \frac{\delta\mathcal{L}}{\delta\partial_\mu\phi}\right] \\ + \epsilon\chi \frac{\delta\mathcal{L}}{\delta\partial_i\phi} n_i(x,t)\Big|_{x\in\partial V} \end{aligned}, \tag{3.3}$$

where $n(x,t)$ is the unit vector field normal to ∂V. Since $\chi(x,t)$ is arbitrary and may take any value on ∂V, one gets both the *Local Ward Identities* or Euler Lagrange equations, when χ vanishes on ∂V,

$$D_{EL}\mathcal{L} \equiv \frac{\partial\mathcal{L}}{\delta\phi} - \partial_\mu \frac{\partial\mathcal{L}}{\delta\partial_\mu\phi} = 0, \tag{3.4}$$

as well as the *Boundary Ward Identities*

$$\sum_i n_i \frac{\delta\mathcal{L}}{\delta\partial_i\phi}(x,t)\Big|_{x\in\partial V} = 0. \tag{3.5}$$

It is not difficult to see that boundary terms like a four divergence $\partial^\mu K_\mu$ do not contribute to the LWI since for localized variations of the fields the variation of ∂K can be integrated out

$$\delta \int_{VT} \partial^\mu K_\mu dx = \epsilon \int_{VT} \partial^\mu \left[\frac{\delta K_\mu}{\delta\phi}\chi + \frac{\delta K_\mu}{\delta\partial_\nu\phi}\partial_\nu\chi\right] dx\ dt = 0,$$

(because χ vanishes together with its derivatives on the boundary $\partial V \times \partial T$). A four divergence term does, however, affect the BWI, since, for variations which do not vanish on the space boundary, one gets in eq. (3.3) a non-vanishing contribution

$$\begin{aligned} \frac{\delta\partial K}{\delta\partial_i\phi} n_i &= \frac{\delta}{\delta\partial_i\phi}\left(\frac{\delta K_\mu}{\delta\phi}\partial_\mu\phi + \frac{\delta K_\mu}{\delta\partial_\nu\phi}\partial_\mu\partial_\nu\phi\right) n_i \\ &= \left(\frac{\delta K_i}{\delta\phi} + \frac{\delta K_\mu}{\delta\phi\delta\partial_i\phi}\partial^\mu\phi\right) n_i \end{aligned} \tag{3.6}$$

(in the last equality we have used the antisymmetry of $\delta K_\mu/\delta\partial_\nu\phi$, which follows from the condition that ∂K should not contain derivatives of the fields higher than the first).

Now, when the LWI define equations of motion for which the Cauchy problem is well posed as a hyperbolic problem, the propagation of the fields has an essentially local structure so that (for large V) it is independent from the boundary conditions and *a fortiori* from the BWI. In this case, the BWI do not play any relevant rôle in the solution of the dynamical problem and one may as well consider only the weaker form of the variational principle, without any loss. This is not the case when there are constraints leading to long range interactions; as we will see, it is through the BWI that a coupling to the boundary and variables at infinity enter in the game to reproduce the same features of the Hamiltonian approach.

In conclusion, boundary terms like e.g. a four divergence term do in general affect the time evolution of the fields. The rôle of the four- divergence terms with respect to the symmetry properties of the action, the definition and conservation of the current etc. will be discussed later.

Example: *The Schwinger model.* To provide a simple illustration of the effect of the BWI, we consider the bosonized massless Schwinger model in the axial gauge. The Lagrangean is

$$\mathcal{L} = \frac{1}{2}\partial_\mu\phi\partial^\mu\phi + \frac{1}{2}(\partial_1 A_0)^2 + g\partial_1\phi A_0 \tag{3.7}$$

(and clearly symmetric under chiral transformations $\phi \to \phi + c$). The LWI are

$$\begin{aligned} \Box\phi + g\partial_1\phi A_0 &= 0\ , \\ \partial_1^2 A_0 - g\partial_1\phi &= 0\ , \end{aligned} \tag{3.8}$$

and the BWI are

$$\begin{aligned} -\partial_1\phi + gA_0 &= 0, \quad \text{on } \partial V, \\ \partial_1 A_0 &= 0, \quad \text{on } \partial V. \end{aligned} \tag{3.9}$$

The second eq. (3.8) implies

$$\partial_1 A_0 - g\phi = c, \qquad \partial_1 c = 0, \tag{3.10}$$

and, by eqs. (3.9),

$$c = -g\phi_B,$$

with ϕ_B the value of ϕ on the space boundary. Hence eq. (3.10) reads

$$\partial_1 A_0 - g(\phi - \phi_B) = 0$$

and the first of eqs. (3.8) becomes

$$\Box\phi + g^2(\phi - \phi_B) = 0. \tag{3.11}$$

In the limit $V \to \infty$, for the solutions having a limit for $|x| \to \infty, \phi_B \to \phi_\infty \equiv \lim_{V\to\infty}(1/V)\int_V \phi(x)dx$ and one gets equations of motion involving the variable at infinity ϕ_∞.

On the other hand, if one adds to $\mathcal{L}$ the (chirally non-symmetric) boundary term

$$g\partial^\nu[\epsilon_{\mu\nu}(\phi - \theta)A_\mu] \tag{3.12}$$

with θ an arbitrary constant, one gets the same EL equations (3.8), but different BWI

$$\partial_1\phi = 0, \qquad \partial_1 A_0 - g(\phi - \theta) = 0, \quad \text{on } \partial V.$$

Thus

$$\partial_1 A_0 - g\phi = c = -g\theta$$

and one gets a different dynamics for ϕ

$$\Box\phi + g^2(\phi - \theta) = 0. \tag{3.13}$$

One may regard the two alternatives as corresponding to different choices of boundary conditions, namely i) free or Neumann (or periodic) boundary conditions for ϕ, in the first case, so that the variations of the fields are allowed to be arbitrary, the value of ϕ_B is not fixed a priori and its value affects the time evolution of ϕ, also in the thermodynamical limit (clearly such boundary conditions do not break chiral symmetry); ii) boundary conditions, e.g. Dirichlet b.c., which fix ϕ_B in the second case, $\phi_B = 0$, so that the variations of the fields at the space boundary must vanish (clearly such b.c. break chiral symmetry).

ii) The Quantum Case: The Functional Integral Approach

The structures discussed in the previous paragraph are no peculiarity of the classical case and actually, with suitable modifications, go through to the quantum case. The bridge is provided by the functional integral approach to QFT. For the sake of simplicity we start by discussing the QFT models which can be ultravioletly regularized on a lattice; we also consider the system in a space time volume V and take the thermodynamical limit at the end.

Formally, the euclidean correlation functions of the fields can be obtained through the functional integral [28] [29]

$$\langle\phi(x_1)\ldots\phi(x_n)\rangle_V = Z_V^{-1}\int d\mu_V(\phi)\phi(x_1)\ldots\phi(x_n) \tag{3.14}$$

where

$$d\mu_V(\phi) = d\mu_{0V}(\phi)e^{\int_V \mathcal{L}dx}, \tag{3.15}$$

($d\mu_{0V}(\phi)$ corresponds to the formal expression $\mathcal{D}\phi = \Pi_{x\in V} d\phi_x$) is a measure over the euclidean field configurations (or *paths*),

$$\mathcal{L} = \mathcal{L}(\phi, \partial\phi)$$

is the euclidean Lagrangean function and $Z_V = \int d\mu_V(\phi)$. To give a precise meaning to the above expression, it is convenient to UV regularize the theory on a lattice (say of lattice spacing $a = 1/K$), so that $d\mu_{0V}$ is then the measure which treats each point of the lattice independently

$$d\mu_{0V}(\phi) = \prod_{i\in V} d\phi_i\ .$$

The action integral A_V becomes a sum over the lattice points, but for simplicity we will keep the continuum notation. Furthermore, on a lattice exp $(-A_V)$ is an integrable function of the fields, so that the above expression (3.14) has a meaning. The ultraviolet limit of $d\mu_V(\phi)$ has to be taken at the level of the correlation functions. The parameters which enter in $\mathcal{L} = \varnothing_{\text{free}} + \varnothing_{\text{int}}$ have to be considered as functions of the UV cut-off K, chosen in such a way that they yield finite correlation functions in the limit $K \to \infty$ (for perturbatively renormalizable theories such functions are expressed as formal power series of the renormalized coupling constant). Thus, $\mathcal{L}$ has to be understood as containing all the relevant counter terms.

From the above functional integral formulation one may derive Local and Boundary Ward Identities. To this purpose one considers an infinitesimal change of the variables in the field configuration space

$$\phi(x) \to \phi'(x) \equiv \phi(x) + \epsilon\chi(x).$$

Putting

$$d\mu'_V(\phi') \equiv d\mu_V(\phi) \tag{3.16}$$

we define

$$\delta d\mu_V(\phi) \equiv d\mu'_V(\phi) - d\mu_V(\phi).$$

Now, if the points $x_1, \ldots, x_n$ in the correlation functions (3.14) do not belong to the support of x, we have

$$\phi'(x_1)\ldots\phi'(x_n) = \phi(x_1)\ldots\phi(x_n)$$

and

$$\begin{aligned}\int d\mu_V(\phi)\phi(x_1)\ldots\phi(x_n) &= \int d\mu'_V(\phi)\phi(x_1)\ldots\phi(x_n)\\ &= \int d\mu'_V(\phi')\phi'(x_1)\ldots\phi'(x_n)\\ &= \int d\mu'_V(\phi)\phi(x_1)\ldots\phi(x_n)\end{aligned}$$

i.e.

$$\int \delta d\mu_V(\phi)\phi(x_1)\ldots\phi(x_n) = 0. \tag{3.17}$$

On the other hand: i) $\delta d\mu_{0V}(\phi) = 0$ since $d\mu_{0V}$ treats the lattice points as independent and on each lattice point such change of variables leaves $d\mu_{0V}$ invariant; ii) furthermore

$$\delta e^{-A_V} = \epsilon\int_V \left[\frac{\delta\mathcal{L}}{\delta\phi}\chi + \frac{\delta\mathcal{L}}{\delta\partial_\mu\phi}\partial_\mu\chi\right] dx e^{-A_V}$$

and if χ has compact support $\subset V$, the second term can be integrated by parts, yielding

$$\delta d\mu_V = d\mu_V\epsilon\int_V \chi(x)D_{EL}\mathcal{L}dx.$$

Eq. (3.7) then reads

$$\epsilon\int d\mu_V(\phi)\int_V \chi(x)dxD_{EL}\mathcal{L}\phi(x_1)\ldots\phi(x_n) = 0.$$

Since $\chi(x)$ is arbitrary, apart from the condition that $x_1, \ldots, x_n \notin$ supp χ, we have

$$\langle\phi(x_1)\ldots\phi(x_j)D_{EL}\mathcal{L}(x)\phi(x_{j+i})\ldots\phi(x_n)\rangle_V = 0. \tag{3.18}$$

In removing the UV cut-off, c-number subtractions or normal ordering may be necessary to define polynomials of the fields at a point and give a meaning to the Euler-Lagrange equations in the UV limit. We will understand that such a procedure has been performed in eq. (3.18) and use the same notation for the renormalized operator $D_{EL}\mathcal{L}(x)$.

By the condition on supp χ, eq. (3.18) holds when the euclidean times $x_{01},\ldots,x_{0n}$ are all different from x_0. Hence, by a backward Wick rotation (or by the OS reconstruction theorem [30]), the corresponding T-ordered correlation functions in Minkowski space vanish and by the time-slice axiom or by analyticity (the Reeh-Schlieder theorem [31]) one gets that the matrix elements of $D_{EL}\mathcal{L}$ on a dense set of states vanish. In this way one obtains the ***Local Ward Identities.*** The Local Ward Identities include in particular the standard Ward-Takahashi identities (which are obtained by variations of the fields corresponding to symmetry transformations). They play a crucial rôle in controlling and partly fixing the UV behaviour of the theory. As it is clear from the previous discussion, to control the infrared structure of QFT models, like gauge theories, the LWI are not enough and one needs additional identities, the BWI. To our knowledge they have not been systematically discussed in the literature and in our opinion they are relevant as the LWI. The derivation of the ***Boundary Ward Identities*** in the quantum case can be done along a pattern similar to that used above by considering variations of the fields which do not vanish on the space boundary. Proceeding as in the classical case (with the minor modifications required by the functional integral as above) , one gets

$$\langle\phi(x_1)\ldots\phi(x_j)\sum_i n_i\frac{\partial\mathcal{L}}{\delta\partial_i\phi}(x)\Big|_{x\in\partial V_S}\phi(x_{j+i})\ldots\phi(x_n)\rangle_V=0 \tag{3.19}$$

(V_S denotes the space volume) and the vanishing of the matrix elements of the boundary operator $(\sum n_i\delta\mathcal{L}/\delta\partial_i\phi)_{x\in\partial V_S}$ in Minkowski space.

Local and Boundary WI can also be derived in the case in which the UV regularization of the measure is not that used previously (where one uses the "flat" measure $d\mu_{0V}$ as a reference measure and UV cuts off the entire Lagrangean). We shall discuss the case in which the UV cut-off K is only in the interaction

$$d\mu_{VK}=d\mu^V_{\text{free}}\exp\left(\int_V\mathcal{L}^K_{\text{int}}dx\right), \tag{3.20}$$

$d\mu^V_{\text{free}}$ being the free measure corresponding to $\mathcal{L}_{\text{free}}$, formally

$$d\mu^V_{\text{free}}=d\mu_{0V}e^{\int_V\mathcal{L}_{\text{free}}dx},$$

the UV cut-off being removed in $\mathcal{L}_{\text{free}}$. This procedure is close to the perturbative approach.

The variation of $d\mu_{VK}$ under the change of variables $\phi(x)\to\phi(x)+\epsilon\chi(x)$ is then given by

$$\begin{aligned}\delta d\mu_{VK}=&(\delta d\mu^V_{\text{free}})\,exp\left(\int_V\mathcal{L}^K_{\text{int}}dx\right)\\&+\epsilon\int_V\left[\frac{\delta\mathcal{L}^K_{\text{int}}}{\delta\phi}\chi+\frac{\delta\mathcal{L}^K_{\text{int}}}{\delta\partial_\mu\phi}\partial_\mu\chi\right]dx\,d\mu_{VK},\end{aligned} \tag{3.21}$$

$$\delta d\mu^{V}_{\text{free}} = d\mu^{V}_{\text{free}} J = d\mu_{\text{free}} \left(1 + \epsilon \int dx \chi D\phi\right), \tag{3.22}$$

where J denotes the Jacobean of the field transformation, which in the free case can be computed and found to be $(1 + \epsilon \int dx\chi D\phi)$, with D^{-1} the kernel defined by the free two-point function. Formally

$$\delta d\mu_{\text{free}} = d\mu_{\text{free}} \left(\epsilon \int \left[\frac{\delta \mathcal{L}_{\text{free}}}{\delta \phi}\chi + \frac{\delta \mathcal{L}_{\text{free}}}{\delta \partial_\mu \phi}\partial_\mu \chi\right] dx\right).$$

By combining (3.21) and (3.22) and proceeding as before one gets in a similar way the LWI and the BWI.

Remark 1. One may wonder about the asymmetry between space and time in the BWI, already present in the classical Hamilton variational principle. As a matter of fact, a covariant QFT yields euclidean correlation functions with no space and time asymmetry. In fact, in the above derivation of the BWI one may as well consider variations with χ of compact support in space, but not in time. The argument goes through up to the analogue of eq. (3.19), and if the theory is covariant, by the Reeh-Schlieder theorem, it is possible to localize a dense set of states in an arbitrarily small bounded region of space time, so that one gets the vanishing of the matrix elements of $\delta\mathcal{L}/\delta\partial_0\phi|_{x_0 \in V_T}$ on a dense set of states. Hence, one gets the covariant BWI

$$\frac{\delta \mathcal{L}}{\delta \partial_\mu \phi(x)} n^\mu(x)\Big|_{x\in\partial V} = 0. \tag{3.23}$$

However, for theories which are not covariant one can in general only use the analyticity in time, which is guaranteed by the existence of the Hamiltonian. In this case, it is possible to localize a dense set of states in space time regions with the time variable in a bounded interval (time-slice axiom) and therefore one may only get the BWI of the form (3.19). (Space time symmetric BWI still hold on the vacuum state.)

The above functional integral discussion may be regarded as a way of justifying the validity of the classical action principle (in the $\hbar \to 0$ limit), merely as a consequence of the fact that: i) the functional integral is invariant under change of variables in the configuration space (eq. (3.17)); ii) unless the theory has Euclidean covariance properties one can obtain only those equations (i.e. BWI) which follow from variations of the configurations which vanish at the time boundary (Hamilton principle).

4. The Schwinger Model and the θ Angle Problem

Semiclassical non-perturbative calculations [1] [2] [3] have shown that, in QCD with massless fermions, the decomposition into pure phases of the correlation functions, obtained from the functional integral, is realized by taking a coherent superposition of topologically distinct sectors, each characterized by a fixed topological number $\nu \equiv (g^2/32\pi^2) \int d^4x F_{\mu\nu} \tilde{F}^{\mu\nu}$, each weighted by the phase $\exp i\nu\theta$. The so obtained quantum *phases* are then labelled by the free parameter θ (θ-vacua), and characterized by chiral symmetry breaking. The occurrence of the phase $\exp i\nu\theta$ is essentially equivalent to the addition of the topological term

$$\mathcal{L}_\theta = \theta(g^2/32\pi^2) F\tilde{F}$$

to the "naive" QCD Lagrangean and it is responsible for a strong CP violation for massive fermions (unless $\theta = 0, \pi$). The experimental data indicate that θ must be extremely small and one would like to have a natural explanation for $\theta = 0$. As emphasized also recently by Jackiw [31] , the vanishing of θ appears a *fine tuning* deep open problem of QCD.

In this section we would like to discuss a possible solution by realizing the very delicate removal of the infrared cut-off and by a careful handling of the boundary terms [32]. As already emphasized in the previous sections, in the presence of long range interactions, the boundary terms may give rise to volume effects and be equivalent to a coupling to an external field (seizing of the vacuum). As we have explicitly seen in the case of Heisenberg-like models (sect. 2), such effective external field coupling arising from boundary terms can either be treated as a boundary variable (or variable at infinity) which is free to "align to the direction" of an additional external field, or be frozen in advance and therefore be in general mismatched with respect to the direction of an additional external field. The two possibilities can e.g. be realized by a suitable choice of the boundary conditions. We argue that a similar structure may occur in QCD, where the vacuum angle θ can be left free to get aligned to the direction of an additional external field like the fermion mass term M, namely

$$\theta = \theta_m \equiv \arg \det M,$$

so that there is no strong CP violation. As we will see this possibility is related to the choice of chirally symmetric boundary conditions.

This is simply and clearly illustrated by the massive Schwinger model, with topological term in the bosonized form (in the axial gauge), provided the boundary terms are carefully handled, in the removal of the infrared cut-off [32]. We start with this model and discuss the QCD case later. We first discuss the massless Schwinger model, since there the relation between the occurrence of a variable at infinity playing the rôle of the θ angle and the careful removal of the infrared cut-off, in the functional integral, without neglecting boundary terms, is particularly transparent.

i) The Massless Schwinger Model

The Euclidean Lagrangean is

$$\mathcal{L} = -\frac{1}{2}(\partial_1 A_0)^2 + i\,(e/\sqrt{\pi})\, A_0 \partial_1 \phi - \frac{1}{2}(\partial_0 \phi)^2 - \frac{1}{2}(\partial_1 \phi)^2 + i\frac{e}{2\pi}\theta \partial_1 A_0 \ , \quad (4.1)$$

where the last term $-i(e/2\pi)\theta E$, $E = -\partial_1 A_0$, is the topological term, i.e. the two-dimensional analogue of the θ-term $ie\theta\tilde{F}F$ in four dimensions.

As we shall see, independently of a fermion mass term, the functional measure exhibits characteristic features of mean field interactions, which completely change the rôle of the topological term. As discussed in sect. 2, in order to properly treat the occurrence of boundary effects and display the coupling to variables at infinity, we adopt the strategy of leaving the decomposition into pure phases at the very end, i.e. after the infinite volume limit on the correlation functions defined by the functional integral. The resulting picture appears more clearly if we put the system on a square lattice V of finite size $2L+1$, so that we can work with free boundary conditions. By exploiting the invariance of the functional integral under variations of the fields, as discussed in sect. 3, we get the Local and Boundary WI. Combining the two kinds of WI we get

$$\Delta\phi_{ij} - \frac{e^2}{\pi}\left(\phi_{ij} - \phi_{Bj} - \frac{\theta}{2\sqrt{\pi}}\right) = 0, \tag{4.2}$$

$$E_{(ij)} = i\left(\frac{e}{\sqrt{\pi}}\right)\left(\phi_{ij} - \phi_{Bj} - \frac{\theta}{2\sqrt{\pi}}\right) = 0, \tag{4.3}$$

where Δ is the discrete Laplacean, ϕ_{ij} is the lattice version of ϕ, i,j are the lattice site indexes in $[-L,L]\times[-L,L]$, the first being the space index,

$$\phi_{Bj} \equiv \frac{1}{2}\left(\phi_{L_j} + \phi_{-L_j}\right), \tag{4.4}$$

E is the electric field and the following lattice version of the interaction has been used

$$\int A_0\partial_1\phi d^2x \to \sum_{i,j} A_0(\alpha_{ij})\left(\phi_{ij+e_1} - \phi_{ij}\right),$$

α_{ij} being the link in the temporal direction with lower end point (ij), and e_1 a unit vector in the space direction.

Eqs. (4.2) (4.3) are the euclidean lattice version of eqs. (3.10') (3.11) including the contribution of the θ term; they are the Lagrangean analogue of eq. (2.10) with the boundary terms now localized on the boundary of the euclidean space time.

The removal of the infrared cut-off ($L \to \infty$) in the correlation functions of $\exp(i\alpha\phi)$ and E can be done through an analysis of the spectrum of the operators which define the action as a quadratic form [33]. Thus, one can show that

i) the infinite volume limit exists for the correlation functions of $\exp(i\phi(f))$, $\partial\phi(f)$ and E,

ii) on such infinite volume correlation functions

$$\lim_{V\to\infty} e^{i\alpha\phi_V} \equiv \lim_{V\to\infty} \exp\left[i\alpha\frac{1}{V}\sum_{i,j\in V}\phi_{ij}\right], \qquad \alpha \in \mathbb{R}$$

exists and it defines a strongly continuous unitary group $\exp(i\alpha\phi_\infty)$,

iii) on the correlation function of $\exp(i\alpha\phi), \alpha \in \mathbb{R}, E, \partial\phi$

$$\lim_{V\to\infty} (\phi_{Bj} - \phi_{ij}) = \phi_\infty - \frac{\theta}{2\sqrt{\pi}} - \phi_{ij}, \tag{4.5}$$

iv) in the limit $L \to \infty$, ϕ converges to $\phi_\infty + \chi$, with χ a free field of mass $|e|/\sqrt{\pi}$ and vanishing mean value (more precisely $\exp(i\alpha\phi) \to \exp(i\alpha\phi_\infty)$ $\exp(i\alpha\chi)$), and eqs. (4.2) (4.3) take the following form in such a limit

$$\Delta\phi - \frac{e^2}{\pi}(\phi - \phi_\infty) = 0, \tag{4.6}$$

$$E = i\left(\frac{e}{\sqrt{\pi}}\right)(\phi - \phi_\infty). \tag{4.7}$$

It is worthwhile to remark that the Lagrangean parameter θ has disappeared from the equations of motion (and its rôle has been taken by the variable at infinity ϕ_∞). Since pure phases are labelled only by $\langle\phi_\infty\rangle = \langle\phi\rangle \equiv \theta_{\text{vac}}/2\sqrt{\pi}$, there is no dependence in the theory on the Lagrangean angle θ. The careful handling of the boundary terms and the chiral symmetry of the Lagrangean are crucial for this result. Freely integrating by parts, without keeping the boundary terms, or equivalently imposing vanishing boundary conditions on ϕ, $\phi_B = 0$ (chiral symmetry breaking boundary conditions), would eliminate the coupling with variables on the boundary (see eqs. (4.2) (4.3)) and lead to eqs. (4.6) (4.7) with ϕ_∞ replaced by $\theta/2\sqrt{\pi}$. In fact in this case the limit (4.5) would give

$$\langle\phi_\infty\rangle = \langle\phi\rangle = \frac{\theta}{2\sqrt{\pi}}\ , \tag{4.8}$$

and the θ angle labels the pure phases. In this way, one would recover the picture of Callan, Dashen and Gross [2].

The rôle played by the boundary conditions in the breaking of chiral symmetry and the very strong relation with the spin models discussed in sect. 2 can be clearly seen by realizing that with free boundary conditions the action can be written as

$$A = A_{\text{loc}} + i\left(\frac{e}{\sqrt{\pi}}\right)\int \partial_0(A_1\phi)d^2x + i\left(\frac{e}{2\pi}\right)\theta\int E d^2x \tag{4.9}$$

with A_{loc} giving rise to Ward Identities without boundary terms. A_{loc} can be viewed as obtained from A with $\theta = 0$ and $\phi = 0$ on the boundary (chiral breaking b.c.). In the limit $L \to \infty$, the value of ϕ on the boundary can essentially be replaced by a single variable ϕ_B, which by the previous results converges to $\phi_\infty - \theta/2\sqrt{\pi}$. Then, the above action becomes

$$A = A_{\text{loc}} + i\left(\frac{e}{\sqrt{\pi}}\right)\phi_B\int E d^2x + i\left(\frac{e}{2\pi}\right)\theta\int E d^2x \tag{4.10}$$

$$\cong A_{\text{loc}} + ie\left(\frac{e}{\sqrt{\pi}}\right)\phi_\infty\int E d^2x\ . \tag{4.11}$$

This is the same as the Callan, Dashen and Gross action [2] with the topological term replaced by $(ie/\sqrt{\pi})\,\phi_\infty E$. It is clear from eq. (4.10) that one gets the same correlation functions either by reducing the theory, obtained by the action (4.9), into pure phases labelled by $\langle\phi_\infty\rangle = \langle\phi\rangle \equiv \theta_{\rm vac}/2\sqrt{\pi}$ (**strategy I**), or by using the Callan, Dashen and Gross action

$$A^{CDG} = A_{\rm loc} + ie\left(\frac{\theta^{CDG}}{2\pi}\right)\int Ed^2x \tag{4.12}$$

and by treating θ^{CDG} as a free parameter (**strategy II**).

ii) The Massive Schwinger Model

The analogies with the spin models of sect. 2, suggest that the two strategies above, namely those based on the actions (4.9) and (4.12) are no longer equivalent if an external field is introduced, since the volume effects induced by the boundary terms are quite different in the two cases. This is therefore what one expects if the fermion mass term

$$m^2 : \cos\left(2\sqrt{\pi}\phi - \theta_m\right) : \tag{4.13}$$

is introduced. Actually, one can argue that one gets the same correlation functions if one uses either the action (4.9) plus the mass term (4.13) with $\theta = \theta_m = 0$ or the action (4.12) plus the mass term (4.13) with $\theta^{CDG} = \theta_m = 0$. In fact, in the first case $\langle \exp(i\phi_B)\rangle$ has the same symmetries of the potential, i.e. ϕ_B essentially converges to one of the minima of $\cos(2\sqrt{\pi}\phi)$, which are all equivalent to $\phi_B = 0$ for the correlation functions of $\exp(2i\sqrt{\pi}\phi)$, E and $\partial\phi$; this implies that the term $\phi_B E$ is irrelevant and the two actions are equivalent. On the other hand, if $\theta_m \neq 0$ and $\theta = 0 = \theta^{CDG}$ the two actions have different symmetry properties. In the strategy I, one can obtain the case $\theta = 0,\ \theta_m \neq 0$ from the case $\theta = 0,\ \theta_m = 0$ by a chiral rotation since the action (4.9) is chirally symmetric. Such correlation functions also coincide with those obtained by using the action (4.12), if the parameter θ^{CDG} is fine tuned to be equal to θ_m; in fact they are obtained by a chiral rotation from the case $\theta^{CDG} = 0 = \theta_m$ (since in strategy II, the action (4.2) is not symmetric under chiral transformations one has to change θ^{CDG} [1] [2]). The equivalence of the two strategies in this case also follows by comparing (4.11) and (4.12) with $\langle\phi_\infty\rangle = \theta_m/2\sqrt{\pi}$. In strategy II, however, with $\theta^{CDG} \neq \theta_m$ *the correlation functions depend on* θ^{CDG} *and* θ_m [1] [2] [30] .

Moreover, one can argue that in the general case $\theta \neq 0$, $\theta_m \neq 0$, ϕ_B essentially converges to $(\theta_m - \theta)/2\sqrt{\pi}$ so that the action (4.10) is actually independent of the Lagrangean parameter θ and therefore *the corresponding correlation functions are independent of* θ *and coincide with those of the case* $\theta = 0$, $\theta_m \neq 0$.

In conclusion, by adopting strategy I, i.e. by using a chirally symmetric action (apart from the fermion mass term) and by imposing chirally symmetric boundary conditions one is automatically led to a theory in which the θ term is not effective (one can put $\theta = 0$) and the correlation functions are invariant under the CP symmetry defined by the fermion mass term. Such correlation functions actually coincide with those obtained with strategy II by fine-tuning $\theta^{CDG} = \theta_m$, but no fine-tuning is needed in strategy I, where the disappearance of the CP violation is a consequence of the chiral symmetry of the action in a finite volume and of the boundary condition. In this respect, such mechanism can be regarded as natural in the technical sense with the chiral symmetry in finite volume playing the rôle of the Peccei-Quinn symmetry; here no extra symmetry is needed.

The above *natural* mechanism for the disappearance of the strong CP violation can be explicitly controlled in the simple case in which the mass term is replaced by $m^2\left(\phi - \theta_m/2\sqrt{\pi}\right)^2/2$. By following strategy I, one can prove that eq. (4.5) is unchanged, but now ϕ_∞ can only take the value $\theta_m/2\sqrt{\pi}$ and eqs. (4.2) (4.3) become

$$\Delta\phi - \left(\frac{e^2}{\pi} + m^2\right)\left(\phi - \frac{\theta_m}{2\sqrt{\pi}}\right) = 0, \tag{4.15}$$

$$E = \left(\frac{ie}{\sqrt{\pi}}\right)\left(\phi - \frac{\theta_m}{2\sqrt{\pi}}\right). \tag{4.16}$$

As anticipated, there is no dependence on θ, no strong CP violation and the unique phase is characterized by $\langle\phi\rangle = \langle\phi_\infty\rangle = \theta_m/2\sqrt{\pi}$. Again, a quite different result [34] [35] would be obtained by using the action (4.12) with θ^{CDG} a free parameter, or equivalently by using the action (4.9) with vanishing boundary conditions on $\phi, \phi_B = 0$ (chiral symmetry breaking boundary conditions!). In fact, in both cases, we get the following equation for E

$$E = i\left(\frac{e}{\sqrt{\pi}}\right)\left(\phi - \phi_B - \frac{\theta}{2\sqrt{\pi}}\right).$$

However, in the limit $L \to \infty$, strategy I yields $\phi_B \to \phi_\infty - \theta/2\sqrt{\pi}$ and

$$\langle E\rangle = 0, \tag{4.17}$$

whereas the other strategy, corresponding to choosing $\phi_B = 0$, yields

$$\langle E\rangle = i\left(\frac{e}{\sqrt{\pi}}\right)\left(\langle\phi\rangle - \frac{\theta}{2\sqrt{\pi}}\right)$$

and

$$\langle\phi\rangle = \left(\frac{1}{2\sqrt{\pi}}\right)\frac{\left(\theta + \frac{m^2\pi}{e^2}\theta_m\right)}{\left(1 + \frac{m^2\pi}{e^2}\right)}\ , \tag{4.18}$$

which give

$$\langle E\rangle = \frac{im^2}{2e}\frac{(\theta_m - \theta)}{\left(1 + \frac{m^2}{e^2}\pi\right)}$$

Thus, unless $\theta^{CDG} \equiv \theta = \theta_m$, there is a non-vanishing background electric field and CP violation.

It is worthwhile to remark that the solution of the CP problem can be argued quite generally to follow from the chiral symmetry of the action (4.9); in fact for $\theta_m = \theta = 0$ the theory is symmetric under the CP transformation defined by the mass term, $CP_0 \equiv CP_{\theta_m=0}$, as in the standard case with $\theta^{CDG} = \theta_m = 0$. Then, the chiral invariance of the action, apart from the mass term, allows to go from $\theta_m = 0$ to $\theta_m \neq 0$ by a chiral transformation $\alpha(\theta_m)$ and therefore the corresponding correlation functions, for $\theta = 0,\ \theta_m \neq 0$, are symmetric under

$$CP_{\theta_m} \equiv \alpha^{-1}(\theta_m) CP_0 \alpha(\theta_m).$$

iii) Boundary Terms and the Screening of the Topological Term

To see in more detail the mechanism by which the boundary terms restore chiral symmetry and lead to a CP invariant theory it is instructive to control the thermodynamical limit of ϕ and ϕ_B. To this purpose we consider the continuum limit of the Local and Boundary WI in the temporal gauge

$$\Delta\phi - \tilde{e}^2 \left(\phi - \phi_B - \tilde{\theta}\right) - m^2 \left(\phi - \tilde{\theta}_m\right) = 0, \tag{4.19}$$

$$E = i\tilde{e} \left(\phi - \phi_B - \tilde{\theta}\right), \tag{4.20}$$

$$\tilde{e} \equiv e/\sqrt{\pi},\ \tilde{\theta} \equiv \theta/2\sqrt{\pi},\ E = -\partial_0 A_1,\ \phi_B \equiv \frac{1}{2}\left[\phi(x_1, T) + \phi(x_1, -T)\right]$$

$$E = -i\tilde{e}\tilde{\theta}, \qquad \text{on } \partial\mathrm{T}, \tag{4.21}$$

$$\partial_0\phi - i\tilde{e}A_1 = 0, \qquad \text{on } \partial\mathrm{T}, \tag{4.22}$$

$$\partial_1\phi = 0, \qquad \text{on } \partial\mathrm{V_S}, \tag{4.23}$$

where $\partial T, \partial V_S$ denote the temporal and the space boundary. Since A_1 is undetermined by a function of x_1, one has to completely fix the gauge by adding a gauge fixing condition on A_1; we will choose

$$A_1(x_1, T) = -A_1(x_1, -T). \tag{4.24}$$

From eq. (4.20) and the definition of E we get

$$A_1(x_1, x_0) = -\int_0^{x_0} dx_0' i\tilde{e} \left(\phi - \phi_B - \tilde{\theta}\right) + f(x_1) \tag{4.25}$$

and by eq. (4.24)

$$f(x_1) = \frac{1}{2}\int_{-T}^{T} dx_0' i\tilde{e}\epsilon(x_0')(\phi - \phi_B). \tag{4.26}$$

Then, the boundary condition (4.22) takes the following form

$$\partial_0\phi(x_1, \pm T) \mp \frac{\tilde{e}^2}{2}\int_{-T}^{T} dx_0'(\phi - \phi_B) = \mp\tilde{e}^2 T\tilde{\theta}. \tag{4.27}$$

Thanks to eq. (4.20) and the definition of ϕ_B, the boundary condition (4.21) is equivalent to

$$\phi(x_1,T) = \phi(x_1,-T). \tag{4.28}$$

For simplicity we put $\theta_m = 0$. Eq. (4.19) can conveniently be written in the following form

$$L_1\phi + L_2\phi = -\tilde{e}^2\tilde{\theta} \tag{4.29}$$

with

$$L_1\phi = \partial_0^2\phi + \frac{\tilde{e}^2}{2}(\phi(x_1,T) + \phi(x_1,-T)) - (\tilde{e}^2 + m^2)\phi$$

$$L_2\phi = \partial_1^2\phi \quad .$$

One can prove that eq. (4.29) plus the boundary conditions (4.23) , (4.27), (4.28) has a unique solution of class C^2 in the interior of $V_S \times T$ and C^1 on the boundary.

Such a solution can be found by Fourier expansion in terms of eigenfunctions $\phi_n(x_1)$ of L_2, satisfying (4.23)

$$\phi(x_1,x_0) = \sum_n c_n(x_0)\phi_n(x_1), \qquad \phi_n(x_1) = \cos\frac{\pi n x_1}{L} \ . \tag{4.30}$$

Then, eq.(4.29) becomes

$$\sum_n \{-n^2 c_n + L_1 c_n\}\, \phi_n(x_1) = -\tilde{e}^2\tilde{\theta}$$

and since the r.h.s. is independent of x_1, one has

$$L_1 c_n(x_0) = n^2 c_n(x_0), \qquad n \neq 0, \tag{4.31}$$

$$L_1 c_n(x_0) = -\tilde{e}^2\tilde{\theta}, \qquad n = 0. \tag{4.32}$$

The boundary condition (4.27) reads

$$\sum_n \left[\partial_0 c_n(\pm T) \mp \frac{\tilde{e}^2}{2}\int_{-T}^{T} dx_0(c_n - c_{nB})\right] \phi_n(x_1) = \mp\tilde{e}^2\tilde{\theta}\, T$$

and again the independence of the r.h.s. from x_1 gives

$$\partial_0 c_n(\pm T) \mp \frac{\tilde{e}^2}{2}\int_{-T}^{T} dx_0(c_n - c_{nB}) = 0, \quad n \neq 0, \tag{4.33}$$

$$\partial_0 c_0(\pm T) \mp \frac{\tilde{e}^2}{2}\int_{-T}^{T} dx_0(c_0 - c_{0B}) = \mp\tilde{e}^2\tilde{\theta}\, T. \tag{4.34}$$

For $n \neq 0$ we have an eigenvalue problem for an operator, L_1, which can be shown to be negative

$$\int_{-T}^{T} dx_0 \phi L_1 \phi = -\int_{-T}^{T} \left[(\partial_0\phi)^2 + m^2\phi^2 + \tilde{e}^2(\phi - \phi_B)^2\right] dx_0 < 0$$

as a consequence of the definition of L_1 and of the boundary conditions (4.27) (4.28). Thus, one must have

$$c_n = 0, \qquad n \neq 0.$$

The general solution of eq. (4.32), satisfying the boundary condition (4.28) is

$$c_0(x_0) = A\cosh(\omega_0 x_0) + c_0^1, \tag{4.35}$$

where $\omega_0 \equiv (\tilde{e}^2 + m^2)^{1/2}$, c_0^1 is a particular solution of the inhomogeneous equation:

$$c_0^1 = \left(\tilde{\theta} + c_B\right) \frac{\tilde{e}^2}{(\tilde{e}^2 + m^2)} \,. \tag{4.36}$$

The constants A and c_B are determined by the condition $c_B = c_0(T)$ and by the boundary condition (4.34). The result is

$$A = -\tilde{\theta}(\cosh(\omega_0 T))^{-1} \left\{ \frac{\tilde{e}^2 \omega_0 T}{\tilde{e}^2 \omega_0 T + m^2 \tanh(\omega_0 T)} \right\} \tag{4.37}$$

$$\tilde{\theta} + c_B = -\tilde{\theta}\left(\tilde{e}^2 + m^2\right) \frac{\tanh(\omega_0 T)}{\tilde{e}^2 \omega_0 T + m^2 \tanh(\omega_0 T)}. \tag{4.38}$$

The solution in the case $\theta_m \neq 0$ is obtained by a chiral transformation, i.e. by a shift of $\tilde{\theta}_m : c_0(x_0) \rightarrow c_0(x_0) + \tilde{\theta}_m$.

Since in solving the above Ward Identities we have not isolated a region of space on which a dense set of states can be localized, the so obtained solution describes the matrix elements on the ground states.

From eqs. (4.35)-(4.38) we learn that for large volumes ($T \rightarrow \infty$) the expectation value of ϕ is essentially equal to $\tilde{\theta}_m$ everywhere, except a region near the boundary of size $1/\omega_0$

$$\lim_{T\rightarrow\infty} \langle \phi \rangle_T = \tilde{\theta}_m$$

Furthermore, the value of ϕ at the boundary, ϕ_B, is

$$\langle \phi_B \rangle_T = -\tilde{\theta} + \tilde{\theta}_m + \mathcal{O}\left(\frac{1}{\omega_0 T}\right).$$

Thus, for large T, the *boundary term* $i\tilde{e}\phi_B E$ in the action (4.10)*completely screens off the topological term* $i\tilde{e}\tilde{\theta}E$ *and effectively replaces it by* $i\tilde{e}\tilde{\theta}_m E$. This means that the action

$$A = A_{\rm loc} + i\tilde{e}\int \phi_B E d^2x + i\tilde{e}\tilde{\theta}\int E d^2x,$$

where $A_{\rm loc}$ is the local action (without boundary terms) corresponding to the Callan, Dashen and Gross-action A^{CDG} with $\theta^{CDG} = 0$, is effectively replaced by

$$A = A_{\rm loc} + i\tilde{e}\tilde{\theta}_m \int E d^2x$$

i.e. by the CDG action with $\theta^{CDG} = \theta_m$. One may say that the topological term induces a condensation of the field ϕ (i.e. a fermionic condensation) on the boundary, which completely screens such a topological term.

5. Fermionic Integration, Boundary Conditions, and Chiral Symmetry

A crucial issue of the above discussion is that the model is defined in a finite volume by a chirally symmetric action (eq. (4.9)). As a consequence of this property, the introduction of a mass term is equivalent to the introduction of an external field coupling, in a theory with spontaneously broken chiral symmetry; this leads immediately to the alinement of the order parameter $\langle\phi\rangle$ to $\theta_m/2\sqrt{\pi}$ and to the disappearance of the CP violating parameter.

The second point is that even if, when the mass term is present, the phase is unique, nevertheless the boundary conditions are not irrelevant since there is a long range coupling with the boundary. The existence of this coupling can be argued *quite generally* as a consequence of the local chiral anomaly and of the invariance of the action under global chiral transformations. In fact, the chiral anomaly implies a mass gap in the Goldstone spectrum associated to chiral breaking; this in turn requires that the effective Hamiltonian is not symmetric, which is compatible with the symmetry of the original finite volume Hamiltonian only if there are long range interactions with the boundary [4-9,11].

By the above remarks, we are thus lead to discuss the chiral properties of the action (more precisely of the functional measure) in presence of chiral anomalies, in comparison with the standard approach [1] [2] [25] [35] [36-38]. Most of the standard wisdom has been obtained by studying the fermion integration in presence of an external (euclidean) gauge field and it is clearly illustrated in Fujikawa's approach [38].

It is convenient to distinguish two types of problems. One is the analysis of the properties of the (euclidean) fermionic correlation function in an external gauge field, formally given by

$$\langle\psi_1\ldots\bar{\psi}_n\rangle_A = (\det \not{D}_A)^{-1}\int \mathcal{D}\psi\mathcal{D}\bar{\psi}\exp\left[\int \mathcal{L}_{\psi A}d^4x\right]\psi_1\ldots\bar{\psi}_n \qquad (5.1)$$

where $\not{D}_A \equiv \gamma_\mu(\partial_\mu + ieA_\mu), \mathcal{D}\psi, \mathcal{D}\bar{\psi}$ denote the Berezin fermionic integration, $\mathcal{L}_{\psi A}$ is the fermionic part of the Lagrangean and the normalization is that corresponding to the fermionic partition function in an external gauge field, $Z_A = \det \not{D}_A$.

The other problem is the contribution to the effective action for the gluons, given by the fermionic integration, i.e. $\det \not{D}_A$

$$\langle A_1\ldots A_n\rangle = Z^{-1}\int \mathcal{D}A\exp\left[\int \mathcal{L}_A d^4x\right]\det \not{D}_A A_1\ldots A_n \quad .$$

i) Fermionic Correlation Functions in an External Field

To better define the correlation functions (5.1) one adopts an ultraviolet regularization, typically by cutting the high eigenvalues of $\not{D}_A$ (see e.g. [38]). Much of the attention in the literature has been paid to the UV regularization; actually a crucial preliminary step is to put the system in a finite volume V and to fix boundary conditions which define $\not{D}_A$ as an antiadjoint operator ($\gamma_\mu^\dagger = \gamma_\mu$). Only then one can deal with the eigenvalues of $\not{D}_A$ (see e.g. [39]). The thermodynamical limit is supposed to be taken on the correlation functions only at the end.

Furthermore, a technical complication arises due to the occurrence of zero modes. In fact, for finite volume and UV cutoff, the occurrence of zero modes yields $\det \not{D}_A = 0$ and the fermion integration in (5.1) vanishes unless the number n of pairs $\psi, \bar{\psi}$ in the integrand is larger or equal to the number k of zero modes ($u_i, i = \bar{i}_1, \ldots, \bar{i}_k$) and in this case it is equal to

$$\sum_{i_1,\ldots,i_n}{}' \prod_{i \neq i_1,\ldots,i_n} \lambda_i \left(u_{i_1} \ldots u_{i_n}^\dagger \right)_a \tag{5.2}$$

where Σ' means a summation with $i_m \neq i_j$, for $m \neq j$, and $i_1 = \bar{i}_1, \ldots, i_k = \bar{i}_k$, λ_i denotes the eigenvalues of $\not{D}_A$, u_i the corresponding eigenfunctions and $(\;)_a$ antisymmetrization. The standard convention is to replace $\det \not{D}_A$ by $\det' \not{D}_A$ defined by the removal of the zero modes and to define

$$\langle \psi_1 \ldots \bar{\psi}_n \rangle'_A = (\det{}' \not{D}_A)^{-1} (5.2) \tag{5.3}$$

when $n \geq k$, and zero otherwise.

We find it more convenient to add a fermion mass term, which thanks to its hermiticity, precludes the existence of zero modes for the new Dirac operator $\not{D}_{A,m} \equiv \not{D}_A + m$. One can show that when $m \to 0$, for $n \geq k$,

$$\det(\not{D}_A + m) \langle \psi_1 \ldots \bar{\psi}_n \rangle_A \to \det{}' \not{D}_A \sum_{i_1,\ldots,i_n}{}' (u_{i_1} \ldots u_{i_n}^\dagger)_a \left(\frac{1}{\lambda_{i_1} \ldots \lambda_{i_n}} \right)' \tag{5.4}$$

where $(\ldots)'$ means that for each mode the weight has to be taken $1/\lambda_{i_j}$ if $\lambda_{i_j} \neq 0$ and 1 if $\lambda_{i_j} = 0$.

More generally one can take a mass term M of the form

$$M = m(\cos\theta_m + i\gamma_5 \sin\theta_m). \tag{5.5}$$

For $m \neq 0$, also in this case zero modes cannot exist. The operator $\not{D}_A + M$ is not a normal operator in general; however the corresponding fermionic quadratic form can be written in diagonal form. In fact, by considering boundary conditions such that γ_5 maps the domain of $\not{D}_A + M$ into itself (γ_5 invariant b.c.), if u_{λ_0} denotes an eigenfunction of $\not{D}_A$ corresponding to the eigenvalue λ_0, one has

$$\gamma_5 u_{\lambda_0} = u_{-\lambda_0}, \qquad \text{for } \lambda_0 \neq 0 \tag{5.6}$$

(since $\not{D}_A$ is antihermitean, $\bar{\lambda}_0 = -\lambda_0$). Therefore for $\lambda_0 \neq 0$ the spectrum of $\not{D}_A$ is symmetric. For $\lambda_0 = 0$, one may use a basis which diagonalizes γ_5 and in general one gets $n_\pm$ modes with positive/negative chirality. Since M does not connect blocks $(\lambda_0, \bar{\lambda}_0)$ corresponding to different eigenvalues λ_0, the operator $\not{D}_A + M$ can be brought to block diagonal form, each block being labelled by the pair $(\lambda_0, \bar{\lambda}_0)$, for $\lambda_0 \neq 0$, plus the block corresponding to the $n_\pm$ modes, which become eigenfunctions of $\not{D}_A + M$ with eigenvalues $me^{\pm i\theta_m}$. We can now expand $\psi(x)$ into eigenfunctions u_n of $\not{D}_A + M$

$$\psi(x) = \sum_n c_n u_n(x), \qquad (\not{D}_A + M)u_n(x) = \lambda_n u_n,$$

and $\bar{\psi}(x)$ into eigenfunctions v_n of $(\not{D}_A + M)^\dagger$

$$\bar{\psi}(x) = \sum_n \bar{c}_n v_n^\dagger(x), \qquad (\not{D}_A + M)^\dagger v_n = \bar{\lambda}_n v_n.$$

The fermionic quadratic form then takes the diagonal form

$$\sum_n \lambda_n \bar{c}_n c_n, \qquad \lambda_{-n} = \bar{\lambda}_n,$$

and the fermionic integration can be done by using Berezin calculus. In this way one gets well defined fermionic functions, for finite volume; e.g.

$$\langle \bar{\psi}(x)\psi(y)\rangle_A = \sum_{\lambda_i} \frac{1}{\lambda_i} v_i^\dagger(x) u_i(y). \tag{5.7}$$

If $\theta_m = 0, v_i$ is also an eigenfunction of $\not{D}_A + M$ with eigenvalue λ_i and therefore $v_i = u_i$. In the following we will mostly consider this case, for simplicity. The r.h.s. of eq. (5.7) is the eigenfunction expansion of the resolvent of $\not{D}_A$, with m playing the rôle of the resolvent parameter. Such expansion exists and the sum over λ_i converges in norm as an operator in L^2, since $(\not{D}_A + m)^{-1}$ is a compact operator, namely the λ_i's have finite multiplicity and have only infinity as accumulation point, for bounded A_μ and for a large class of boundary conditions. For simplicity this case is considered in the following.

After this preliminaries we can state

Proposition 5.1. Once the boundary conditions have been fixed, the fermionic correlation functions in an external field are independent of the UV regularization.

We will show that the same fermionic correlation functions are obtained if instead of cutting the high eigenvalues of $\not{D}_A + m$, as used in deriving eq. (5.7), one introduces a standard UV cutoff in the interaction

$$\not{A} \to \not{A}_\Lambda \equiv P_\Lambda \not{A} P_\Lambda \tag{5.8}$$

where P_Λ is the projection onto the eigenspace of the free fermionic modes with eigenvalues $|\lambda_i^0| < \Lambda$. It is enough to show that $(\not{D}_{A_\Lambda} + m)^{-1}$ converges in norm to $(\not{D}_A + m)^{-1}$.

To this purpose we consider the Neumann expansion

$$(\not{D}_{A_\Lambda} + m)^{-1} = (\not{\partial} + m)^{-1/2} \left[\sum_m ((\not{\partial} + m)^{-1/2} \not{A}_\Lambda (\not{\partial} + m)^{-1/2})^n \right] (\not{\partial} + m)^{-1/2} \tag{5.9}$$

which is norm convergent uniformly in Λ for m large enough, since

$$\| (\not{\partial}+m)^{-1/2} \not{A}_\Lambda(\not{\partial}+m)^{-1/2} \| \leq \| (\not{\partial}+m)^{-1/2} \|^2 \|\not{A}_\Lambda \| \leq \frac{1}{m} \|\not{A} \| \equiv K/m.$$

By the same argument, also the analogous Neumann expansion for $(\not{D}_A+m)^{-1}$ is norm convergent for m large enough and therefore if the partial sums in (4.9) converge in norm to the corresponding partial sums for $(\not{D}_A+m)^{-1}$, when $\Lambda \to \infty$,so does the total series. In fact,

$$\begin{aligned}
&\| [(\not{\partial}+m)^{-1/2} \not{A}_\Lambda(\not{\partial}+m)^{-1/2}]^n - \left[(\not{\partial}+m)^{-1/2} \not{A}(\not{\partial}+m)^{-1/2}\right]^n \| \\
&\leq \| (\not{\partial}+m)^{-1/2}(\not{A}_\Lambda- \not{A})(\not{\partial}+m)^{-1/2} \| \, n \left(\frac{K}{m}\right)^{n-1} \\
&\leq 2\Lambda^{-1/2}(K/m^{1/2})n \left(\frac{K}{m}\right)^{n-1}
\end{aligned}$$

since $P_\Lambda \not{A} P_\Lambda - \not{A} = (P_\Lambda - 1)AP_\Lambda + A(P_\Lambda - 1)$ and

$$\| (\not{\partial}+m)^{-1/2}(1-P_\Lambda) \| \leq \Lambda^{-1/2}.$$

Finally, both $(\not{D}_{A_\Lambda}+m)^{-1}$ and $(\not{D}_A+m)^{-1}$ are analytic functions of m for $\Re m > 0$, and the analytic functions $(\not{D}_{A_\Lambda}+m)^{-1}$, which are all bounded in norm by $(\Re m)^{-1}$, converge in norm to $(\not{D}_A+m)^{-1}$ for $\Re m > K$; then they converge in norm to $(\not{D}_A+m)^{-1}$ for $\Re m > 0$, (the same argument holds for $\Re m < 0$.

In conclusion, the norm convergence of the resolvents implies the norm convergence of the correlation functions as convolution operators in L^2, hence weakly in L^2 and therefore as distributions in the two variables and in L^2 in each variable.

The independence of the fermionic correlation functions (in an external gauge field) from the UV regularization of the functional measure implies that also the equations obeyed by the currents, which are defined through a point splitting procedure, in terms of the correlation functions, have a content independent of the UV regularization. It should be mentioned that the UV regularization of the functional measure can also be used to define compound field operators and therefore different UV regularizations may lead to different currents; for example the gauge invariant UV regularization based on cutting the large eigenvalues of $\not{D}_A$ leads to the construction of the anomalous gauge invariant chiral current, whereas the chiral invariant UV regularization, based on cutting the large eigenvalues of the free Dirac operator, leads to a chiral current which is conserved in the limit of massless fermions (but not gauge invariant). In any case, however, even if the equations obeyed by the so obtained different currents are different, they must be equivalent, i.e. have the same content, since the underlying correlation functions are the same. The anomaly for the gauge invariant chiral current is equivalent to the conservation equation for the gauge non-invariant chiral current.

The above arguments also show that the possible breaking of chiral symmetry, being decided by the correlation functions, cannot be a consequence of the UV regularization, for example of the gauge invariant regularization, as sometimes stated in the literature. The possible lack of chiral symmetry of the functional measure in finite volume can, and it actually does, only depend on the boundary conditions, as we will discuss below.

The mechanism is already clearly displayed by the massless Schwinger model discussed in the previous section, where the anomaly of the gauge invariant chiral current $j^5_\mu = (1/\sqrt{\pi})\partial_\mu \phi$, being independent of the boundary conditions, does not imply the lack of chiral symmetry of the correlation functions. As we have discussed above, the chiral symmetry depends on the boundary conditions; it holds for chirally symmetric b.c. (strategy I) and its breaking shows up only at the level of the decomposition into pure phases, but it does not hold if one chooses b.c. which break chiral symmetry (strategy II).

ii) Boundary Conditions and Chiral Symmetry.

The chiral symmetry of the functional measure for finite volume V is equivalent i) for $M = 0$ to the invariance of the correlation functions (for finite volume) under rigid chiral transformations

$$\psi \to e^{i\alpha\gamma_5}\psi, \qquad A_\mu \to A_\mu, \tag{5.10}$$

and ii) for $M \neq 0$ to the invariance of the correlation functions when the above transformation is accompanied by

$$\theta_m \to \theta_m + 2\alpha. \tag{5.11}$$

To investigate the chiral symmetry of the functional measure it is convenient to discuss separately the fermionic correlation functions in an external gauge field and the effective measure $\det(\not{D}_A + M)$ (or $\det' \not{D}_A$) for A_μ. We start by discussing the first ones in the case $M \neq 0$. By using chiral transformations we get

$$\not{D}_A + M_{\theta_m} = e^{i\theta_m\gamma_5/2}(\not{D}_A + M_{\theta_m=0})e^{i\theta_m\gamma_5/2} \tag{5.12}$$

and therefore

$$(\not{D}_A + M_{\theta_m})^{-1} = e^{-i\theta_m\gamma_5/2}(\not{D}_A + M_0)^{-1}e^{-i\theta_m\gamma_5/2}\ . \tag{5.13}$$

Thus, the fermionic correlation functions in finite volume, being related to the eigenfunction expansion of $(\not{D}_A + M_{\theta_m})^{-1}$ (e.g. eq. (5.7)), are covariant under the chiral transformation (5.10), (5.11).

Less obvious is the behavior of $\det(\not{D}_A + M)$ under chiral transformations also because a regularization is needed to define the determinant. Choosing a regularization which cuts the large eigenvalues in a symmetric way ($|\lambda_n| < \Lambda$), one gets

$$\det(\not{D}_A + M_{\theta_m}) = \prod_{\substack{|\lambda_n|<\Lambda \\ \lambda_{0n}\neq 0}} [\det(\not{D}_A + M_{\theta_m})]_{\lambda_n} \cdot [\det(\not{D}_A + M_{\theta_m})]_0$$

where the subscript λ_n denotes the restriction to the block corresponding to the pairs $\lambda_n, \bar{\lambda}_n$ for $\lambda_{n0} \neq 0$ (λ_{n0} denotes the eigenvalue when $M = 0$), and the subscript zero denotes the restriction to the space generated by the $n_\pm$ modes with $\lambda_{n0} = 0$. Clearly

$$[\det(\not{D}_A + M_{\theta_m})]_0 = e^{i\theta_m(n_+ - n_-)} m^{(n_+ + n_-)} \tag{5.14}$$

$$[\det(\not{D}_A + M_{\theta_m})]_{\lambda_n} = |\lambda_n|^2 \tag{5.15}$$

so that $\det(\not{D}_A + M_{\theta_m})$ is chiral invariant iff $n_+ = n_-$. The same result can be obtained by a ζ function regularization which treat λ_n and $\bar{\lambda}_n$ symmetrically, e.g.

$$\zeta'(s) = -\sum_n \ln \lambda_n e^{s(\ln \lambda_n + \ln \bar{\lambda}_n)} =$$

$$= -[(n_+ + n_-)\ln m + i(n_+ - n_-)\theta_m]\, e^{-s \ln m^2} - \sum_{n>0} \ln |\lambda_n|^2 e^{-s \ln |\lambda_n|^2}.$$

Since $|\lambda_n|^2$ is independent of θ_m, the above function is independent of θ_m and therefore chiral symmetric if $n_+ = n_-$. We have then to discuss the zero modes of $\not{D}_A$. Clearly, the number of zero modes depends on the external field, but also on the b.c.. The index theorem relation

$$n_+ - n_- = \nu \equiv -(1/32\pi^2) Tr \int {}^*F F dx \tag{5.16}$$

holds (apart from corrections which become irrelevant in the infinite volume limit) if the so-called spectral b.c. of Atiyah, Patodi and Singer (APS) [40] are imposed. Such b.c. give rise therefore to a measure which is not chirally invariant, in the presence of gauge fields with $\nu \neq 0$.

It is important to note that the dependence of $n_+ - n_-$ on A_μ only arises through the dependence on A_μ of the b.c.. In fact, if the b.c. are independent of A_μ (and γ_5 invariant) then, since eigenvalues and eigenvectors are continuous in the external field with the uniform topology and for $\lambda \neq 0$ the spectrum is always symmetric, $n_+ - n_-$ is independent of the gauge field. Examples of b.c. independent of A_μ are the periodic or antiperiodic b.c., for which, by the above argument, one gets $n_+ = n_-$ and a functional measure (for finite volume) invariant under global chiral transformations. This is not the case of the APS b.c. ([36][41][42]), since they change discontinuously when the topological number varies.

Moreover, one can find b.c. which yield $n_+ - n_-$ = independent of A_μ (and therefore a chirally invariant functional measure) without losing gauge invariance. One can actually modify the APS b.c. in such a way that they still yield:

1) antiself-adjointness of $\not{D}_A$,
2) a discrete spectrum for $\not{D}_A$ with finite multiplicity and having only $\lambda = \infty$ as accumulation point,
3) γ_5 invariant domain of $\not{D}_A$,

4) gauge invariance of $\not{D}_A$, namely if A' is related to A by a globally defined gauge transformation $\mathcal{V}$, then $\not{D}_{A'} = \mathcal{V}\mathcal{D}_A\mathcal{V}^{-1}$,
5) CP invariance of the domain of $\not{D}_A$ as in the APS case,
but give
6) $n_+ - n_- =$ independent of A.

To see this, as in the APS case in finite volume V, one considers gauge fields which are pure gauge around the boundary and one writes the euclidean Dirac operator in the neighbourhood of the (closed) boundary, defined by $r = 0$, in the form $i\gamma_r(\partial_r + B_A)$, where the subscript r denotes the component in the direction normal to the boundary. Furthermore, one can assume that A_μ has no normal component [36]. Since B_A is a differential operator defined on a compact manifold, its spectrum is discrete and one may introduce the subspace $H^A_{<0}(H^A_{\geq 0})$ of states generated by eigenstates of the restriction $B^+_A(B^-_A)$ of B_A to positive (negative) chirality states. The APS b.c. then require

$$\psi|_{r=0} \in H^A_{<0} \quad \text{if } \gamma_5\psi = \psi; \qquad \psi|_{r=0} \in \mathrm{H}^A_{\geq 0} \text{ if } \gamma_5\psi = -\psi.$$

The above definition of $H^A_{<0}$ implies that if on the boundary the field configuration A and $A = 0$ are connected by the gauge transformation $U(A)$

$$A_\mu|_{r=0} = U(A)\partial_\mu U(A)^{-1}|_{r=0}$$

then, one has

$$H^A_{<0} = U(A)H^{A=0}_{<0}$$

and similarly for $H^A_{\geq 0}$.

The modified b.c. are defined by requiring

$$\psi|_{r=0} \in K^A_{<0} \equiv U(A)U^{-1}_{\nu(A)}H^{A=0}_{<0}, \text{ if } \gamma_5\psi = \psi,$$

where $\nu(A)$ is the topological number of A and $U_{\nu(A)}$ is the gauge transformation which on the boundary connects $A = 0$ to a fixed reference configuration $A^{(\nu)}$ of the sector labelled by $\nu(A)$. Similarly one defines $K^A_{\geq 0}$ and requires $\psi|_{r=0} \in K^A_{\geq 0}$ if $\gamma_5\psi = -\psi$. The idea is that for each sector $[\nu]$ one chooses a representative configuration $A^{(\nu)}$ and the b.c. corresponding to it are the APS b.c. for the free Dirac operator ($A = 0$); within each sector then on the boundary different gauge configurations are connected by the gauge transformation $U(A)U^{-1}_{\nu(A)}$.

The APS b.c. are singled out if, in addition to the gauge invariance condition 4), one further requires that if A_μ is a pure gauge around the boundary $A|_{r=0} = U\partial U^{-1}|_{r=0}$, (with U not globally defined on V if $\nu \neq 0$), then the corresponding b.c. is obtained from that for $A = 0$ by acting with U. For the modified APS b.c. the gauge invariance is required only if two configurations of A_μ are related by a gauge transformation over the whole volume.

Property 4) follows from the gauge covariance of K^A since gauge transformations $\mathcal{V}$ globally defined on V do not change $\nu(A)$.

Properties 1)-3) follow as in the APS case, since by a globally defined gauge transformation one is reduced on the boundary to a gauge configuration, $A^{(\nu)}$, for which the b.c. are those of APS case with $A = 0$; then 1)-3) hold for $\not{D}_{A=0}$ and therefore also for $\not{D}_{A\neq 0}$ since A is a bounded perturbation.

Property 5) holds as in the APS case if the set of representative configurations $\{A^{(\nu)}\}$ is chosen to be CP stable.

Finally, property (6) follows from the remark that as before one is reduced on the boundary to the representative configurations and for them the b.c. are independent of A and, once the b.c. are fixed the spectrum of $\not{D}_A$ is continuous in A; this and the symmetry of the $\lambda \neq 0$ spectrum under γ_5 therefore implies that $n_+ - n_-$ must be independent of A, and therefore $n_+ = n_-$ if it is so for $A = 0$. As it will be clear from the following discussion, the analogue of Strategy I discussed in the massive Schwinger model above, is the choice of boundary conditions which yield $n_+ = n_-$. Actually, since what is relevant is $\det(\not{D}_A + M)/Z$, the property $n_+ - n_- =$ independent of A gives the same results as the case $n_+ = n_-$. From the lesson of the massive Schwinger model we have strong reasons to believe that in this case the topological term gets screened off by a fermion condensation on the boundary. However, for the moment we will put $\theta = 0$ (we will comment below on the naturality of this choice).

In this case, the generic correlation functions in finite volume V are given by

$$\langle A_1 \ldots \psi_j \ldots A_k \ldots \bar{\psi}_n \rangle^V_{M_{\theta_m}} = \int d\mu_V(A) A_1 \ldots A_k \ldots \\ \det(\not{D}_A + M_{\theta_m})_V \langle \psi_j \ldots \bar{\psi}_n \rangle^V_A \tag{5.17}$$

where $d\mu_V(A)$ is the measure over the gauge field configurations, formally

$$d\mu_V(A) = \prod_{x \in V} dA_x \exp \int_V \mathcal{L}_A dx.$$

As a consequence of eq. (5.13), for γ_5 invariant boundary conditions, the fermionic correlation functions in external gauge fields $\langle \psi_j \ldots \bar{\psi}_n \rangle^V_A$ transform covariantly under chiral transformations. Moreover by the above choice of the boundary conditions yielding $n_+ = n_-$, $\det(\not{D}_A + M_{\theta_m})_V$ is independent from θ_m and therefore chirally invariant. The resulting (finite volume) measure on A is therefore invariant under rigid chiral transformations and the correlation functions (5.17) transform covariantly under chiral transformations:

$$\langle A_1 \ldots \psi_j \ldots A_k \ldots \bar{\psi}_n \rangle^V_{M_{\theta_m}} = \\ \langle A_1 \ldots e^{-i\theta_m \gamma_5/2} \psi_j \ldots A_k \ldots \bar{\psi}_n e^{-i\theta_m \gamma_5/2} \rangle^V_{M_{\theta_m = 0}} . \tag{5.18}$$

The above choice of b.c. giving $n_+ = n_-$ also yields the CP symmetry of the correlation functions (5.18). In fact, if CP_m denotes the CP symmetry defined by M_{θ_m},

$$(\not{D}_A + M_{\theta_m})^{CP_m} = e^{i\theta_m \gamma_5/2} (\not{D}_A + M_0)^{CP_0} e^{i\theta_m \gamma_5/2}$$

and the spectrum of $(\not{D}_A + M_0)^{CP_0}$ is obtained from that of $\not{D}_A + M_0$ by changing each eigenvalue λ_n into its complex conjugate $\bar{\lambda}_n$, so that the CP_m symmetry of the fermionic correlation function in an external gauge field follows from their chiral covariance (for γ_5 invariant b.c.). It remains to discuss the CP transformation properties of $\det(\not{D}_A + M_{\theta_m})$. For $\theta_m = 0$ the symmetry follows easily from eqs. (5.14), (5.15), for γ_5 invariant boundary conditions. However, for arbitrary $\theta_m \neq 0$ the spectrum of $\not{D}_A + M_{\theta_m}$ is not symmetric under $\lambda_n \to \bar{\lambda}_n$, the determinant is not real (with a complex phase depending on θ_m), and therefore not CP_m symmetric, unless $n_+ = n_-$.

We can now comment on the naturality of $\theta = 0$; since, for $n_+ = n_-$, $\det(\not{D}_A + M_{\theta_m})$ is real and independent of θ_m, the effective action for A_μ is CP symmetric at each order in the fermion loop expansion and the renormalization of $\det(\not{D}_A + M_{\theta_m})$ cannot require counter-terms of the type $i\theta\tilde{F}F$ since the determinant is real and the θ term represents a complex contribution. The presence of weak interactions does not change this property since again, for $n_+ = n_-, \det(\not{D}_A + M_{\theta_m})$ is real independently of θ_m and therefore at each order in the expansion in the weak coupling constant g_W.

We can now try to compare the above results with the standard wisdom [1], [2], [25], which usually makes reference to the formulation in the infinite volume limit. From a rigorous point of view this limit is problematic and quantities like $\det(\not{D}_A + M)$ lose their meaning. We will then consider the case of large V, with the understanding that the actual limit in the formulas below should be taken at the level of the correlation functions.

In the limit $V \to \infty$, the fermionic correlation functions in an external gauge field become independent of the boundary conditions, since, for $V \to \infty$, the operator $\not{D}_A + M$ is unique and therefore so is also $(\not{D}_A + M)^{-1}$, which defines the fermionic correlation functions. Thus, in the infinite volume limit, the fermionic correlation functions in external gauge fields are the same as those obtained in the standard strategy (CDG). On the other hand, $\det(\not{D}_A + M)^V$ is a global quantity, and it crucially depends on the boundary conditions also when V becomes large. By denoting by $\det(\not{D}_A + M_{\theta_m})^{\rm sym}$ the determinant obtained with chirally symmetric boundary conditions, for large V

$$\det(\not{D}_A + M_{\theta_m})^{\rm sym} = \det(\not{D}_A + M_0)^{\rm sym} \simeq \det(\not{D}_A + M_{\theta_m})^{CDG} e^{-i\theta_m(n_+ - n_-)}, \tag{5.19}$$

where the superscript CDG denotes the determinant obtained by using the APS chiral breaking boundary conditions. Hence, in the limit of large V

$$\begin{aligned}\langle A_1 \ldots \psi_j \ldots A_k \ldots \bar{\psi}_n \rangle^{\rm sym}_{M_{\theta_m}} &\simeq \int d\mu(A) det(\not{D}_A + M_{\theta_m})^{CDG} \\ &\qquad A_1 \ldots A_k \ldots e^{-i\theta_m(n_+ - n_-)} \langle \psi_j \ldots \bar{\psi}_n \rangle_{M_{\theta_m}} \\ &= \Big(\int d\mu(A) e^{-i\theta(n_+ - n_-)} \det(\not{D}_A + M_{\theta_m})^{CDG} \\ &\qquad A_1 \ldots A_k \ldots \langle \psi_j \ldots \bar{\psi}_n \rangle_{M_{\theta_m}} \Big)_{\theta = \theta_m} \cdot \\ &= \langle A_1 \ldots \psi_j \ldots A_k \bar{\psi}_n \rangle^{CDG}_{M_{\theta_m}, \theta = \theta_m}\end{aligned} \tag{5.20}$$

It is worthwhile to remark that, even if the correlation functions obtained by using chirally symmetric b.c. in the infinite volume limit coincide with those of the standard picture with $\theta = \theta_m$, the perturbative renormalization of the two theories may face different problems, since in the first case the vanishing of the CP violating parameter is guaranteed by the chiral symmetry of the finite volume measure (equivalently by the Boundary Ward Identities), whereas in the second case the equality $\theta = \theta_m$ is not assured by a symmetry (the finite volume measure is not chirally symmetric) nor by the BWI, and therefore requires a fine tuning. The use of chirally symmetric boundary conditions (gives rise to BWI which) assure the vanishing of the CP violation parameter at each step of the renormalization procedure. Roughly, it is like comparing a model with a given symmetry and a model obtained by adding and subtracting a symmetry breaking term

$$\begin{aligned} \mathcal{L}_{\text{sym}} &= \mathcal{L}_{\text{sym}} + \theta \mathcal{L}_{\text{br}} - \theta_m \mathcal{L}_{\text{br}}|_{\theta=\theta_m} \\ &\equiv \mathcal{L}'_\theta - \theta_m \mathcal{L}_{\text{br}}|_{\theta=\theta_m}. \end{aligned}$$

A perturbative expansion based on $\mathcal{L}'$ with $\mathcal{L}_{\text{br}}$ as a perturbation, *without recognizing* the symmetry in question would require a fine tuning of the parameters order by order, in order to restore the original symmetry.

In the QCD case, if $\theta_m \equiv \arg\det M_{\theta_m} \neq 0$ and $\theta = 0$, the correlation functions of the gauge fields are CP_m symmetric at the tree level, but in the standard strategy (chiral breaking b.c.), such symmetry is not stable under quantum corrections, since $\det(\not{D}_A + M_{\theta_m})$ is complex (and so is also its fermion loop expansion) and its phase must be compensated by a θ term with a "fine tuning" $\theta = \theta_m$ (even if θ_m is not renormalized by strong interactions). In fact, *after* taking the sum over fermion loops, i.e. by considering the effective action for the gauge fields

$$(\det(\not{D}_A + M_{\theta_m}))^{CDG} e^{-i\theta(n_+ - n_-)}|_{\theta=\theta_m}$$

is independent of θ_m, real and CP_m symmetric.

However, such restoration of symmetry through a fine tuning is not stable under the addition of the weak interactions, since in this case, if one puts $\theta = \theta_m^{\text{tree}}$, where the subscript tree means the value at the tree level in the weak interactions, one does not get a CP_m symmetric expansion in the weak coupling constant g_W. CP_m symmetry can be obtained only if, order by order in g_W one performs a fine tuning $\theta = \theta_m$ (lack of naturality). Instead, with the strategy based on chirally symmetric b.c., $\det(\not{D}_A + M_{\theta m})$ is real, independent of θ_m, and therefore CP symmetric at each order of the weak expansion of θ_m. It is clear from the above discussion that the local chiral anomaly, which is independent of the existence of zero modes [37], and therefore of b.c., does not imply the lack of global chiral invariance of the functional measure (and of the correlation functions) in a finite volume. The chiral non invariance of the correlation functions is rather related to the choice of the b.c. and therefore the mechanism is typical of spontaneous symmetry breaking in a situation, however, where b.c. give rise to volume effects and may be equivalent to the addition of an explicit symmetry breaking term in the local Lagrangean (*seizing of the vacuum*). This is not in contradiction with the local chiral variations of the functional measure

$$d\mu \rightarrow d\mu \exp\{-i(g^2/16\pi^2)Tr \int \alpha(x)^* FF dx\}. \tag{5.21}$$

E.g. for periodic b.c., fermion field configurations live on a torus, and symmetry under global chiral transformations can be understood as a consequence of the vanishing of the total topological charge, which enters in eq. (5.21) with $\alpha(x) = 1$, due to a δ-function contribution resulting from gauge fields discontinuity on the torus across the boundary. Working with periodic b.c. is therefore similar to restricting the gauge field configuration to those with $\nu = 0$. This is in fact the approach of Nielsen and Schroer [37] for the massless Schwinger model, where it leads to correlation functions which, after decomposition into pure phases, coincide with those obtained by integrating over all gauge field configurations.

From the above discussion it follows that the use of a chirally invariant action (apart from the mass term) for the bosonized Schwinger model is equivalent to a fermion integration with b.c. leading to a symmetric functional measure. Similarly, the APS b.c. correspond to the use of the action A^{CDG}, eq. (4.12), in the bosonized Schwinger model. In this case, the index theorem implies that a change of the phase in the zero mode eigenfunctions (equivalent to a chiral transformation) is the same as changing θ^{CDG} in A^{CDG}.

The origin of the long range coupling with the boundary (a basic feature of the analysis of sect. 4) can be understood in terms of fermion integration by noticing that, with finite volume and e.g. periodic b.c., the picture of zero modes changes completely: the zero modes, concentrated around the localizations of the instantons, of the standard (APS or compactified) approach are replaced by a space of eigenfunctions (with eigenvalues close to zero) with an equal number of right and left dimensions. In general, each such eigenfunction will be localized partly inside the volume, around the instantons, and partly near the boundary, where periodic boundary conditions produce (on the torus) a non-trivial topological density. Such strong correlations between the interior and the boundary is expected to give rise to long range coupling with the boundary. From this point of view, the APS b.c., or the compactification of the euclidean space to a sphere, have the effect of eliminating such boundary coupling; this strategy is followed by Callan, Dashen and Gross [2] and by 't Hooft [30,25] and it is clearly equivalent to the other (based on periodic b.c.) as far as local Ward identities are considered. On the other hand, the identification of the pure phases, requiring a careful handling of the thermodynamical limit, crucially depends on the choice of the boundary conditions.

Finally, it may be instructive to compare the two strategies in the limit of massless fermions. Once the thermodynamical limit has been taken, the limit $m \rightarrow 0$ in strategy I yields pure phases labelled by the parameter θ_m ("the direction" along which the external field goes to zero); they are exactly the phases obtained in strategy II with $\theta = \theta_m$. However, the limits $V \rightarrow \infty$ and $m \rightarrow 0$ do not commute. In strategy I, letting $m \rightarrow 0$ before $V \rightarrow \infty$ yields a mixed phase (with chirally symmetric correlation functions); the decomposition into pure phases yields the phases labelled by the vacuum parameter θ_{vac}. The same result is obtained with strategy II (with $m \rightarrow 0$ before $V \rightarrow \infty$), where one gets a pure phase labelled by a vacuum angle $\theta = \theta_{\mathrm{Lagrangean}}$. The picture is very close to that discussed for the spin models and the Schwinger model of sects. 2, 4. The two strategies are equivalent in the absence of an external field (or of a fermion mass) but they cease to be so after the addition of an external field.

References

1 R. Jackiw and C. Rebbi: Phys. Rev. Lett. **37**, 172 (1976)

2 C. Callan, R. Dashen and D. Gross: Phys. Lett. **63B**, 334 (1976)

3 J. Kogut and L. Susskind: Phys. Rev. **D11**, 3594 (1975)

4 G. Morchio and F. Strocchi: J. Math. Phys. **28**, 1912 (1987)

5 G. Morchio and F. Strocchi: *Infrared problem, Higgs phenomenon and long range interactions*, in *Fundamental Problems of Gauge Field Theory*, Erice School 1985, G. Velo and A.S. Wightman dir. (Plenum Press 1986)

6 G. Morchio and F. Strocchi: Comm. Math. Phys. **111**, 593 (1987)

7 G. Morchio and F. Strocchi: Ann. Phys. (N.Y.) **170**, 310 (1986)

8 F. Strocchi: *Long range dynamics and spontaneous symmetry breaking in many-body systems*, lectures at the Workshop *Fractals, Quasicrystals, Knots and Algebraic Quantum Mechanics* , Maratea 1987, A. Amann et al. eds. (Kluwer Academic Publ. 1988)

9 G. Morchio and F. Strocchi: *Removal of the infrared cutoff, seizing of the vacuum and symmetry breaking in many-body and in gauge theories*, invited talk at the *IX Int. Conf. on Mathematical Physics* , Swansea 1988, B. Simon et al. eds. (Adam Hilger Publ. 1989)

10 R. Haag: Nuovo Cim. **25**, 1078 (1962); see also W. Thirring and A. Wehrl: Comm. Math. Phys. **4**, 303 (1967)

11 G. Morchio and F. Strocchi: Comm. Math. Phys. **99**, 153 (1985)

12 D. Ruelle: *Statistical Mechanics: Rigorous Results* (W.A. Benjamin, N.Y. 1969)

13 C. Domb and M.S. Green eds.: *Phase Transition and Critical Phenomena*, Vol. I (Academic Press 1971)

14 For a review see e.g. F. Strocchi: *Elements of Quantum Mechanics of Infinite Systems* (World Scientific, Singapore 1985)

15 G. Morchio and F. Strocchi: J. Math. Phys. **28**, 622 (1987)

16 J. Goldstone, A. Salam and S. Weinberg: Phys. Rev. **127**, 965 (1962)
J.A. Swieca: *Goldstone theorem and related topics*, in *Cargese Lectures in Physics* , Vol. 4, D. Kastler ed. (Gordon and Breach, New York 1970)

17 J.A. Swieca: Comm. Math. Phys. **4**, 1 (1967)

18 J. Schwinger: Phys. Rev. **128**, 2425 (1962)
J. Lowenstein and A. Swieca: Ann. Phys. (N.Y.) **68**, 172 (1971)

19 S. Coleman: Phys. Rev **D11**, 2088 (1975)
S. Mandelstam: Phys. Rev. **D11**, 3026 (1975)

20 A.L. Carey and S.N.M. Ruijsenaars: Acta Appl. Math. **10**, 1 (1987)

21 S. Coleman, R. Jackiw and L. Susskind: Ann. Phys. (N.Y.) **93**, 267 (1975)

22 O.E. Lanford and D. Ruelle: Comm. Math. Phys. **13**, 194 (1969)

23 T.W. Kibble: *Proc. Int. Conf. Elementary Particles* (Oxford Univ. Press 1965)
G.S. Guralnik, C.R. Haagen and T.W. Kibble: In *Advances in Particle Physics*, Vol. 2, R.L. Cook and R.E. Marshak eds. (Interscience 1968)

24 R.F. Streater: *Goldstone's theorem and related topics*, in *Many Degrees of Freedom in Field Theory*, L. Streit ed. (Plenum Press 1978)

25 G. 't Hooft: *How instantons solve the U(1) problem*, Phys. Reports **142**, 357 (1986)

26 E.L. Hill: Rev. Mod. Phys. **23**, 253 (1951)

27 C. Lanczos: *The Variational Principles of Mechanics* (University of Toronto Press 1949)

28 L.D. Faddeev: *Introduction to Functional Methods*, in *Methods in Field Theory*, Les Houches 1975, R. Balian et al. eds. (North-Holland 1976)
P. Ramond: *Field Theory. A Modern Primer* (Benjamin-Cummings 1981)

29 J. Glimm and A. Jaffe: *Functional Integral Methods in Quantum Field Theory*, in *New Developments in Quantum Field Theory and Statistical Mechanics*, H. Levy et al. eds., Cargese 1976 (Plenum Press 1977); *Quantum Physics. A Functional Integral Point of View* (Springer 1987)
30 G. 't Hooft: Phys. Rev. Letters **37**, 8 (1976)
31 R. Jackiw: *Topological investigations of quantized gauge theories*, in *Relativity, Groups and Topology II*, B.R. De Witt and R. Stora eds. (North Holland 1984)
32 G. Morchio and F. Strocchi: *Infrared structures in QFT models and the θ angle problem*, invited talk at the Conference *Selected Topics in Quantum Field Theory and Mathematical Physics* , Liblice, June 1989, edited by J. Niederle and J. Fisher (World Scientific)
33 G. Morchio: to be published
34 S. Coleman, R. Jackiw and L. Susskind: Ann. Phys. (N.Y.) **93**, 267 (1975)
S. Coleman: Ann. Phys. (N.Y.) **101**, 239 (1976)
35 S. Coleman: In *The Whys of Subnuclear Physics*, A. Zichichi ed. (Plenum Press 1979)
36 B. Schroer: *Topological Methods for Gauge Theories*, Schladming Lectures 1978, in *Facts and Prospects of Gauge Theories* , P. Urban ed. (Springer Verlag 1978)
37 N.K. Nielsen and B. Schroer: Nucl. Phys. **B127**, 493 (1977)
38 K. Fujikawa: Phys. Rev. **D21**, 2848 (1980)
39 A. Andrianov, L. Bonora and R. Gamboa-Saravi: Phys. Rev. **D26**, 2821 (1982)
40 M.A Atiyah, V.K. Patodi and I.M. Singer: Math. Proc. Camb. Phil. Soc. **77**, 43 (1975)
41 M. Hortacsu, K.D. Rothe and B. Schroer: Nucl. Phys. **B171**, 530 (1980)
42 M. Ninominiya and C.I. Tan: Nucl. Phys. **B257**, 199 (1985)

Two-Dimensional Nonlinear Sigma Models: Orthodoxy and Heresy [1]

A. Patrascioiu[1], *E. Seiler*[2]

[1] Physics Department and Center for the Study of Complex Systems
University of Arizona, Tucson, AZ 85721, U.S.A.
[2] Max-Planck-Institut für Physik und Astrophysik
– Werner-Heisenberg-Institut für Physik –
P.O. Box 40 12 12, D – 8000 München, Fed. Rep. of Germany

Abstract: It is generally believed that two-dimensional nonlinear σ models can be used as toy models for four-dimensional gauge theories. In particular it is believed that both classes of field theories show the same characteristic difference depending on the nature of their symmetry group: If it is abelian they have a massless phase at weak coupling, whereas the nonabelian models show spontaneous mass generation as well as asymptotic freedom. In these lectures we consider the so-called $O(N)$ models in two dimensions. We give a critical review of the traditional arguments advanced to support the above-mentioned beliefs ("the orthodoxy") and we outline the strategy of a proof for our "heretical" view that all those models have a massles low temperature phase in their lattice versions, which implies that possible continuum limits cannot be asymptotically free.

1. Introduction

In these lectures the nonlinear σ models will be understood as continuum limits of classical ferromagnets. For definiteness we will limit ourselves to the discussion of the $O(N)$ vector models which are defined as follows: To each point x in (a finite subset of) the square lattice we associate a classical spin $S(x)$ which is a unit vector in $\mathbb{R}^N$. The statistical distribution of the spins is determined by the Hamiltonian

$$H = -\sum_{\langle xy\rangle} S(x)\cdot S(y) \tag{1}$$

through the Gibbs measure

$$d\mu_\Lambda = Z_\Lambda^{-1} e^{-\beta H_\Lambda} \prod_x \delta(1-S(x)^2)d^D S(x) \tag{2}$$

As usual it is understood that one wants to take the thermodynamic limit $\Lambda \to \mathbb{Z}^2$.

[1] Lectures given by E. Seiler at XXIX. Internationale Universitätswochen für Kernphysik 1990, Schladming, Austria, March 1990.

A euclidean quantum field theory is then obtained by moving into a critical point at which the correlation length ξ diverges, and rescaling. That means that the continuum correlation functions are for instance given by

$$\lim_{\beta \to \beta_c} \langle S_{n_1}(x_1\xi) \cdots S_{n_k}(x_k\xi) \rangle Z_\beta^{-k/2} \tag{3}$$

where the wave function renormalization $Z(\beta)$ has to be chosen appropriately to obtain a nontrivial limit.

Construction of the continuum limit requires therefore a good enough understanding of the critical behavior of the theory on the lattice. This will be the main issue of these lectures.

We first recall what is rigorously known about these models.

***Facts*:**

(1) $\underline{N=1}$ (Ising model): There exists an ordered phase at large β showing spontaneous symmetry breaking. The reason is that defects (misaligned neighboring spins) cost a certain minimum amount of energy.

(2) $\underline{N \geq 2}$ ($N = 2$: plane rotator, $N = 3$: classical Heisenberg model): There is no spontaneous breaking of symmetry at any β ("Mermin-Wagner theorem" [1,2,3,4]). The reason is that soft spin waves cost arbitrarily little energy but have sufficient entropy to disorder the system.

(3) $\underline{N=2}$ (plane rotator): There is a "Kosterlitz-Thouless" transition at $\beta = \beta_c$ that manifests itself as follows: For $\beta < \beta_c$ the system shows exponential clustering, i.e. there are constants ξ and $C(A,B)$ such that for any two observables A and B (bounded functions of finitely many spins)

$$|\langle A\tau^n(B)\rangle - \langle A\rangle\langle B\rangle| \leq C(A,B)\exp(-n/\xi) \tag{4}$$

where $\tau^n(B)$ denotes the translate of B by n lattice units. For $\beta \geq \beta_c$, on the other hand, this is not so and there is only algebraic clustering, for instance

$$|\langle S(x)S(y)\rangle| \geq a|x-y|^{-\eta} \tag{5}$$

for $|x-y| \geq 1$.

In this low temperature phase, often called a spin wave phase, the system behaves essentially like a massless Gaussian, or more precisely

$$\begin{pmatrix} S_1 \\ S_2 \end{pmatrix} \approx \begin{pmatrix} \cos \alpha\Phi \\ \sin \alpha\Phi \end{pmatrix} \tag{6}$$

where Φ is a massless Gaussian field determined by the Hamiltonian - $H_\Lambda = \frac{1}{2}\sum_{\langle xy\rangle}(\Phi(x) - \Phi(y))^2$.

2. Beliefs

Before 1975 it was widely believed that the $O(N)$ models show for all $N \geq 2$ a phase transition to some kind of spin wave phase.

After 1975 the present orthodoxy caught hold which says that for $N > 2$ there is exponential clustering (a mass gap in the language of field theory) at all $\beta < \infty$, with an inverse correlation length (mass gap) behaving as

$$m \equiv \xi^{-1} \approx c\beta^a e^{b\beta} \tag{7}$$

This change in the conventional wisdom was mostly brought about by the influential papers of Polyakov [5] and Brézin and Zinn-Justin [6].

Generally there are three reasons that are advanced in support of this belief:

(1) *Instantons*:
The $O(3)$ model on $\mathbb{R}^2$, i.e. in the continuum has local minima of the Hamiltonian, i.e. finite energy solutions of the classical field equations, called instantons which can be given in terms of rational functions

$$\mathbb{R}^2 \cong \mathbb{C} \ni z \longmapsto w = c \prod_{i=1}^{k} \frac{z - a_i}{z - b_i} \in \mathbb{C}^* \cong S_2 \tag{8}$$

if we use a complex variable z as a coordinate on $\mathbb{R}^2$ and another one, w on the target manifold S_2 via stereographic projection. In this interpretation (8) and their complex conjugates become the announced finite energy solutions of the field equations (and there are no others).

The idea is now that these minima of the energy in some sense "dominate" the Gibbs measure and that a reasonable approximation is obtained by considering only Gaussian fluctuations around those minima. This leads to the so-called instanton gas picture in which the $O(3)$ model is replaced by a Coulomb gas right at the first collapse point [7,8], which is equivalent to a massive free fermion field.

(2) *Large N*:
In the limit $N \to \infty$, keeping $\tilde{\beta} = \beta/N$ fixed, the $O(N)$ model goes over into the exactly soluble "spherical model" of Berlin and Kac ([9]), which has a mass gap for all $\tilde{\beta}$.

There are even some rigorous results based on this observation, in particular A.Kupiainen showed that for any β there is an $N_0(\beta)$ such that the $O(N)$ model shows exponential clustering for $N > N_0(\beta)$ [10].

(3) *Perturbative Renormalization Group*:
For $N \geq 3$ the $O(N)$ models, when analyzed in the framework of formal perturbation theory (PT), show the property of asymptotic freedom (AF). This can be described as follows:

Consider the two point function of the spin, expanded to second order in $1/\beta$ according to the rules of formal perturbation theory:

$$\begin{aligned} G_\beta(x) &\equiv \langle S(0) \cdot S(x) \rangle_\beta \\ &\approx 1 - \frac{N-1}{2\beta} \ln|x| - \frac{N-1}{2\beta^2}\left((\ln|x|)^2 - \frac{1}{2} \ln|x| \right) + \mathcal{O}(\beta^{-3}) \end{aligned} \tag{9}$$

(see for instance [11]).

If we now try to adjust $\Delta\beta$ in such a way that

$$\frac{G_\beta(x)}{G_\beta(x)} \approx \frac{G_{\beta+\Delta\beta}(2x)}{G_{\beta+\Delta\beta}(2y)} \tag{10}$$

to order $1/\beta^2$ (as would be necessary to obtain the continuum limit), we obtain:

$$\Delta\beta = \frac{N-2}{2\pi}\ln 2 + \mathcal{O}(\frac{1}{\beta}) \tag{11}$$

In other words, there is a Renormalization Group (RG) transformation of β

$$R(\beta) = \beta + \Delta\beta \tag{12}$$

that seems to offset essentially the effect of refining the lattice by a factor of 2. Since obviously

$$R^n(\beta) \to \infty$$

as $n \to \infty$, and since $1/\beta = g$ is normally considered the coupling constant of the model, one says that there is asymptotic freedom because the smaller the lattice cutoff, the smaller the coupling has to be.

The quantity $\Delta\beta$ is directly related to the Callan-Symanzik β function by

$$\beta(g) = -g^2\Delta\beta/\ln 2 \tag{13}$$

If, on the other hand, in a model the two point function for large distances behaves like a power

$$G_\beta(x) \sim |x|^{-\eta} \tag{14}$$

then it follows that $\Delta\beta$ as defined before, and by (13) therefore the Callan-Symanzik β function vanishes. This is the situation in the $O(2)$ model according to Fact(3) above.

It seems that many people find the following inversion of this argument plausible (though it is not a strict logical consequence): If a model has asymptotic freedom in the sense explained above, it will show exponential clustering at arbitrarily large β. If one accepts in addition the perturbative calculation of $\Delta\beta$ outlined above, one is led to the belief stated in the beginning of this section.

3. Critique

Let us now have a critical look at the three arguments given above.

(1) ***Instantons***:
It has been known for many years that the fluctuations about the instantons are beset by uncontrollable infrared divergencies [12]; in particular it is not at all justified to think of them as small, so the whole calculation loses its credibility.

Another problem is that the instanton gas by its very nature creates a conflict with clustering: Since as "dominating" configurations one takes either holomorphic or antiholomorphic maps from the plane to the sphere, the sign of the

Jacobian of the map is constant and one has introduced by hand "orientational long range order" which would lead to spontaneous breakdown of the subgroup of reflections contained in $O(3)$ [13]. By analysis of the Fortuin-Kasteleyn representation described below, it is not hard to see that this impossible.

(2) *Large N*:
The large N arguments have something to say about making N large at fixed β. They say nothing about the question of interest, namely what happens at fixed N for large β. In particular there is no conflict whatsoever between the fact that the spherical model has no phase transition and a possible transition for finite N at a $\beta_c(N)$ that goes to ∞ faster than N.

(3) *Perturbation Theory*:
If one is not content with just playing a game, the minimal goal of PT is to produce an asymptotic expansion to a theory that makes sense beyond a formal expansion. The expansion parameter in our case is $g = 1/\beta$ (i.e. the bare coupling) but equivalently one can use any other parameter that has itself an asymptotic expansion in g (for instance a renormalized coupling constant).

For a **finite system** such as our ferromagnets on a finite lattice, such an asymptotic expansion of expectation values is easily obtained by Laplace's method, since for $\beta \to \infty$ the Gibbs measure becomes concentrated on the completely ordered configurations and the fluctuations away from them can be controlled. We note that the expansion coefficients are fully determined by the derivatives of the Hamiltonian H_Λ at its minimum.

For an **infinite system** of the type considered here ($2D$ $O(N)$ model) PT is a priori suspect, as has been stressed in [14]. The reason is that one is again trying to expand around a completely ordered state, whereas by the Mermin-Wagner theorem (Fact(2) above) there are always arbitrarily large fluctuations away from this state, since the system is disordered (no spontaneous symmetry breaking). So there is never anything small in which to expand.

Many people object to this view by pointing to the fact that in expectation values of $O(N)$ invariant observables the infrared divergencies cancel order by order in PT [11,15]. The underlying philosophy seems to be:

" If I get a finite answer, it must be correct".

But this is definitely a dangerous principle to follow as the following examples show.

Example 1 ([16]):
Consider the internal energy of the $1D$ $O(N)$ model with periodic boundary conditions on a finite lattice (chain of length L):

$$\langle (S(n) - S(n+1))^2 \rangle (\beta, L) = \frac{a_1(L)}{\beta} + \frac{a_2(L)}{\beta^2} + \mathcal{O}(\beta^{-3}) \qquad (15)$$

On the other hand it is not difficult to calculate directly in the infinite volume limit

$$\langle (S(n) - S(n+1))^2 \rangle (\beta, \infty) = \frac{b_1}{\beta} + \frac{b_2}{\beta^2} + \mathcal{O}(\beta^{-3}) \qquad (16)$$

Now taking the infinite volume limit in (15) term by term one obtains

$$a_1(\infty) = b_1 = \frac{N-1}{2} \tag{17}$$

$$a_2(\infty) = \frac{(N-1)(N-8)}{48} \tag{18a}$$

whereas b_2 (16) has the value

$$b_2 = \frac{(N-1)(N-3)}{8} \tag{18b}$$

So here PT up this order has a finite thermodynamic limit, but it is wrong.

Example 2 ([17]):
Consider the following "truncated sphere model" in $2D$: the model is defined like the $O(3)$ model, except that the spin is not allowed to take values in the polar caps defined by $|S_3| \geq 1 - \epsilon$. To set up PT for this model one should expand around an ordered state where the spin points in an equatorial direction (those directions are in fact favored for entropic reasons). Since the derivatives of the Hamiltonian at this configuration are totally oblivious of the constraint, PT turns out to be identical to PT for the $O(3)$ model. So one would also calculate perturbatively the same Callan-Symanzik β function as for the $O(3)$ model

$$\beta_{\text{PT}}(g) = -g^2/2\pi \tag{19}$$

and conclude that the model is asymptotically free. On the other hand by a simple yet rigorous argument based on Ginibre's correlation inequalities, in [17] it is shown that this model has only algebraic clustering at large β. This shows, as explained above, that $\Delta\beta = 0$, i.e. the true β function vanishes

$$\beta_{\text{exact}}(g) = 0 \tag{20}$$

What has gone wrong with PT in those two examples? For fixed finite L Laplace's method shows that PT gives an asymptotic expansion in $1/\beta$. But it is not **uniformly asymptotic** in L. Precisely:

$$\langle F\rangle(L,\beta) = \sum_{n=0}^{K} \frac{a_n(L)}{\beta^n} + R_K(\beta, L) \tag{21}$$

where

$$\beta^K R_K(\beta, L) \to 0 \tag{22}$$

but **not**

$$|\beta^K R_K(\beta, L)| \leq const \tag{23}$$

uniformly in L. In the $1D$ example on can see this explicitly by numerically determining $R_K(\beta, L)$.

The perturbative Renormalization Group leads also to intrinsic inconsistencies if taken seriously [17]. We will explain this in a simple form by considering the distribution of the block spin variable

$$\bar{S}_\Lambda = \frac{1}{|\Lambda|} \sum_{x \in \Lambda} S(x) \tag{24}$$

Binder [18] suggested to study the following quantity

$$U_L(\beta) = -\frac{\langle \bar{S}_\Lambda \cdot \bar{S}_\Lambda \bar{S}_\Lambda \cdot \bar{S}_\Lambda \rangle^T}{\langle \bar{S}_\Lambda \rangle^2} = \frac{2}{N} - \frac{\langle \bar{S}_\Lambda^2 ; \bar{S}_\Lambda^2 \rangle}{\langle \bar{S}_\Lambda^2 \rangle^2} \tag{25}$$

(similar quantities were studied in the probabilistic Renormalization Group pioneered by Jona-Lasinio et al [19]). $U_L(\beta)$ measures in some sense how much the distribution of the block spin deviates from a Gaussian.

If distant spins are nearly independent, by the central limit theorem block spin will be asymptotically Gaussian for large L (see for instance Malyshev [20]). This means that U_L will go to 0 as $L \to \infty$. So the quantity $U_L(\beta)$ measures how non-Gaussian the fluctuations of the block spin $\bar{S}_\Lambda$ are.

If the system is in a pure phase showing long range order (l.r.o.), on the other hand, by ergodicity the block spin will go to a nonfluctuating random number equal to its expectation value (the magnetization); this means that $U_L(\beta)$ will converge to $2/N$ for $L \to \infty$ in that case.

If the system is critical, one expects larger than normal fluctuations and therefore that $U_L(\beta)$ has a limit as L goes to ∞ that lies between the extreme values 0 and $2/N$. $U_L(\beta)$ should be an asymptotic renormalization group invariant for large L.

One can in turn use $U_L(\beta)$ to determine the RG flow of β by requiring that $U_L(\beta)$ is invariant under it (this is Binder's "phenomenological Renormalization Group"). As above this means that one is determining a $\Delta\beta$ by the requirement

$$U_L(\beta) \approx U_{2L}(\beta + \Delta\beta) \tag{26}$$

Again one can try to use the perturbative expansion of $U_L(\beta)$ to determine this RG flow. Luckily P.Weisz ([21]) has calculated the asymptotic expansion for $U_L(\beta)$ up to order $1/\beta^3$:

$$U_L(\beta) = \frac{2}{N} - \frac{a(L)}{\beta^2} - \frac{b(L)}{\beta^3} + \mathcal{O}(\beta^{-4}) \tag{27}$$

The asymptotic behavior of the coefficients for large L is

$$\lim_{L\to\infty} a(L) = a_\infty \quad (\approx .004) \tag{28}$$

$$\lim_{L\to\infty} \frac{b(L)}{\ln L} = b_\infty = \frac{N-2}{\pi} a_\infty \tag{29}$$

From this one can again derive the perturbative expression for $\Delta\beta$ if one assumes that $\ln L << \beta$. So all seems well, especially since it is generally believed that PT is trustworthy under the same condition.

But not all is well if one looks more closely. If one takes the PT expression (27) to plot the level lines of $U_L(\beta)$ in the plane with vertical coordinate β and horizontal coordinate $\ln L$ one discovers that the level lines curve to the right, i.e. invariably leave the region where $\ln L << \beta$ and where optimistic people would expect PT to be trustworthy. The perturbative RG ("asymptotic scaling") on the other hand would require that the level lines become asymptotically linear with slope $\frac{N-2}{2\pi}\ln 2$. This conflict can also be seen by noting that the perturbative expression (27) is manifestly not invariant under the perturbative RG (12) (including higher order terms does not help). In fact, if we let β vary with L as suggested by the perturbative RG, $U_L(\beta)$ will go to the unphysical value $2/N$ corresponding to l.r.o.!

The situation can be summed up as follows:

"If PT can be trusted, it cannot be trusted in the long run."

There is, however, another possibility that is certainly not in conflict with either any rigorous results, nor the perturbative calculations where they are trustworthy. It could be that the true level lines of $U_L(\beta)$ become asymptotically horizontal for large L. An analytic example of a function that has both the right PT coefficients and goes to a nontrivial constant as $L \to \infty$ is the following:

$$U_L(\beta) = \frac{2}{N} - \frac{a}{\beta^2}\left(2 - \exp\left(-\frac{N-2}{\pi\beta}\ln L\right)\right) \tag{30}$$

This would mean that there is a line of fixed points of the simple minded RG defined by the level lines of $U_L(\beta)$ and to consider this possibility is clear heresy.

So far we have found some problems with the orthodoxy, but not yet produced any positive evidence for this heretical possibility. In the next section we will go much further and present a strategy by which we think this heresy can actually be proven, if one works hard enough.

4. Heresy: Strategy for a Proof

Before explaining our strategy (which is described in [22]) in some detail, we have to collect some information that is both rigorous and in principle well known, though not everywhere.

4.1 The FK Representation

In 1969 Fortuin and Kasteleyn [23] found a way of rewriting the Ising model as a model of correlated bond percolation. Edwards and Sokal [24] extended this model to one whose configurations are described in terms of Ising spins $\sigma_x \in \{-1, 1\}$ associated to the sites x of a lattice and bond occupation variables $n_{xy} \in \{0, 1\}$ associated to the bonds $\langle xy\rangle$. The configurations are distributed according to the following probabilities:

$$P\Big(\{\sigma_x\}, \{n_{xy}\}\Big) = Z_\Lambda^{-1} \prod_{\langle xy\rangle} \Big\{\delta_{\sigma_x\sigma_y} n_{xy}(1-q) + (1-n_{xy})\Big\} \tag{31}$$

where

$$q = \exp(-2\beta) \tag{32}$$

Summing this over the bond occupation numbers n_{xy} gives

$$P\Big(\{\sigma_x\}\Big) = Z_\Lambda^{-1} \prod_{\langle xy\rangle} \Big\{\delta_{\sigma_x\sigma_y} + q(1-\delta_{\sigma_x\sigma_y})\Big\} \tag{33}$$

which is easily recognized as the standard Gibbs measure of the Ising model. On the other hand summing over the Ising spins we obtain

$$P\Big(\{n_{xy}\}\Big) = 2^{N_c} \prod_{\langle xy\rangle} \Big\{n_{xy}(1-q) + (1-n_{xy})q\Big\} \tag{34}$$

where N_c denotes the number of connected clusters. (34) is the correlated percolation measure of Fortuin and Kasteleyn [23]. Of interest are also the following conditional probabilities following from (31):

$$P\Big(\{\sigma_x\} \mid \{n_{xy}\Big) = \prod_C \Big\{\frac{1}{2}\prod_{x\in C}(1+\sigma_x) + \frac{1}{2}\prod_{x\in C}(1-\sigma_x)\Big\} \tag{35}$$

where the product runs over the connected clusters.

In words: Given the bond occupation numbers n_{xy}, the Ising spins are distributed as follows: All sites belonging to a connected cluster C carry the same spin, whereas the spins belonging to different clusters are chosen independently to be $+1$ or -1 with probability $1/2$. Furthermore we get

$$\begin{aligned} P\Big(\{n_{xy}\} \mid \{\sigma_x\}\Big) = Z_\Lambda^{-1} \prod_{\langle xy\rangle} \Big\{\delta_{\sigma_x\sigma_y}\Big[n_{xy}(1-q) \\ + (1-n_{xy})q\Big] + \Big(1-\delta_{\sigma_x\sigma_y}\Big)\Big(1-n_{xy}\Big)\Big\} \end{aligned} \tag{36}$$

In words: Given the Ising spins, a link $\langle xy\rangle$ is inactive (i.e. $n_{xy}=0$) if $\sigma_x \neq \sigma_y$. If $\sigma_x = \sigma_y$ the link is activated (i.e. $n_{xy}=1$) with probability $1-\exp(-2\beta)$.

These conditional probabilities are the basis of the so-called Swendsen-Wang algorithm [25] , which alternates assigning spins according to (35) with activating bonds according to (36). It is not hard to see that this procedure, if iterated infinitely often, produces an ensemble of configurations of the extended model distributed according to (31). U.Wolff [26] proposed (and used) the following modification: Instead of identifying the clusters and assigning Ising spins to them at random, he just picks one cluster attached to a randomly chosen site and flips its spin, before reassigning active bonds according to (36) again.

The importance of these Monte-Carlo algorithms lies in the fact that they give a much faster approach to equilibrium, especially in a critical region. Wolff's single-cluster algorithm is even more efficient than the Swendsen-Wang algorithm.

The representation (31) and the algorithms based on it can easily be generalized to models with space dependent couplings β_{xy} in place of a constant β, replacing (32) by

$$q_{xy} = \exp(-2\beta_{xy}) \tag{37}$$

Even mixed ferro-and antiferromagnetic couplings can be accommodated – one only has to replace q_{xy} by $\exp(-2|\beta_{xy}|)$ and $\delta_{\sigma_x\sigma_y}$ by $1-\delta_{\sigma_x\sigma_y}$ for the antiferromagnetic bonds.

Generalization to O(N):

The most important aspect of Wolffs paper [26] is his generalization of this type of cluster algorithm from the Ising model to the $O(N)$ models. We will describe how his algorithm can be understood by a generalization of the FK representation (see [27]); this representation will also be crucial for our strategy to prove the existence of a spin wave phase. The generalization is extremely simple: Choose a unit vector $e \in S_{N-1}$ and write

$$e \cdot S(x) \equiv \sigma_x |e \cdot S(x)| \tag{38}$$

where $\sigma_x = \mathrm{sgn}(e \cdot S(x))$ is now an Ising variable. This allows us to think of the $O(N)$ model as an Ising model with fluctuating coupling constants. The FK representation (31) can be applied as before, with q replaced by

$$q_{xy} = \exp(-2\beta|S(x)\cdot e \quad S(y)\cdot e|) \tag{39}$$

and of course the q's themselves are random variables with a distribution induced by the Gibbs measure, so that the full probability measure for the system of spins and link variables becomes

$$P(\{n_{xy}\},\{S(x)\}) = Z_\Lambda^{-1} \prod_{\langle xy\rangle} \Big\{\delta_{\sigma_x\sigma_y}[n_{xy}(1-q_{xy})$$
$$+(1-n_{xy})q_{xy}] + (1-\delta_{\sigma_x\sigma_y})(1-n_{xy}\Big\} \times \prod_{\langle xy\rangle} e^{\beta S(x)\cdot S(x)} \tag{40}$$

Wolff's algorithm for the $O(N)$ model is the following:

(0) Choose any starting configuration of $O(N)$ spins.
(1) Choose a unit vector $e \in S_{N-1}$ with uniform probability.
(2) If the projections $e \cdot S(x)$ and $e \cdot S(y)$ have the same sign, assign the link $\langle xy\rangle$ the variable $n_{\langle xy\rangle} = +1$ with probability q_{xy} given in (30), otherwise assign 0.
(3) Construct the connected cluster attached to a randomly chosen site x_0, and on that cluster flip the components parallel to e of all spins and repeat the steps until time runs out.

It can be checked easily that this algorithm corresponds to an ergodic stochastic process that has (40) as its unique equilibrium measure.

This algorithm represents a technological breakthrough in the numerical study of the $O(N)$ models comparable to the invention of the internal combustion machine: It increases the range by a factor of about 20, if measured by the correlation lengths that can reliably be simulated.

For us the main interest is in the new way of thinking about the system that is suggested by the algorithm and the underlying FK representation. We want to mention, however, that Wolff [28], using his algorithm, for the $O(3)$ model at $\beta = 1.4, 1.5, 1.6, 1.7, 1.8$ and 1.9 found values of the correlation length ξ

$$\xi^{-1} = -\lim_{|x|\to\infty} \frac{1}{|x|} \ln\langle S(0) \cdot S(x)\rangle \tag{41}$$

and the magnetic susceptibility

$$\chi = \sum_x \langle S(0) \cdot S(x)\rangle \tag{42}$$

that do not show the behavior predicted by the perturbative RG ("asymptotic scaling"), but show a faster increase (that can in fact be fitted with a power law singularity with a critical point at a finite β_c). Of course this kind of numerical evidence can never be conclusive, because it may always happen that the behavior changes again at larger β that are still inaccessible to numerical simulation.

Returning to analytic considerations, let us discuss the thermodynamic significance of the FK clusters produced by the probability measures (31) and (40). For the Ising model this is very simple, because we have

<u>*Ising*</u>:

$$\langle FK\rangle = \chi \tag{43}$$

where, by brutal abuse of notation, we denote by $\langle FK\rangle$ the expected size of the connected clusters ("FK clusters") in the FK representation.

(43) follows from the fact that the two-point function of the spins maybe computed by averaging only over those configurations where the two points belong to the same FK cluster (because for given bond variables the spins on different clusters are independent), and it is contained already in the work of Fortuin and Kasteleyn [23] (see also [26]).

(43) implies that in the disordered high temperature phase $\langle FK\rangle$ is finite whereas in the ordered phase and already at the critical point it diverges.

For the $O(N)$ models ($N \geq 2$) we can see similarly

<u>$O(N)$</u>:

$$\langle FK\rangle = \sum_x \langle \mathrm{sgn} S_N(0)\, \mathrm{sgn} S_N(x)\rangle \tag{44}$$

i.e. $\langle FK\rangle$ is the susceptibility of the imbedded Ising spins. Since by the Mermin-Wagner theorem there is no long range order, divergence of $\langle FK\rangle$ implies that the system does not show exponential clustering:

If $\langle FK\rangle = \infty$ the system has no mass gap

Wolff [28] found in addition empirically that over a large range of β values

$$\chi \approx 1.333\langle FK\rangle \tag{45}$$

Finally we want to mention the relation of the FK clusters to the more directly observable "hemispherical clusters" (maximal connected subsets of the lattice in which the spin points in a certain hemisphere): By their definition, the FK clusters maybe obtained from the hemispherical (H) clusters by "dilution", i.e. removal of bonds with probability q_{xy}.

4.2 Interlude on Percolation Theory

Percolation models are the simplest models of statistical mechanics showing nontrivial phase structure and critical behavior. In percolation models the sites or bonds of a lattice are "activated" according to a certain probability distribution and one studies the connected clusters of active sites or bonds, respectively. In the first case one talks about *site*, in the second about *bond percolation*. If the activation probabilities for different sites or bonds are independent, the model is called *Bernoulli* percolation.

The typical question asked in these models is: " Are there infinite clusters?"

If the answer is yes, one also says that there is percolation in the system.

By the use of the generalized FK representation we have transformed the $O(N)$ models into percolation models, and as we have seen, the question of the existence of a spin wave phase is equivalent to the question whether the FK clusters have divergent mean size at large β.

Before developing our argument that this in fact so, we need some results from the literature.

Theorem 1 (Coniglio et al. [29]): In any two-dimensional Ising ferromagnet with a Gibbs measure that is invariant under the lattice symmetries, in a pure phase there is at most one percolating cluster of +1's or −1's.

The intuitive reason is that in $2D$ two infinite clusters would get into each others way, but the proof is by no means trivial.

On a square lattice we can, in addition to the ordinary notion of connectedness based on the lattice links, also consider * *connectedness* that uses also the diagonals of the elementary squares. This leads the notion of * *clusters* , * *percolation* etc.

The following is a result of fundamental importance for our problem:

Theorem 2 (Russo [30]): Let μ be a translation invariant probability measure on $\{-1,1\}^{\mathbb{Z}^2}$. If neither the + sites percolate nor the − sites * percolate, the mean cluster sizes $\langle C_+\rangle$ of +1's and $\langle C_-\rangle$ of −1's both diverge with μ-probability 1.

Corollary: If neither the clusters of +1's nor the ones of −1's percolate, both their mean sizes are infinite.

The intuitive reason is that to prevent both types of percolation, each site has to be surrounded by infinitely many clusters of arbitrary size of both types.

From the corollary we can learn something about the $O(N)$ models: Let us denote by $\langle H\rangle$ the expected size of the clusters where the spin points in a certain hemisphere, and by $\langle H^c\rangle$ the same for the complementary hemisphere. Then either

(1) both types of clusters show * percolation

or

(2) $\langle H\rangle = \infty$ and $\langle H^c\rangle = \infty$

At high temperature (small β), the first alternative holds on the square lattice (the percolation threshold for Bernoulli $*$ percolation, i.e. $\beta = 0$ is about .4). We will argue that at large β the other alternative holds.

On the triangular lattice the situation is simpler, because there is no difference between ordinary and $*$ percolation. Since by the Mermin-Wagner theorem there cannot be ordinary percolation, on the triangular lattice it is always alternative (2) that applies.

4.3 The H Clusters of the $O(N)$ Model on the Square Lattice at Large β

By ergodicity of the lattice translations it is enough to consider a single configuration on the infinite lattice and replace ensemble averages by averages over translations.

At low temperature the spin configurations will be **smooth** in the sense that two neighboring spins will be very close with overwhelming probability.

Beyond this somewhat vague statement there are the following precise results:

Theorem 3 (Georgii [31]): The bonds satisfying $|S(x) - S(y)| \leq \epsilon$ will percolate for large enough β, no matter how small ϵ.

Theorem 4 (Bricmont and Fontaine [32]): The probability of having a cluster of bonds satisfying $|S(x) - S(y)| \geq \epsilon$ of size $\geq C$ is bounded by $\exp(-\beta a C/\epsilon)$ for some constant a.

For this reason we feel justified in first simplifying the situation by considering configurations in the continuum $\mathbb{R}^2$ instead of on the lattice $\mathbb{Z}^2$, satisfying uniform Lipschitz continuity and Mermin-Wagner in the sense that space averages satisfy $O(N)$ invariance.

We first consider $O(2)$ and want to argue that not only the H clusters do not percolate, but even the clusters corresponding to the complement A^c of an arbitrarily small open connected part A of the circle do not percolate. By Russo (Theorem 2) this means that the clusters corresponding to A will form clusters (rings) of divergent mean size.

Assume the contrary, namely that there is percolation of A^c. Then A forms islands in the "sea" of A^c. Next consider the reflected image A^r of A, which for small A will be disjoint from A. So there are also islands of A^r which do not touch those of A. This means that C, the complement of $A \cup A^r$ still percolates. But C consists of two disconnected pieces D and D^r. By continuity the regions where the spin is in D and the ones where it is in D^r cannot touch, so either D or D^r has to percolate (by Theorem 1 they cannot both percolate). But this contradicts the symmetry condition embodying the Mermin-Wagner theorem.

q.e.d.

Let us now turn to the case $N \geq 3$, where the situation is not quite as simple to analyze. For a Lipschitz continuous configuration $*$ percolation is a where two H clusters and two H^c clusters meet in a single point (like the "4 corners" where the states of Utah, Colorado, New Mexico and Arizona meet). Such points are characterized by $S \cdot e = 0$ and $\nabla S \cdot e = 0$ which is clearly an overdetermination. More importantly, such " 4 corners" are unstable under $O(N)$ transformations that change the reference vector e. This means that the H and H^c cannot percolate, or equivalently that the equator does not percolate.

But to return from the continuum to the lattice we need a little more, namely that also a small enough equatorial strip

$$S_\epsilon \equiv \left\{ S \in S_{N-1} \mid |S \cdot e| \leq \epsilon \right\} \tag{46}$$

does not percolate, provided β is large enough. To see this one has to use $O(N)$ invariance; If one such strip percolates, all its $O(N)$ rotated versions also have to percolate. One can see that this, together with the principle of noncoexistence of disjoint infinite clusters embodied in Theorem 1, cannot be satisfied. If the equatorial strip S_ϵ does not percolate, the "reduced hemispheres"

$$H_\epsilon \equiv \left\{ S \in S_{N-1} \mid S \cdot e > \epsilon \right\} \tag{47}$$

and

$$-H_\epsilon \equiv \left\{ S \in S_{N-1} \mid S \cdot e < -\epsilon \right\} \tag{48}$$

cannot form islands, and since by continuity they do not touch, also $S_\epsilon \cup H_\epsilon$ cannot percolate. This means that H_ϵ and its complement have to form rings of arbitrary size and

$$\langle H_\epsilon \rangle = \infty \tag{49}$$

To return from the continuum to the lattice and to relax the strict Lipschitz continuity does now not pose any real problems; the defects where something may go wrong are extremely rare and do not suffice to break up the large clusters.

We may also appeal to the sacred principle of universality, replacing the square lattice by the triangular one and imposing the strict Lipschitz condition

$$|S(x) - S(y)| \leq \delta \qquad \text{for } |x - y| = 1 \tag{50}$$

According to universality these modification do not affect the critical behavior of the models.

4.4 From H to FK Clusters

The dilution process that reduces the H clusters to the FK clusters is most severe near the equator since there the cutting probability q_{xy} can be near 1. But if we consider instead the smaller H_ϵ clusters, the cutting probability becomes arbitrarily small uniformly for large β. Also increasing β can only make the H_ϵ clusters more robust. One should expect that except possibly at "critical points", where they first arise, such clusters (rings) of divergent mean size posses a certain stability against dilution with very small probability. We have conducted a detailed numerical study of the stability of such clusters against dilution with small probability [33] and confirmed the stability principle just formulated.

So we feel justified in claiming that at large enough β $\langle FK \rangle = \infty$, which implies masslessness (no exponential clustering).

Finally we want to remark that there is an alternative strategy to prove the existence of a soft phase:

A sufficiently fat equatorial strip will always percolate for the following reason: At $\beta = 0$ this is so (Bernoulli percolation has a percolation threshold of about .6). Increasing β has to make such an equatorial strip even more solid. The complement, consisting of the two polar caps, will always form islands of finite mean size.

On the percolating equatorial strip we have essentially Richard's truncated sphere model and by a simple correlation inequality (as in [17] we can bound the two point-function of our model by the one of an $O(2)$ model on the percolating set. The strategy by which Fröhlich and Spencer proved the existence of a spin wave phase in $O(2)$ [34] becomes applicable: They transform the system into a Coulomb gas; the holes that are present in our case become conducting regions. Fröhlich and Spencer showed that at low temperature the Coulomb gas forms neutral "molecules" and does not screen. The conducting holes will not change the situation qualitatively but only affect the dielectric constant.

5. Conclusions

In conclusion we can say the following:

— There are good reasons to question the orthodoxy

— There is a strategy to prove our heretical point of view (with details remaining to be filled in).

References

1 D.Mermin: J.Math.Phys. **8**, 1061 (1967)
2 R.L.Dobrushin and S.B.Shlosman: Commun.Math.Phys. **42**, 31 (1975)
3 C.Pfister: Commun.Math.Phys. **79**, 181 (1981)
4 J.Fröhlich and C.Pfister: Commun.Math.Phys. **81**, 277 (1981)
5 A.M.Polyakov: Phys.Lett **59B**, 79 (1975)
6 E.Brézin and J.Zinn-Justin: Phys.Rev. **B14**, 3110 (1976)
7 V.A.Fateev, I.V.Frolov, A.S.Schwarz: Nucl.Phys. **B154**, 1 (1979)
8 B.Berg and M.Lüscher: Commun.Math.Phys. **69**, 57 (1979)

9 T.H.Berlin and M.Kac: Phys.Rev. **86**, 821 (1952)
10 A.Kupiainen: Commun.Math.Phys. **73**, 273 (1980)
11 S.Elitzur: Nucl.Phys. **B212**, 501 (1983)
12 A.Patrascioiu and A.Rouet: Lett. Nuovo Cim. **35**, 107 (1982); Nucl.Phys. **B214**, 481 (1983)
13 Y.Iwasaki: Phys.Lett. **104B**, 458 (1981)
14 A.Patrascioiu: Phys.Rev.Lett. **54**, 1023 (1985)
15 F.David: Phys.Lett. **96B**, 371 (1980); Commun.Math.Phys. **81**, 149 (1981)
16 A.Patrascioiu, E.Seiler, I.O.Stamatescu: Nuovo Cim. **11 D**, 1165 (1989)
17 J.-L.Richard: Phys.Lett. **184B**, 75 (1987)
18 K.Binder: Z.Phys. **B43**, 119 (1981)
19 G.Jona-Lasinio: Nuovo Cim. **26B**, 99 (1975)99;
for a review see M.Cassandro and G.Jona-Lasinio: Adv.Phys. **27**, 913 (1978)
20 V.A.Malyshev: Soviet Math.Dokl. **16**, 1141 (1975)
21 P.Weisz: private communication
22 A.Patrascioiu and E.Seiler: *Failure of Perturbation Theory in Nonabelian Models*, University of Arizona preprint, February 1990
23 P.W.Kasteleyn and C.M.Fortuin: J.Phys.Soc.Jpn. **26** (Suppl.), 11 (1969);
C.M.Fortuin and P.W.Kasteleyn: Physica (Utrecht) **57**, 536 (1972)
24 R.G.Edwards and A.D.Sokal: Phys.Rev. **D38**, 2009 (1988)
25 R.H.Swendsen and J.-S.Wang: Phys.Rev.Lett. **58**, 86 (1987)
26 U.Wolff: Phys.Rev.Lett. **62**, 361 (1989)
27 A.Patrascioiu: *Employing the Ising Representation to Implement Nonlocal Monte-Carlo Updating in $O(N)$ Models*, University of Arizona preprint 1989
28 U.Wolff: *Asymptotic Freedom and Mass Generation in the $O(3)$ Nonlinear σ Model*, preprint DESY 89-021
29 A.Coniglio, C.R.Nappi, F.Peruggi and L.Russo: Commun.Math.Phys. **51**, 315 (1976)
30 L.Russo: Z. Wahrsch. verw. Gebiete **42**, 39 (1975)
31 H.O.Georgii: Commun.Math.Phys. **81**, 455 (1981)
32 Bricmont and Fontaine: Commun.Math.Phys. **87**, 417 (1982)
33 A.Patrascioiu and E.Seiler: *Critical Behavior in a Model of Correlated Percolation*, Max-Planck-Institut preprint MPI-PAE/PTh15/90
34 J.Fröhlich and T.Spencer: Commun.Math.Phys. **81**, 527 (1981)

Gauge-Independence of Anomalies

W. Kummer

Institut für Theoretische Physik, Technische Universität Wien
Wiedner Hauptstrasse 8–10, A – 1040 Wien, Austria

Abstract: Experience with string theories suggests a new look at some old problems regarding anomalies. The extended BRS-technique is an excellent tool to control the dependence on gauge-fixing of anomalies. These methods are explained together with an extension, so as to include 'external' anomalies as well.

1. Introduction

The history of anomalies started more than 20 years ago with the discovery of the chiral anomaly by Adler, Bell and Jackiw [1]. It was soon recognized that the breakdown of a classical symmetry by quantum mechanical corrections represents a complication which may afflict a wide range of field theories, spoiling their internal consistency. On the other hand, precisely the historically oldest example, the chiral anomaly mentioned above, proved to possess remarkable physical significance. By spontaneous symmetry breaking of the chiral symmetry in quantum chromodynamics (QCD), considering the pion as the corresponding Goldstone boson, it was possible to relate the chiral anomaly to an observable process, namely the decay of π_0 into two photons. Comparison with experiments, taking due account of a proper factor 3 for colors and of the charges of the quarks, provided one of the corner stones of our present belief that QCD is correct.

This apparent contradiction with respect to the statement that anomalies are disasters for quantum field theories, is resolved by the observation that e.g. the chiral anomaly is related to an 'external' (rigid) symmetry of the QCD action, affecting the Fermi fields of righthanded and lefthanded chirality (in the chiral limit of vanishing quark masses), while the 'internal' (gauge-symmetry) of QCD is perfectly nonanomalous.

In gauge theories quantization always implies fixing the gauge. Obviously any physical observable must be gauge-independent. In the case of the chiral anomaly the latter is essentially determined by the one-loop triangle diagram of (chiral) Fermion lines only. No reasonable gauge choice influences the propagator of those Fermions and thus the gauge-(in)dependence for chiral anomalies never did attract much interest.

Fields and Particles Editors: H. Mitter · W. Schweiger

On the other hand, gauge independence plays an important role within modern renormalization methods for gauge theories, based upon the BRST-formalism [2]. Within a certain regularization scheme, the Slavnov-Taylor (ST) identity for the generating functional Γ of 1-p-i-vertex functions experiences a certain modification $\mathcal{A}$ which is determined by the action principle [3]. A consistency condition [4] for $\mathcal{A}$ in this approach reduces the renormalization to the so called cohomology problem of a nilpotent BRS-operator. Trivial solutions contain the usual local counterterms, whereas any residual nontrivial solution indicates the presence of an anomaly. The trick to extend BRS-transformations by including gauge parameters [5], allows a very effective 'weeding out' of those terms in the would-be anomaly $\mathcal{A}$ which depend on the parameters of the chosen gauge. For linear gauges it has been shown by a very straightforward argument that all gauge-dependent contributions in $\mathcal{A}$ may be absorbed in local counter-terms. Thus, the cohomology problem mentioned above can be simplified greatly, especially for more complicated gauge-theories like supersymmetric ones [5],[6]. Still, as long as 'internal' anomalies are regarded as diseases, to be cured for useful 'physical' field theories, their gauge-independence does not seem too important. New interesting types of anomalies have been encountered with the advent of string theories [7], where ordinary gauge transformations are replaced by diffeomorphisms plus Weyl-scaling on the two-dimensional world sheet. Treating the bosonic string according to the BRS-technique as a 2-dimensional field theory [8] the BRS-anomaly vanishes at the critical dimension of ordinary space time. Because for the most popular gauge choices (and choices of counter terms) this BRS-anomaly is pushed into the conformal sector, it is dubbed 'conformal' anomaly. Already for the bosonic string, yet another anomaly obstructs the conservation of the number of Faddeev-Popov ghosts ('ghost current anomaly'). Also here, if standard gauge choices are generalized so as to include derivative conditions like the harmonic or the deDonder-gauge,it was found that the ghost current anomaly may move away (completely or in part) from the ghost current [9], but not from the theory [10,11].

Of course, powerful methods to treat anomalies in Ward-identities related to global symmetries are known for a long time [3,4,6,12]. They are usually based upon the introduction of some additional external gauge field for each external global current. The anomalies arising from a gauge-variation with respect to those external gauge fields are investigated for a generating functional with the <u>sources</u> of quantized (internal) fields <u>turned off</u>. On the other hand, internal anomalies and gauge-independence of the theory are studied separately within the BRS-renormalization program for generating functionals depending on such sources [2–5].

It is clear that both approaches must be combined, if we aim at a simple proof of gauge independence for certain external anomalies as well.

The lectures are organized as follows: In section 2 we try to elucidate the difference between internal and external symmetries for the case of the Yang Mills (YM) theory of QCD and chiral symmetry. Then (section 3) path-integral quantization is performed for an arbitrary gauge theory, whereby a very compact notation is introduced. The 'extended' BRS-technique is the subject of section 4, where general nonlinear gauge-fixing is assumed. In this manner the gauge-dependence of internal anomalies can be treated. In section 5 we describe the difficulties encountered, when one tries to discuss the anomaly of some external symmetry together with the consequences of renormalization, related to the internal (gauge-)symmetry. Section 6 explains the peculiar example of the ghost-anomaly of the bosonic string, where gauge-dependence turned out very explicitly to be a central issue. In section 7 we introduce external gauge fields and ghosts in order to gauge global symmetries as well for a general gauge theory. As expected intuitively, this works if the global symmetry commutes with the BRS-symmetry, after suitable 'gaugification' of the BRS-operation itself, if necessary. Both may be incorporated into one symmetry-extended BRS approach, yielding one comprehensive ST-identity and one consistency equation for the extended anomaly. The gauge-independence in this case follows as well. These identities contain the whole renormalization problem including the Ward-identities for external global symmetries and the consistency conditions. In the applications (section 8) examples like chiral symmetry and its supersymmetric (SUSY) generalizations (including superconformal symmetry) are discussed. The appendix A illustrates the difficulties of a noncommuting external symmetry for a simplified toy-model.

2. Gauge-Invariance, Gauge-Dependence, External Symmetry

In order to obtain a feeling for the true content of the formal developments to be encountered in section 3, we discuss first the well-known example of an ordinary nonabelian gauge-theory (QCD) and chiral symmetry.

The gauge invariant action $\delta L_{\rm inv} = 0$ needs a gauge breaking action L to become amenable to quantization:

$$L = L_{\rm inv} + L_{\rm gb} \tag{2.1}$$

For ordinary gauge theories like QCD with a Lie-algebra valued gauge field depending on generators T^a with structure constants f_{abc}

$$A_\mu = A_\mu^a T^a$$

$$[T^a, T^b] = i f_{abc} T^c,$$

and with a (Dirac) matter field Ψ the infinitesimal gauge transformation reads

$$\begin{aligned} \delta A_\mu &= [D_\mu, \delta\omega] \\ \delta\Psi &= \mathrm{i}\delta\omega\Psi. \end{aligned} \tag{2.2}$$

These fields, together with the gauge covariant derivative

$$D_\mu = \partial_\mu - igA_\mu \tag{2.3}$$

appear in (we oppress consistently differentials in space-time integrals)

$$L_{\text{inv}} = -\frac{Tr}{4}\int_x F_{\mu\nu}F^{\mu\nu} + \int_x \bar{\Psi}(i\not{D} - m)\Psi \ , \tag{2.4}$$

where

$$F_{\mu\nu} = g^{-1}[D_\mu, D_\nu].$$

D_μ acting on Ψ is understood with generators appropriate for the representation of Ψ. L_{gb} consists of the gauge-breaking term

$$L_{\text{gb}}^{(1)} = Tr\int_x \left(BF_\mu A^\mu - \frac{\alpha}{2}B^2\right) \tag{2.5}$$

with an Lie-algebra-valued auxiliary field B for a linear gauge $F_\mu A^\mu$ = fixed ($F_\mu = \partial_\mu$ for the covariant gauge, $F_\mu = n_\mu$ for an axial gauge etc.). In the following we shall mostly consider homogeneous gauges $F_\mu A^\mu = 0$ i.e. $\alpha = 0$ in (2.5). (2.5) must be accompanied by the Faddeev-Popov (F.P.) action

$$L_{\text{gb}}^{(2)} = -Tr\int_x \bar{b}\,\delta(F_\mu A^\mu)\Big|_{\delta w = c} = -Tr\int \bar{b}F_\mu[D^\mu, c] \ , \tag{2.6}$$

containing the right (c) and left ($\bar{\text{b}}$) anticommuting FP-ghost. Despite breaking of the local gauge-symmetry by L_{gb}, (2.1) still permits a global BRS-symmetry [2] with an anticommuting global infinitesimal parameter $\delta\lambda$

$$\begin{aligned} \delta A_\mu &= \delta\lambda[D_\mu, c] = \delta\lambda(sA_\mu) \\ \delta\Psi &= \delta\lambda igc\Psi = \delta\lambda(s\Psi) \\ \delta c &= \delta\lambda igc^2 = \delta\lambda(sc) \\ \delta\bar{b} &= \delta\lambda B = \delta\lambda(s\bar{b}) \\ \delta B &= 0 = \delta\lambda(sB) \end{aligned} \tag{2.7}$$

which has the 'nilpotence' property $s^2 = 0$. In terms of the BRS-operation s, L_{gb} may be rewritten as a BRS-variation:

$$L_{\text{gb}} = Tr\int s\left(\bar{b}F_\mu A^\mu\right) \ . \tag{2.8}$$

By inspection of (2.1) with (2.4), one realizes that L possesses the chiral symmetry

$$\Psi \to e^{i\alpha\gamma_5}\Psi \tag{2.9}$$

or

$$\Psi_\pm = e^{\pm i\alpha}\Psi_\pm \tag{2.10}$$

of the Dirac fields with right, resp. left-handed chirality. This symmetry is exact at $m \to 0$, otherwise it is broken by a ('soft') term proportional to m. Classically Noether's theorem attaches a (softly broken) conservation law

$$\partial_\mu\left(\bar{\Psi}\gamma^\mu\gamma_5\Psi\right) = 2m\bar{\Psi}\gamma_5\Psi \tag{2.11}$$

to this symmetry. (2.7) and (2.9) entail symmetry relations for certain quantities X of the quantized theory:

$$W^{\mathrm{int}}(X) = 0 \tag{2.12}$$

$$W^{\mathrm{ext}}(X) = 0 \; . \tag{2.13}$$

Breakdown at the quantum level is signalled by the appearance of an 'internal' anomaly at the r.h.s. of (2.12), resp. (2.13). As emphasized in the introduction, a genuine internal anomaly, i.e. one that cannot be eliminated by renormalization, is not tolerable in field theory. However, an external anomaly may have even physical consequences: For the chiral symmetry in the limit $m \to 0$ an anomaly in (2.13) can be accounted for at the level of an anomalous conservation law (2.11) by [1]

$$\begin{gathered} \partial_\mu \left(\bar{q}\gamma^\mu \gamma_5 q \right) = c f_{\mu\nu} \tilde{f}^{\mu\nu} \\ c = 3 \sum_{i=1}^{3} \frac{e_i^2}{4\pi^2} \end{gathered} \tag{2.14}$$

where in QCD the anomalous term for the axial current of quarks with charge e_i contains the photon field-tensor $f_{\mu\nu}$ and its dual $\tilde{f}_{\mu\nu}$. Using the fact that chiral symmetry is spontaneously broken in Nature with the pion taking the role of the Goldstone boson, it can be shown that g_π, the decay constant of $\pi_0 \to 2\gamma$, is proportional to the anomaly coefficient c ($g_\pi = -2c/f_\pi, f_\pi =$ pion decay constant). The observed value of g_π agrees perfectly with this prediction from the anomaly, where the factor 3 for the 3 colors is crucial.

Some trivial, but necessary remarks are the following: Calling p the collection of gauge parameters ($\alpha, n_\mu, \ldots$ in our example), for a physical observable X (S-matrix-element or quantity derived thereof) it must be obviously possible to prove $\frac{\partial X}{\partial p} = 0$. This notion of gauge-independence is distinct from gauge-invariance, although for certain cases more or less obvious relations exist. Of course, gauge-invariance does not in general imply gauge-independence (e.g. $Tr(F_{\mu\nu}n^\nu)^2$ with a fixed 'gauge-direction' n^μ is gauge-dependent but gauge-invariant), also $\frac{\partial X}{\partial p} = 0$ or $\delta_{\mathrm{gauge}} X = 0$ do not allow the conclusion that $X =$ physical, although they are necessary conditions for a physically observable X. If an external anomaly leads to physically observable effects as in the example of chiral symmetry of QCD, c/f_π must be gauge-independent. Since f_π appears by itself in the decay amplitude of $\pi \to \mu\nu$ and thus must be gauge-independent, it can be expected that a proof for $\partial c/\partial p = 0$ exists.

3. Quantization

From now on we shall employ a comprehensive notation $\varphi_i = (A_\mu, \Psi, \ldots)$ for the fields affected by the ('internal') gauge-transformation. φ_i may also cover the metric field in Einstein-relativity, the spin-connection field etc. The fields in an action like (2.1) also include the FP ghosts $c, \bar{b}$ and the Lagrange-multiplier fields like B:

$$\phi_A = (\varphi_i, c, \bar{b}, B) \tag{3.1}$$

The subset (φ_i, c) sometimes will be denoted by ϕ_a. The (anti-)commuting properties of the fields in (3.1) are duly taken into account by the definition

$$\phi_A \phi_B = (-1)^{AB} \phi_B \phi_A \tag{3.2}$$

where the grading A is denoted by the same symbol as the corresponding index. $A = 0(1)$ for (anti)-commuting fields. For the functional differentiation of anticommuting fields we shall always assume left derivatives:

$$\frac{\delta}{\delta \phi_A} F(G(\phi)) = \int_{x'} \frac{\delta G'}{\delta \phi_A} \frac{\delta F}{\delta G'}. \tag{3.3}$$

Quantization of a field theory proceeds from the generating functional in terms of a path-integral, assuming that this is correct for the theories and gauge conditions under consideration:

$$Z(j) = \int_{(\phi)} \exp i \left[L_0 + \int_x j_A \phi_A \right] . \tag{3.4}$$

Also here the path integral measure has not been written explicitly. From (3.4) the Green's functions with n legs

$$G_{A_1 \cdots A_n}(x_1, \cdots, x_n) = \left[\frac{(-i)^n}{Z} \frac{\delta^n Z}{\delta j_{A_i}(x_1) \cdots \delta j_{A_n}(x_n)} \right]_{j=0} \tag{3.5}$$

can be derived. In terms of the generating functional Z_c of connected Green's functions

$$Z_c = -i \ln Z , \tag{3.6}$$

the Legendre transform

$$\underline{\phi}_A(j) = \delta Z_c / \delta j_A \tag{3.7}$$

towards a generating functional Γ for 1-p-i vertices

$$\Gamma(\underline{\phi}) = Z_c - \int_x j_A \delta Z_c / \delta j_A \tag{3.8}$$

implies

$$\delta \Gamma / \delta \underline{\phi}_A = (-1)^{A+1} j_A, \tag{3.9}$$

whereas for other variables y (fields, sources, gauge parameters etc.)

$$\delta \Gamma(\underline{\phi}, y) / \delta y = \delta Z_c(j(\underline{\phi}), y) / \delta y . \tag{3.10}$$

The 1-p-i vertices follow from $\Gamma(\underline{\phi})$:

$$\Gamma_{A_1 \cdots A_n} = \left[\delta^n \Gamma / \delta \underline{\phi}_{A_1} \cdots \delta \underline{\phi}_{A_n} \right]_{\underline{\phi}=0} . \tag{3.11}$$

4. Extended BRS-Identity, Internal Anomaly

Since we are interested in the dependence on gauge-parameters p as well, it is useful to include a BRS transform for p[5]

$$sp = z \; , \qquad sz = 0 \tag{4.1}$$

in an 'extended' BRS-transformation which consists of $(s\phi_A)$ <u>and</u> (4.1). Commuting gauge parameters p are assumed to lead to z with grading 1. For simplicity we only consider <u>one</u> gauge parameter p in the following ($z^2 = 0$), the generalization being easy. The BRS-operation on the fields

$$s = \int_x (s\phi_A)\,\delta/\delta\phi_A \tag{4.2}$$

and for the gauge-parameters

$$d = z\,\partial/\partial p \tag{4.3}$$

anticommute by construction

$$\{s, d\} = 0 \tag{4.4}$$

so that because of $d^2 = z^2\partial^2/\partial p^2 = 0$

$$(s + d)^2 = 0 \; . \tag{4.5}$$

This suggests an 'extended' gauge-breaking action [5]

$$L_{\text{gb}} = (s + d)\int \bar{b}\left(K(\varphi, p) - \frac{\alpha}{2}B\right) \tag{4.6}$$

instead of a simple BRS variation like (2.8). $K(\varphi, p) =$ fixed denotes a general <u>nonlinear</u> gauge-condition, depending on a gauge parameter p. In the following we consider homogeneous gauges ($\alpha = 0$). Ordinary gauge theories allow a covariant nonlinear gauge-condition $K = \partial_\mu A^\mu + \beta A^2$ [13]. Higher powers of A_μ would destroy power-counting renormalizability. This restriction is absent e.g. in supersymmetric (SUSY) Yang-Mills (YM) theory and in twodimensional gravitation (bosonic string) where the respective 'gauge fields' are dimensionless. Our formal treatment covers the most general case.

The relation describing the effect of the (internal) gauge symmetry on Green's functions is obtained by changing the path-integral variables ϕ_A according to a BRS transformation $\delta\phi_A = \delta\lambda(s\phi_A)$ with an infinitesimal global anticommuting parameter $\delta\lambda$. In order to facilitate the subsequent Legendre-transform (cf. section 3) one adds also sources k_A for $(s\phi_A)$ in

$$Z(j, k, z, p) = \int_{(\phi)} \exp i\left(L + \int_x j_A\phi_A\right) \; , \tag{4.7}$$

$$L = L_0 + \int_x k_A(s\phi_A) \, ,$$
$$L_0 = L_{\text{inv}} + L_{\text{gb}} \, . \tag{4.8}$$

Performing this change of variables yields

$$0 = \delta Z = \delta\lambda \int_{(\phi)} \left[(-1)^A \int_x j_A(s\phi_A) - d \int_x s\bar{b}K \right] \exp i(\cdots) \, . \tag{4.9}$$

Thanks to the new source-term, $s\phi_A$ may be replaced by $\delta/i\delta k_A$. In addition

$$d\left(\int s\bar{b}K\right) = d(s+d) \int \bar{b}K = d\left(L + \int_x j_A\phi_A\right)$$

holds, because the p-dependence resides in K alone and $d^2 = 0$. In this manner from (4.9) the 'Slavnov-Taylor-identity' (STI) for Z is obtained:

$$\int_x (-1)^A j_A(x) \frac{\delta Z}{\delta k_A(x)} - z\frac{\partial Z}{\partial p} = 0 \, . \tag{4.10}$$

These formal manipulations are justified only, if a regularization exists which respects the symmetry. We assume this for the moment. Eqs. (3.6 - 3.10) allow an immediate transformation to a 'Lee-identity' (LI) [15] for $\Gamma(\underline{\phi}, k, p, z)$:

$$\mathcal{B}(\Gamma) := \int_x \frac{\delta\Gamma}{\delta\underline{\phi}_A} \frac{\delta\Gamma}{\delta k_A} + d\Gamma = 0 \, . \tag{4.11}$$

Renormalizability means that the symmetry properties, as expressed by (4.11), continue to hold for the renormalized theory whose 1-p-i vertices may be calculated from a (renormalized) action $\Gamma(L^{\text{ren}} - \Delta)$ including local counter terms Δ. The number of possible terms in Δ, determined by physical parameters with masses and coupling constants, must be finite [14]. This means that for the renormalized theory

$$\mathcal{B}(\Gamma(L^{\text{ren}} - \Delta)) = 0 \tag{4.12}$$

must hold.

What happens, if a gauge-invariant regularization (like dimensional regularization for a parity conserving ordinary YM theory) does not exist, or if we decide not to use an existing one ? Obviously a term $\mathcal{A}$, reflecting the breaking of symmetry, will appear on the r.h.s. of (4.10) and (4.11):

$$\mathcal{B}(\Gamma) = \mathcal{A} \neq 0. \tag{4.13}$$

Clearly $\mathcal{A}$ must be at least of $\mathcal{O}(\hbar)$. As shown by Lam, Breitenlohner and Maison [16] for 'renormalizable' gauges, the so called 'action principle' allows the general statement

$$\mathcal{A} = \int_x a(x) + \mathcal{O}(\hbar a) \, , \tag{4.14}$$

where $a(x)$ is a local polynomial in $\underline{\phi}_A, k_A$ and derivatives thereof. Nonlocal terms of $\mathcal{A}$ appear only in the next loop order. $\mathcal{A}$ is a candidate for an 'anomaly'.

Now one defines a 'linearized' operator

$$\mathcal{B}_X := \int_x \left(\frac{\delta X}{\delta \phi_A} \frac{\delta}{\delta k_A} + \frac{\delta X}{\delta k_A} \frac{\delta}{\delta \phi_A} \right) + d \,. \tag{4.15}$$

Using the grading properties of $\underline{\phi}_A$ and k_A — the latter has grading $A+1$ with respect to ϕ_A — it is straightforward to show the identity

$$\mathcal{B}_X \mathcal{B}(X) = 0, \tag{4.16}$$

where $\mathcal{B}(X)$ is defined as in (4.11). Another important identity for (4.15) is

$$(\mathcal{B}_X)^2 = 0 \;\Leftrightarrow \mathcal{B}(X) = 0. \tag{4.17}$$

(4.16) applied to (4.13) yields the 'consistency equation' [4] in its formulation with a 'fermionic' operator $\mathcal{B}_X$:

$$\mathcal{B}_\Gamma \mathcal{A} = 0 \,. \tag{4.18}$$

To lowest nontrivial $\mathcal{O}(\hbar)$

$$\mathcal{B}_\Gamma \mathcal{A} = \mathcal{B}_L \int_x a(x) + \mathcal{O}(\hbar^2) = 0 \tag{4.19}$$

follows, because the (local) action L is the zero loop part of Γ. Since L of (4.8) fulfills $\mathcal{B}(L) = 0$ [17], from (4.17)

$$\mathcal{B}_L^2 = 0 \tag{4.20}$$

is obtained. The solution of (4.19) and (4.20) determines possible solutions of $\mathcal{A}$. This is the celebrated 'cohomology problem' of quantum gauge theories [18], extended so as to include the variation of the gauge parameter [5]. We now expand $\mathcal{B}_L$ and $\mathcal{A}$ in terms of z (remember $z^2 = 0$)

$$\begin{aligned} \mathcal{B}_L &= \mathcal{B}^{(0)} + z\mathcal{B}^{(1)} \\ \mathcal{A} &= \int a(x) = \mathcal{A}^{(0)} + z\mathcal{A}^{(1)}. \end{aligned} \tag{4.21}$$

For $z = 0$ we have the 'unextended' cohomology problem $\mathcal{B}^{(0)}\mathcal{A}^{(0)} = 0$, ${\mathcal{B}^{(0)}}^2 = 0$ whose solution depends crucially on the field theory under consideration. Solutions are known for all 'usual' field theories (YM, SUSY–YM, gravitation etc.). The 'trivial' solution of $\mathcal{A}$ is of the form

$$\mathcal{A}^{(0)} = \mathcal{B}^{(0)}\Delta \approx 0 \tag{4.22}$$

where Δ is still an integral of a local polynomial in the fields and k. The l.h.s. of (4.13) up to $\mathcal{O}(\hbar)$ becomes $\mathcal{B}_L\Gamma_{(1)}$, where $\Gamma_{(1)}$ consists of its zero loop part L and the (nonlocal) part of $\mathcal{O}(\hbar)$. Combining this with the r.h.s. (4.22) yields

$$\begin{aligned} \mathcal{B}^{(0)}\left(\Gamma_{(1)} - \Delta\right) &= \mathcal{B}^0\left(\Gamma_{(1)}(L - \Delta)\right) + \mathcal{O}(\hbar^2) = \\ &= \mathcal{B}\left(\Gamma(L - \Delta)\right)\Big|_{z=0} + \mathcal{O}(\hbar^2) \,. \end{aligned} \tag{4.23}$$

The subtraction of the local counterterm Δ from the action L removes the 'anomaly' $\mathcal{A}$ so that the symmetry content of the theory, as expressed by (4.11) with $z = 0$, to $\mathcal{O}(\hbar)$ remains the same for the theory with appropriate counter-terms Δ. We shall assume that no nontrivial solution of (4.18) exists for the gauge theories considered here so that no genuine anomaly occurs. With $L - \Delta$ instead of L the 'anomaly' $\mathcal{A}^{(0)}$ may now have only terms $\mathcal{O}(\hbar)^2$. Repeating the same argument to this order removes that term, and so on. Clearly, besides these counter-terms for a 'trivial' $\mathcal{A}$, removal of the regularization and the consequential infinities require the introduction of further counter-terms to be added to Δ [19]. This part of the procedure is outside the scope of our present considerations.

Now we turn to the content of (4.19), as far as linear terms in z are concerned:

$$\partial \mathcal{A}^{(0)}/\partial p = -\mathcal{B}^{(1)}\mathcal{A}^{(0)} + \mathcal{B}^{(0)}\mathcal{A}^{(1)}$$

$$\begin{aligned}\mathcal{A}^{(0)} = \mathcal{B}^{(0)} \int_{p_0}^{p} dp'\, \mathcal{A}^{(1)}(p') - \int_{p_0}^{p} dp' \frac{\partial \mathcal{B}^{(0)}}{\partial p'} \int_{p_0}^{p'} dp''\, \mathcal{A}^{(1)}(p'') \\ - \int_{p_0}^{p} dp'\, \mathcal{B}^{(1)}(p')\mathcal{A}^{(0)}(p') \,.\end{aligned} \tag{4.24}$$

From the explicit form of L with (4.7) in $\mathcal{B}_L$ we conclude

$$\mathcal{B}^{(1)} = (-1)^A \int_x \frac{\delta}{\delta \phi_A} \frac{\partial}{\partial p} \int_{x'} (\bar{b}' K') \frac{\delta}{\delta k_A} \tag{4.25}$$

which may be simplified if we take into account the fact that $(s\phi_A)$ with (3.1) in all practical cases (cf. (2.7), $\underline{\phi}_a = (\underline{\varphi}_i, c), \underline{\phi}_3 = \underline{\bar{b}}, \underline{\phi}_4 = \underline{B}$) contains

$$s\underline{\phi}_3 = \underline{\phi}_4 \,, \qquad s\underline{\phi}_4 = 0 \,. \tag{4.26}$$

Hence the source k_4 drops out and k_3 is not needed. Instead of (4.10) we obtain

$$(-1)^a \int_x j^a \frac{\delta Z}{\delta k_a} + (-1)^{(\bar{b})} \int_x j_3 \frac{\delta Z}{\delta j_4} - dZ = 0 \tag{4.27}$$

and the corresponding LI

$$\mathcal{B}(\Gamma) = \int_x \frac{\delta \Gamma}{\delta \underline{\phi}_a} \frac{\delta \Gamma}{\delta k_a} + \int_x \underline{\phi}_4 \frac{\delta \Gamma}{\delta \underline{\phi}_3} + d\Gamma = \mathcal{A} \tag{4.28}$$

in the presence of an 'anomaly' $\mathcal{A}$. Defining a linearized operator as above,

$$\mathcal{B}_X = \int_x \left(\frac{\delta X}{\delta \underline{\phi}_a} \frac{\delta}{\delta k_a} + \frac{\delta X}{\delta k_a} \frac{\delta}{\delta \underline{\phi}_a} + \underline{\phi}_4 \frac{\delta}{\delta \underline{\phi}_3} \right) + d, \tag{4.29}$$

it is clear that $\mathcal{B}^{(1)}$ may acquire only contributions from the first two terms in the integral. Moreover $c = \phi_2$ is absent in $\int \bar{b}K$ so that only the derivative $\delta/\delta\varphi_i$ is relevant in $\mathcal{B}^{(1)}$:

$$\mathcal{B}^{(1)} = (-1)^i \int_x \frac{\delta}{\delta\varphi_i} \frac{\partial}{\partial p} \left(\int_{x'} \underline{\bar{b}}' K' \right) \frac{\delta}{\delta k_i}$$
$$\frac{\partial \mathcal{B}^{(0)}}{\partial p} = \int_x \frac{\delta}{\delta \underline{\phi}_a} \left(\frac{\partial L}{\partial p} \right) \frac{\delta}{\delta k_a} \,. \tag{4.30}$$

If for a special case the explicitly known solution of the cohomology problem obeys $\delta\mathcal{A}^{(0)}/\delta k_i = 0$, we conclude from (4.24) and (4.30) the 'gauge-independence' [20] $\partial\mathcal{A}^{(0)}/\partial p \approx 0$, even for such a nonlinear gauge. This situation occurs in nonlinear YM theories [13]. A weaker, but more generally valid statement, following from (4.30) is e.g. $\partial\mathcal{A}^{(0)}/\partial p|_{\bar{b}=0} \approx 0$. Obviously for some arbitrary gauge theory we cannot expect 'gauge-independence' of $\mathcal{A}$ in a general nonlinear gauge.

In the extensively studied linear gauges $K = F_i\varphi_i$, the result [5] is much simpler. For such a gauge we profit from the socalled 'equation of motion' for $\phi_3 = \bar{b}$ and $\phi_4 = B$. Performing arbitrary variations of δB and $\delta\bar{b}$ in (4.8) produces two further relations for Z

$$\begin{aligned} &F_i \delta Z/\delta j_i + i j_4 Z = 0 \\ &F_i \delta Z/\delta k_i + i j_3 Z + d F_i \delta Z/\delta j_i = 0 \end{aligned} \tag{4.31}$$

which, when Legendre-transformed by (3.6)–(3.10),

$$\begin{aligned} &F_i \underline{\varphi}_i - \delta\Gamma/\delta\underline{B} = 0 \\ &F_i \delta\Gamma/\delta k_i + \delta\Gamma/\delta\underline{\bar{b}} + d F_i \underline{\varphi}_i = 0 \end{aligned} \tag{4.32}$$

lead to a peculiar dependence of Γ on its variables:

$$\begin{aligned} &\Gamma =: \hat{\Gamma} + \int_x \underline{B} F_i \underline{\varphi}_i + d \int_x \underline{\bar{b}} F_i \underline{\varphi}_i \\ &\hat{\Gamma} = \hat{\Gamma}\left(\underline{\phi}_a, \hat{k}_a, p, z\right) \\ &\hat{k}_i = k_i - \underline{\bar{b}} F_i \\ &\hat{k}_2 = k_2 \,. \end{aligned} \tag{4.33}$$

Inserting (4.33) into (4.28),

$$\int_x \frac{\delta\hat{\Gamma}}{\delta\underline{\phi}_a} \frac{\delta\hat{\Gamma}}{\delta\hat{k}_a} + d\hat{\Gamma} = \hat{\mathcal{A}} \tag{4.34}$$

shows that z, the BRS-variation of p, now only resides in d, i.e. $\hat{\mathcal{B}}^{(1)} = 0$ and $\partial\hat{\mathcal{A}}^{(0)}/\delta p \approx 0$ [5].

The methods described in this section remain crucial for the cohomology of $\mathcal{A}^{(0)}$, even if a 'genuine' internal anomaly must not occur in a consistent field theory.

Below we shall deal with the even simpler case of linear BRS-transformations

$$s\phi_A = \phi_B s_{BA} \tag{4.35}$$

with s_{BA} independent of ϕ. Sources k_A are superfluous then, because in the STI all terms linear in ϕ may be directly replaced by $\frac{\delta}{\delta j}$:

$$\int_x (-1)^A j_A s_{BA} \delta Z/\delta j_B - dZ = 0 \tag{4.36}$$

and the LI of Γ (with an anomaly $\mathcal{A}$)

$$(s+d)\Gamma = \mathcal{A} \tag{4.37}$$

simply contains the extended BRS-transformation of Γ. The consistency equation of $\mathcal{A}$ reads now

$$(s+d)\mathcal{A} = 0 \tag{4.38}$$

and again $\frac{\partial \mathcal{A}^{(0)}}{\partial p} \approx 0$.

5. External Symmetries, External Anomalies

As in the example of chiral invariance we now study a global (linear) symmetry with generators $\tau^a (a = 1, \ldots n)$

$$\delta\phi_A = \delta\alpha^a t^a_{BA} \phi_B \tag{5.1}$$

$$\tau^a = \int_x t^a_{BA} \phi_B \delta/\delta\phi_A \tag{5.2}$$

which leaves invariant the classical action ($\tau^a L_{\text{inv}} = \tau^a \int_x \mathcal{L}_{\text{inv}} = 0$), or ($\partial_\mu X = X_{,\mu}$)

$$t^a_{BA}\phi_B \partial\mathcal{L}_{\text{inv}}/\partial\phi_A + t^a_{BA}\phi_{B,\nu}\partial\mathcal{L}_{\text{inv}}/\partial\phi_{A,\nu} = -U^{\mu a}{}_{,\mu} \tag{5.3}$$

for a Lagrangian density depending on at most first derivatives. The total derivative of $U^{\mu a}$ is determined by the specific symmetry operation. It is different from zero e.g. for SUSY transformations. The consequences of (5.3) for the generating functional (4.8) are obtained by considering a <u>local</u> change

$$\tau = \int_x \delta\alpha^a(x) t^a_{BA} \phi_B \delta/\delta\phi_A \tag{5.4}$$

of the variables of integration in the path integral. Using (5.3) the result is (after a partial integration of the term with $\delta\alpha^a{}_{,\nu}$ from $\mathcal{L}_{\text{inv}}$)

$$\begin{aligned} 0 = \int_{(\phi)} \Big\{ \int_x \delta\alpha^a(x) \left(-J^{\mu a}_{(cl),\mu} + j_A t^a_{BA} \phi_B \right) + \\ + \tau \Big[(s+d) \int_x \bar{b} K + \int_x k_A (s\phi_A) \Big] \Big\} \exp\, i(\cdots)\,. \end{aligned} \tag{5.5}$$

Taking an arbitrary $\delta\alpha^a(x)$ (which in the second line is hidden in τ) yields n <u>local</u> Ward-identities, containing the divergence of respective Noether currents

$J^{\mu a}$ whose 'classical' part is

$$J^{\mu a}_{(cl)} = t^a_{BA}\Phi_B \partial\mathcal{L}_{\text{inv}}/\partial\phi_{A,\mu} + U^{\mu a}, \tag{5.6}$$

but which also receive contributions from several terms with derivatives from the fields in the second line of (5.5). Since $s\phi_A$ already contains first order derivatives, any further derivatives in K (like the covariant gauge condition $\partial_\mu A^\mu$) yields second order derivatives. In order to maintain our first order Noether-formalism, such derivatives should be moved upon $\bar{b}$ and B by partial integrations. An immediate Legendre-transform of (5.5) would lead for Γ to an extremely complicated nonlocal relation, quite unlike (4.11) and quite unsuitable for discussing anomalies, local counter terms and renormalization. Instead, one may try to introduce new sources in L for the nonlinear expressions in the fields ϕ_A occurring in (5.5), such as a source for $J^{\mu a}$

$$L \to L + \int_x Q_\mu{}^a J^{\mu a} + \cdots . \tag{5.7}$$

However, those sources yield new contributions in the STI which require yet another set of sources for their BRS variations. The latter produce new additional terms in (5.5), and so on at infinitum. For a consistent renormalization of the identities for Γ for those <u>local</u> Ward identities and the <u>global</u> STI this structural difference is another mayor stumbling block. Also the fact, whether or not s commutes with τ^a is important. This can be illustrated in a toy model with <u>linear</u> BRS-symmetry, a current which is assumed to be BRS-invariant, τ^a-invariant and independent of the gauge parameter, but $[s,\tau] \neq 0$ although even $[s,\tau]L_{\text{gb}} = 0$ (Appendix A). The result

$$\frac{\partial C^a}{\partial p} \approx N^a$$

where N^a is a nonlocal expression, signals trouble, the more so for more realistic theories with noninvariant, gauge-dependent currents [21].

6. Ghost Number Anomaly for the Bosonic String

The bosonic string provides a highly nontrivial illustration for these problems ($\boldsymbol{X}(\sigma,\tau) = (\boldsymbol{X}(x_\alpha)$ are the space-time coordinates)

$$\begin{aligned} L_{\text{inv}} &= \frac{1}{2}\int_x \sqrt{-g}\, g^{\alpha\beta} \boldsymbol{X}_{,\alpha}\boldsymbol{X}_{,\beta} , \\ L_{\text{gb}} &= \int_x [B_{\alpha\beta}K^{\alpha\beta} - \bar{b}_{\alpha\beta}(sK^{\alpha\beta})] = \\ &= s\int_x \bar{b}_{\alpha\beta}K^{\alpha\beta} , \end{aligned} \tag{6.1}$$

whose $\mathcal{L}_{inv}$ is invariant with respect to diffeomorphisms and Weyl-transformations $\bar{g}^{\alpha\beta} = \lambda(x)g^{\alpha\beta}$. The gauge-breaking with a symmetric 2×2 field $B_{\alpha\beta}$ provides the necessary three gauge conditions

$$K^{\alpha\beta}(g, \hat{g}, p) = 0 \tag{6.2}$$

fixing all components of $g^{\alpha\beta}$ in terms of some background metric $\hat{g}^{\alpha\beta}$. The extensive literature on the bosonic string almost exclusively employs the 'standard' gauge [22]

$$K^{(st)\alpha\beta} = g^{\alpha\beta} - \hat{g}^{\alpha\beta}, \tag{6.3}$$

however other gauges known from 4d-gravity are equally possible, e.g. the harmonic gauge ($\hat{\nabla}_\rho$ represents the background-covariant derivatives)

$$K^{(h)\alpha\beta} = n^\alpha \hat{\nabla}_\rho \sqrt{-g}\left(g^{\rho\beta} - \hat{g}^{\rho\beta}\right) + \hat{g}^{\alpha\beta}\left(\sqrt{-g} - \sqrt{-\hat{g}}\right) \tag{6.4}$$

and the deDonder gauge

$$K^{(dD)\alpha\beta} = K^{(h)\alpha\beta}|_{\hat{\nabla}_\rho \to \partial_\rho} \,. \tag{6.5}$$

In order to avoid fields with second derivatives, a derivative in the gauge condition is assumed to be transferred to $\bar{b}_{\alpha\beta}$ (or $B_{\alpha\beta}$) by partial integration. The BRS symmetry of (6.1) is easily checked:

$$\begin{aligned} sg^{\alpha\beta} &= -c^\alpha{}_{,\lambda} g^{\lambda\beta} - c^\beta{}_{,\lambda} g^{\lambda\alpha} + c^\lambda g^{\alpha\beta}{}_{,\lambda} - cg^{\alpha\beta} \\ s\bar{b}_{\alpha\beta} &= B_{\alpha\beta}\,, \qquad sB_{\alpha\beta} = 0 \\ s\left[c^\alpha, c, X\right] &= c^\lambda \left[c^\alpha, c, X\right]_{,\lambda} . \end{aligned} \tag{6.6}$$

An external symmetry of (6.1) is ghost number conservation

$$\tau^{(c)} = \int_x \left(c^\alpha \delta/\delta c^\alpha + c\delta/\delta c - \bar{b}_{\alpha\beta}\delta/\delta\bar{b}_{\alpha\beta}\right) \tag{6.7}$$

with the appropriate Noether current

$$J^{\mu(c)} = c^\alpha \partial\mathcal{L}/\partial c^\alpha{}_{,\mu} + c\partial\mathcal{L}/\partial c_{,\mu} - \bar{b}_{\alpha\beta}\partial\mathcal{L}/\partial\bar{b}_{\alpha\beta,\mu} \,. \tag{6.8}$$

We emphasize that this symmetry, in contrast to the situation considered above, refers to the gauge-breaking part of the Lagrangian density. $J^{\mu(c)}$ not only is gauge-dependent, but also $sJ^{\mu(c)} \neq 0$. Moreover it is wellknown that this current is not conserved at the quantum level for the standard gauge:

$$J^{\mu(c,st)}{}_{,\mu} = -\frac{3}{4\pi}\sqrt{-\hat{g}}R(\hat{g}) \,. \tag{6.9}$$

The anomaly depends on the invariant curvature of the background, the total derivative $\sqrt{-\hat{g}}\,R(\hat{g})$ in 2 dimensions yielding nontrivial ghost number breaking terms for a nontrivial topology, characterized by the genus γ of the world sheet

$$\frac{1}{4\pi}\int_x \sqrt{-\hat{g}}\hat{R} = 2 - 2\gamma. \tag{6.10}$$

The contribution (6.10) is crucial for the construction of strings with vertices and for the field theory of strings. At the critical dimension 26 (with 26 components of X) this external anomaly is the only one for the bosonic string in a flat space-time background. On the other hand, in the harmonic gauge, this anomaly was found to disappear [23]

$$J^{\mu(c,h)}{}_{,\mu} = 0. \tag{6.11}$$

However, as shown by Rebhan and Kraemmer (24) it simply moves to a hitherto 'forgotten' sector of the theory. From the fact that only one loop-effects are essential for this anomaly, it can be shown that the 'effective' action ($h^{\rho\sigma} = g^{\rho\sigma} - \hat{g}^{\rho\sigma}$)

$$L^{eff} = \hat{L}_{\text{inv}} + \int_x \left(B_{\alpha\beta} \hat{K}^{\alpha\beta}_{(\rho\sigma)} h^{\rho\sigma} - \bar{b}_{\alpha\beta} \hat{K}^{\alpha\beta}_{\rho\sigma} \hat{\sigma}^{\rho\sigma} \right) \tag{6.12}$$

with

$$\begin{aligned} \hat{L}_{\text{inv}} &= L_{\text{inv}}|_{g=\hat{g}} \\ \hat{K} &= K|_{g=\hat{g}} \\ \hat{\sigma}^{\rho\sigma} &= (sg^{\rho\sigma})|_{g=\hat{g}} \end{aligned} \tag{6.13}$$

already determines the necessary Feynman rules. This action possesses the additional $U(1)$ symmetry

$$\tau^{Bh} = \int_x \left(h^{\alpha\beta} \delta / \delta h^{\alpha\beta} - B_{\alpha\beta} \delta / \delta B_{\alpha\beta} \right) \tag{6.14}$$

and a corresponding Noether-current $J^{\mu(Bh)}$. As shown in [24], the 'Rebhan-Kraemmer' current, which is the sum of $J^{(c)}$ and $J^{(Bh)}$, has a gauge-independent anomaly

$$J^{\mu(R-K)}{}_{,\mu} = \left(J^{\mu(c)} + J^{\mu(Bh)} \right)_{,\mu} = -\frac{3}{4\pi} \sqrt{-\hat{g}} \hat{R}, \tag{6.15}$$

which was verified explicitly for the harmonic and the deDonder gauge. A proof for arbitrary string gauges by the same authors relied heavily on the fact that the sum of $\tau^{(c)}$ and $\tau^{(Bh)}$ commutes with the residual effective linear BRS symmetry

$$s^{eff} = \int_x \left(B_{\alpha\beta} \delta / \delta \bar{b}_{\alpha\beta} + \hat{\sigma}^{\alpha\beta} \delta / \delta h^{\alpha\beta} \right) \tag{6.16}$$

of (6.12), whereas both transformations by themselves do not exhibit this property.

7. The Symmetry Extended BRS–Technique

For an arbitrary gauge theory (with nonlinear gauge condition) we thus expect the general result that if an external symmetry with an anomalous Ward identity commutes with the (internal) BRS-transformation, the external anomaly should be equivalent to a gauge-independent one. In order to get rid of the ugly current and the other terms in (5.5) we promote the current J^μ to become the source of an external gauge field $\tilde{\varphi}^a_\mu$, replacing derivatives $\partial_\mu \phi$ of the fields, affected by τ, by $\tilde{D}_\mu \phi = (\partial_\mu + \tilde{\varphi}_\mu)\phi$, $\tilde{\varphi}_\mu = \tilde{\varphi}^a_\mu t^a$. This 'gaugification' is a well known trick [4], but it seems to have never been used together with nonvanishing sources for internal (quantized) fields. This is, however, necessary for the discussion of vertices involving the anomaly and an arbitrary number of 'internal' fields. If the external symmetry commutes with s, the gaugified $\tilde{L} = \tilde{L}(\varphi, \tilde{\varphi})$ is gauge invariant for external gauge transformations. Taking the gauge parameter as an (anticommuting) external ghost field $\tilde{u}^a(x)$

$$\tilde{t}\phi_A =: \tilde{t}_A = \tilde{u}^a(x)(\tau^a \phi_A) = \tilde{u}^a t^a_{BA} \phi_B \,, \tag{7.1}$$

the operation $\tilde{t}$ is the restriction to an (external) BRS-symmetry with

$$\tilde{t}\tilde{\varphi}^a_\mu = \left(\tilde{D}_\mu\right)^{ab} \tilde{u}^b \tag{7.2}$$

$$\tilde{t}\tilde{u}^a = -\frac{1}{2} f_{abc} \tilde{u}^b \tilde{u}^c \tag{7.3}$$

so that $\tilde{t}^2 = 0$ on all fields. Since from $[\tau, s] = 0$ we conclude $\{\tilde{t}, s\} = 0$, it is possible to construct a 'symmetry'-extended BRS-operation

$$\hat{s}^2 = (s + \tilde{t})^2 = 0 \tag{7.4}$$

with s given by (4.2) [25] and, summarizing (7.1)–(7.3),

$$\tilde{t} = \int_x \left(\tilde{t}_A(\phi, \tilde{\varphi})\delta/\delta\phi_A + \tilde{t}_a(\tilde{\varphi})\delta/\delta\tilde{\varphi}_a\right), \tag{7.5}$$

where we have introduced the shorthand $\tilde{\varphi}_a$ for $(\tilde{\varphi}^a_\mu, \tilde{u}^b)$. It will be essential that $\tilde{t}_a$ does not depend on ϕ. Imitating the 'extended' BRS-case (4.8) we consider the generating functional

$$\tilde{Z}(j, k, \tilde{\varphi}, z, p) = \int_{(\phi)} \exp\, i \left(\tilde{L} + \int_x j_A \phi_A\right), \tag{7.6}$$

$$\tilde{L} = \tilde{L}_{\text{inv}} + \int_x (s + \tilde{t} + d)\, \bar{b}K + \int_x (s + \tilde{t})\, k_A \phi_A \,. \tag{7.7}$$

Repeating our previous argument in (4.9) for $\delta\phi_A = \delta\lambda(s + \tilde{t})\phi_A$ we must not overlook that $\tilde{L} - \int d(\bar{b}K)$ now is invariant only, if at the same time the external fields $\tilde{\varphi}_a$ are changed:

$$\begin{aligned} \delta\left(\tilde{L} - d\int \bar{b}K\right) &= \delta\lambda\left(s + \tilde{t} - \int_x \tilde{t}_a \frac{\delta}{\delta\tilde{\varphi}_a}\right)\left(\tilde{L} - d\int_x \bar{b}K\right) = \\ &= -\delta\lambda \int_x \tilde{t}_a \delta/\delta\tilde{\varphi}_a \left(\tilde{L} - d\int_x \bar{b}K\right) . \end{aligned} \tag{7.8}$$

Thus, as compared to (4.10) a new term appears in the 'symmetry extended' STI

$$\int_x \left[(-1)^A j_A \delta\tilde{Z}/\delta k_A - \tilde{t}_a(\tilde{\varphi})\delta\tilde{Z}/\delta\tilde{\varphi}_a\right] - d\tilde{Z} = 0 \tag{7.9}$$

and in the 'symmetry extended' LI [26]

$$\tilde{\mathcal{B}}(\tilde{\Gamma}) = \int_x \frac{\delta\tilde{\Gamma}}{\delta\underline{\phi}_A}\frac{\delta\tilde{\Gamma}}{\delta k_A} + \int_x \tilde{t}_a \frac{\delta\tilde{\Gamma}}{\delta\tilde{\varphi}_a} + d\tilde{\Gamma} = \tilde{\mathcal{A}}\,. \tag{7.10}$$

At $\tilde{\varphi} = 0$ (7.9) and (7.10) reduce to (4.10) and (4.11), $\tilde{\mathcal{A}}|_{\tilde{\varphi}=0} = \mathcal{A}$. The anomaly is obtained as a factor of the corresponding FP ghost [18,19]. In the absence of a genuine internal anomaly ($\mathcal{A} \approx 0$), the external anomaly is obtained as

$$\mathcal{C}^a(x) = \left.\frac{\delta\tilde{\mathcal{A}}}{\delta\tilde{u}^a(x)}\right|_{\tilde{u}=0} . \tag{7.11}$$

(7.10) allows a comprehensive treatment of internal and (commuting) external symmetries and their anomalies, simply following the steps after (4.14). The identities (4.16) and (4.17) now hold for $\tilde{\mathcal{B}}(\tilde{X})$ and

$$\tilde{\mathcal{B}}_{\tilde{X}} = \mathcal{B}_{\tilde{X}} + \int_x \tilde{t}_a \hat{\upsilon}/\delta\tilde{\varphi}_a \tag{7.12}$$

so that the comprehensive 'consistency equation' becomes

$$\tilde{\mathcal{B}}_{\tilde{\Gamma}}\tilde{\mathcal{A}} = 0, \tag{7.13}$$

and the result for the 1-loop gauge dependence of $\tilde{\mathcal{A}}^{(0)} = \tilde{\mathcal{A}}|_{z=0}$ is

$$\frac{\partial\tilde{\mathcal{A}}^{(0)}}{\partial p} = -\mathcal{B}^{(1)}_{\tilde{L}}\mathcal{A}^{(0)} + \mathcal{B}^{(0)}_{\tilde{L}}\mathcal{A}^{(1)}\,, \tag{7.14}$$

where $\tilde{\mathcal{B}}^{(i)}$ depend on L_{gb} only which is, in general, unaffected by the 'gaugification' (cf. however [25]). Also the case of a linear BRS-transformation (4.36)–(4.38) can be taken over by simply replacing $s \to s + \tilde{t}$.

If a 'background' field $\hat{\varphi}_\alpha$ (as for the bosonic string in section 6) happens to participate in the external symmetry, this adds another term to $\tilde{t}$,

$$\tilde{t} \to \tilde{t} + \int_x \hat{t}_\alpha \frac{\delta}{\delta\hat{\varphi}_\alpha}. \tag{7.15}$$

For the ghost number of the bosonic string we explicitly considered a symmetry of the gauge-breaking action. Then the gaugification must be extended to

$$\tilde{L}_{gb} = (\tilde{s} + \tilde{t} + d)\int \bar{b}K(\varphi,\tilde{\varphi})\,. \tag{7.16}$$

Here derivatives of φ in the gauge-condition are gaugified — if they are affected by t —, but also the BRS–transformation $\tilde{s}$ with its own derivatives of fields. In fact, $\tilde{t}$ may be dropped in (7.16) in this case, but it survives in the source-term of k_A.

8. Examples

8.1 Chiral Anomaly

In ordinary gauge theories the fields which are subject to gauge transformations (2.2) are A_μ and the Dirac-field Ψ. Powercounting renormalizability only permits nonlinear gauge-conditions at most quadratic in A [13] without Ψ. Hence the external symmetry (2.9), (2.10) leaves L_{gb} trivially invariant. The second term of (2.4) is gaugified by a $U(1)$-gauge field $\tilde{\varphi}_\mu$ as

$$\bar{\Psi} i\partial\!\!\!/ \Psi \to \bar{\Psi}_+ \left(i\partial\!\!\!/ + \tilde{\varphi}\!\!\!/\right) \Psi_+ + \bar{\Psi}_- \left(i\partial\!\!\!/ - \tilde{\varphi}\!\!\!/\right) \Psi_- \,. \tag{8.1}$$

Thus all conditions are fulfilled for the (trivial) gauge-independence of anomaly.

8.2 Horizontal Symmetry

In the presence of scalar fields χ affected by some 'horizontal' symmetry, the situation is essentially as in (8.1), except that χ may now be involved in a renormalizable gauge condition like $\partial_\mu A^\mu + m\chi =$ fixed. Then $tL_{\text{gb}} \neq 0$, but the same conclusion as in (8.1) is reached.

8.3 The Bosonic String

Here L_{gb} is gaugified according to a local $U(1)$–symmetry [27]

$$\begin{aligned} c^\alpha{}_{,\mu} &\to c^\alpha{}_{,\mu} + i\tilde{\varphi}_\mu c^\alpha \\ \bar{b}_{\alpha\beta,\mu} &\to \bar{b}_{\alpha\beta,\mu} - i\tilde{\varphi}_\mu \bar{b}_{\alpha\beta} \end{aligned} \tag{8.2}$$

and similar relations for c, $B_{\alpha\beta}$ and $h^{\alpha\beta}$ (cf. section 6). This creates a gaugified version of the effective BRS–transformation (6.16) ($\hat{\sigma}^{\alpha\beta} \to \tilde{\hat{\sigma}}^{\alpha\beta}$). Still $\{\tilde{s}, \tilde{t}\} = 0$ and the argument of the (symmetry extended) linear BRS-situation applies. — As shown in [27] also a scaling symmetry which involves the background metric may survive the gauge-fixing

$$\tau^{(W)} = \int_x \left(g^{\alpha\beta}\delta/\delta g^{\alpha\beta} + \hat{g}^{\alpha\beta}\delta/\delta\hat{g}^{\alpha\beta}\right). \tag{8.3}$$

The application of the version of (7.15) of our argument in this case shows the vanishing of a corresponding anomaly in all gauges, since it vanishes in the standard gauge.

8.4 The Lorentz Anomaly in Noncovariant Gauges

Noncovariant gauges in principle could cause anomalies [28]. The gaugified external symmetry (Lorentz-invariance) now demands a spin-connection field. Again all requirements are fulfilled for the 'gauge-independent' anomaly, i.e. by interpolation from a covariant gauge, where such an anomaly certainly vanishes, a 'candidate' anomaly may only consist of $\mathcal{B}$-variations of local counterterms.

8.5 Chiral Breaking in SUSY YM Theory (Trivial Case)

In SUSY YM–theory, A_μ is built into a vector superfield ([29], $\theta \not{A}\bar{\theta} = \bar{\theta}^\alpha \sigma^\mu_{\alpha\dot{\beta}} \bar{\theta}^{\dot{\beta}} A_\mu$) for the Weyl spinors θ which are Grassmann variables in the superspace consisting of $(x_\mu, \theta_\alpha, \bar{\theta}_{\dot{\alpha}})$

$$V(x,\theta,\bar{\theta}) = \cdots + \theta \not{A}\bar{\theta} + \frac{1}{2}(\lambda\theta)\bar{\theta}^2 + \cdots \tag{8.4}$$

together with its (Majorana) 'gaugino' λ_α, whereas the fermion Ψ_+ with positive chirality, together with a scalar field S_+ and an auxiliary field Ψ_+ makes up the chiral superfield

$$\phi_+ = e^{-i\theta\not{\partial}\bar{\theta}}(S_+ + \theta\Psi_+ + F_+) \ , \tag{8.5}$$

obeying $\bar{D}_{\dot{\alpha}}\phi_+ = 0$. Here the covariant derivatives of SUSY

$$D_\alpha = (\bar{D}_{\dot{\alpha}})^\dagger = \partial/\partial\theta^\alpha - i(\not{\partial}\bar{\theta})_\alpha \tag{8.6}$$

are used. Similarly $D_\alpha\phi_- = 0$ holds for the chiral superfield with Ψ_- and S_-. The invariant part of the action

$$L = \int d^8x \left[\mathcal{L}_+ + \mathcal{L}_- + cg^{-2}\left(W^2 + \bar{W}^2\right)\right] + L_{\text{gb}} \tag{8.7}$$

with

$$\mathcal{L}_\pm = \phi_\pm^\dagger e^{\pm gV}\phi_\pm \ , \tag{8.8}$$

$$W^2 = W^\alpha W_\alpha = \left(W^2\right)^\dagger \ , \qquad W_\alpha = \bar{D}^2 e^{-gV} D_\alpha e^{gV} \tag{8.9}$$

is SUSY–gauge invariant

$$\delta\phi_+ = i\delta\Lambda_+\phi_+ \ , \qquad \delta\phi_- = i\delta\Lambda_+^\dagger\phi_- \tag{8.10}$$

with the gauge-function $\delta\Lambda_+$ again a chiral superfield, if

$$\delta\left(e^{gV}\right) = i\left(\delta\Lambda_+^\dagger e^{gV} - e^{gV}\delta\Lambda_+\right) \tag{8.11}$$

for Lie–algebra valued fields V and Λ_+ (like A_μ, resp. Ψ in section 2). From (8.11) the quantities $R(V), \bar{R}(V)$ in

$$\delta V = R(V)\delta\Lambda_+ + \bar{R}(V)\delta\Lambda_+^\dagger \tag{8.12}$$

can be computed [29] so that the SUSY–BRS-transformations read

$$
\begin{aligned}
s\phi_+ &= ic_+\phi_+ \\
s\phi_- &= ic_+^\dagger\phi_- \\
sV &= Rc_+ + \bar{R}c_+^\dagger \\
sb_+ &= B_+ \ , \qquad sB_+ = 0
\end{aligned}
\tag{8.13}
$$

With an appropriate (linear) gauge breaking part of L

$$
L_{\text{gb}} = \int d^8x \, s\,(b_+F + \bar{b}_+\bar{F})\, V, \tag{8.14}
$$

$F = 1$ corresponds to the 'covariant' gauge, but even SUSY-noncovariant gauges like the Wess-Zumino gauge can be perfectly accommodated in (8.14), retaining the supergraph formalism [29].

The 'trivial' generalization of chiral symmetry to this case is based upon the observation that (8.7) permits a global chiral phase-transformation

$$
\delta\phi_\pm = \pm i\delta\alpha\phi_\pm \ , \qquad \delta V = 0 \tag{8.15}
$$

which affects all component fields of $\phi_\pm$ (including the scalar field $S_\pm$!) in the same manner. This symmetry trivially commutes with (8.13), gaugification adds a term $\tilde{\varphi}$ to $\pm gV$ in (8.8), where $\tilde{\varphi}$ is an external abelian vector superfield [30]. The anomalous current conservation, analogous to (2.14) and, in fact, containing (2.14) for $\theta = \bar{\theta} = 0$, for the abelian case can be written as [31]

$$
[D^2, \bar{D}^2](\mathcal{L}_+ - \mathcal{L}_-) = \frac{\alpha}{16\pi}\frac{1}{g^2}(D^2W^2 - \bar{D}^2\bar{W}^2) \tag{8.16}
$$

where the SUSY generalization of j_5^μ is $\mathcal{L}_+ - \mathcal{L}_-$. The result of our theorem simply means that (8.16) may be calculated in any gauge of supersymmetry, the difference between e.g. the Wess-Zumino gauge (where all component fields in (8.4) are set equal zero, except A_μ and λ) and the covariant gauge may be absorbed by local counter-terms to the action.

8.6 Superconformal Invariance

If $\phi_\pm$ and V are attributed proper phase-transformations (8.17) (conformal dimensions), together with a so called $\mathcal{R}$-transformation

$$
\mathcal{R} = \theta^\alpha\partial/\partial\theta^\alpha + \bar{\theta}_{\dot{\alpha}}\partial/\partial\bar{\theta}_{\dot{\alpha}} \ , \tag{8.17}
$$

(8.7) also shows superconformal invariance [32]. For us it is essential only that (8.17) affects $\phi_\pm$ <u>and V</u>, and does not commute with (8.13). Hence our argument does not apply. Still, a formally supersymmetric relation like (8.16) can be given [32]

$$
D^\alpha J_{\alpha\dot{\alpha}} = \bar{\mathcal{A}}_{\dot{\alpha}} \tag{8.18}
$$

from which all anomalies of the general set of superconformal Noether currents can be derived — precisely like the anomalous trace of the energy momentum tensor T^{μ}_{μ} in the nonsupersymmetric case. The current $J_{\alpha\dot{\alpha}}$ besides $T_{\mu\nu}$ now also contains j^{μ}_{5} and further currents $Q_{\mu\alpha}$ and $S_{\mu\alpha}$ related to the SUSY transformations. In the same manner in $\bar{\mathcal{A}}_{\dot{\alpha}}$ on the r.h.s. of (8.18) the conformal anomaly and the chiral anomaly are hidden. Our argument was a sufficient one, thus one could argue that still (8.18) may be gauge independent, especially so because gaugifying the superconformal symmetry with an appropriate external superfield $\tilde{\varphi}(L_{\text{inv}}(\phi, V)) \rightarrow \tilde{L}_{\text{inv}}(\phi, V, \tilde{\varphi})$ creates Green's functions of a 'physical' external field $\tilde{\varphi}$ with other legs describing the internal quantized fields. On the mass shell of the latter, these Green's functions should be gauge independent. This is a necessary prerequisite for a physical process. However, for such a proof of gauge independence, the BRS-invariance of the action is a crucial ingredient, also for the SUSY case [29]. The noncommuting superconformal generators in $\tilde{\varphi}$ now prevent $\tilde{L}$ to remain BRS-invariant. Hence these proofs break down as well. These consideration should be seen against the background of a long-standing controversy, regarding the validity of the Adler-Bardeen theorem (nonrenormalization of the chiral anomaly) versus the conformal anomaly (proportional to the β-function) which both appear in the superconformal anomaly $\bar{\mathcal{A}}_{\dot{\alpha}}$ and thus should behave similarly. Although ingenious explanations have been constructed [34], our arguments suggest a simple culprit: In the superconformal case (as contrasted to the conformal one, where commutativity with BRS-transformations is maintained) fixing the SUSY gauge (e.g. to the Wess-Zumino one) is not harmless: It essentially destroys general results (like a SUSY Adler-Bardeen theorem).

Appendix A: A Toy Model for $[s, \tau^a] \neq 0$

This model takes the linear BRS-transformation (4.35) and $[s, \tau]L_{\text{gb}} = 0$, although $[s, \tau] \neq 0$ in general. The action (without the superfluous sources k_A) reads

$$\begin{aligned} \tilde{L} &= L + \int_x Q_\mu J^\mu \\ L &= L_{\text{inv}} + L_{\text{gb}} \end{aligned} \tag{A.1}$$

with a new source term for the Noether current J which is assumed to obey the simplest possible relations

$$sJ = \tau J = \partial J / \partial p = 0 \ . \tag{A.2}$$

Under these special conditions the LI remains as in (4.37),

$$\begin{aligned} (s + d)\tilde{\Gamma} &= \tilde{\mathcal{A}} \\ (s + d)\tilde{\mathcal{A}} &= 0 \quad , \end{aligned} \tag{A.3}$$

whereas the Ward-identity for $\tilde{Z}$, considered as a functional of (A.1),

$$-\partial_\mu \delta \tilde{Z}/\delta Q^a_\mu + j_A(x) t^a_{BA} \delta \tilde{Z}/\delta j_B(x) = 0,$$

in the presence of a candidate anomaly from the global symmetry is transformed easily into

$$\partial_\mu \delta\tilde{\Gamma}/\delta Q^a_\mu + t^a_{BA} \underline{\phi}_B \delta\tilde{\Gamma}/\delta\underline{\phi}_A = \tilde{C}^a. \tag{A.4}$$

In both quantities $\tilde{\Gamma}$ and $\tilde{A}$ only terms of at most $\mathcal{O}(Q)$ are needed and no genuine internal anomaly (at $Q = 0$) should be present:

$$\tilde{\Gamma} = \Gamma + \int_x Q^a_\mu \Gamma^{\mu a}(\underline{\phi}) + \mathcal{O}(Q^2) \tag{A.5}$$

$$\tilde{A} = \int_x Q^a_\mu a^{\mu a}(\underline{\phi}) + \mathcal{O}(Q^2)\,. \tag{A.6}$$

We insert (A.5) into (A.4) at $Q = 0$ ($\tilde{C}^a|_{Q=0} = C^a$),

$$\Gamma^{\mu a}{}_{,\mu} + t^a_{BA}\underline{\phi}_B(x)\frac{\delta\Gamma}{\delta\underline{\phi}_A(x)} = C^a(x)\,, \tag{A.7}$$

and (A.5) together with (A.6) into (A.3):

$$(s+d)\Gamma^{\mu a}(x) = a^{\mu a}(x). \tag{A.8}$$

Applying $(s+d)$ onto (A.7), $\Gamma^{\mu a}$ may be eliminated by (A.8); then

$$a^{\mu a}{}_{,\mu} + \underline{\phi}_B r^a_{BA}\delta\Gamma/\delta\underline{\phi}_A = (s+d)C^a \tag{A.9}$$

simply relates $a^{\mu a}$ to C^a.

$$r^a_{BA} = s_{BC}t^a_{CA} - t^a_{BC}s_{CA} \tag{A.10}$$

is the remnant of commuting s through t (note that $(s+d)\Gamma = 0$). Expanding all quantities y in z as $y = y^{(0)} + zy^{(1)}$, we obtain

$$a^{\mu a(0)}{}_{,\mu} + \underline{\phi}_B r^a_{BA}\frac{\delta\Gamma^{(0)}}{\delta\underline{\phi}_A} = sC^{a(0)}\,, \tag{A.11}$$

$$a^{\mu a(1)}{}_{,\mu} + sC^{a(1)} + \underline{\phi}_B r^a_{BA}\frac{\delta\Gamma^{(1)}}{\delta\underline{\phi}_A} = \frac{\partial C^{a(0)}}{\partial p}\,. \tag{A.12}$$

Both equations have nontrivial content for $\mathcal{O}(\hbar)$ or higher. The gauge-dependent part $\hat{C}^a$ of $C^{(0)a}$, according to (A.12) consists of three terms:

1. The (local) divergence $a^{\mu(1)}{}_{,\mu}$ which may be absorbed by renormalization of $J^\mu \to J^\mu + \Delta^\mu$,
2. a BRS variation which according to (A.11) does not affect $a^{\mu(0)}{}_{,\mu}$,
3. the term with r which contains the, in general, nonlocal $\Gamma^{(1)}$. Hence for $r \neq 0$ the counter-term $C^{a(0)}$ is not equivalent to a gauge-independent one, even in this extremely simplified situation.

References

1 K. Johnson: Phys.Lett. **5**, 353 (1963)
S.L. Adler: Phys.Rev. **117**, 2426 (1969)
J.Bell and R. Jackiw: Nuovo Cim. **60A**, 47 (1969)
W. Bardeen: Phys.Rev. **182**, 1517 (1969)
S.-Y. Pi and S. Shei: Phys.Rev. **D11**, 2946 (1975)

2 C. Becchi, A. Rouet, R. Stora: Phys.Lett. **52B**, 344 (1974); Comm.Math.Phys. **42**, 127 (1975); Am.Phys. **98**, 287 (1976)
J.Zinn-Justin, in *Trends in Elementary Particle Theory*, ed. by H. Rollnik, K. Dietz (Lecture Notes in Physics, Vol. 37, Springer 1975)

3 Y.M. Lam: Phys.Rev. **D6**, 2145 (1972); **D7**, 2943 (1973)

4 J. Wess and B. Zumino: Phys.Lett. **37B**, 95 (1971)

5 O. Piguet and K. Sibold: Nucl.Phys. **B248**, 301, 336 (1984); **249**, 396 (1985). These authors streamlined earlier related ideas of H. Kluberg-Stern and J.B. Zuber: Phys.Rev. **D12**, 482 (1975)

6 W. Kummer and M. Schweda: Phys.Lett. **141B**, 363 (1984)
W. Kummer, H. Mistelberger, P.Schaller, M.Schweda and T. Kreuzberger: Nucl.Phys. **B281**, 411 (1987)
W. Kummer, M. Mistelberger, P. Schaller and M. Schweda: Zs. f. Phys. **40C**, 91 (1988)
W. Kummer, M. Mistelberger and P. Schaller: Zs.f.Phys. **40C**, 103 (1988)

7 L. Brink, P. di Vecchia and P.S. Howe: Phys.Lett. **B65**, 369 (1976)
S. Deser and B. Zumino: Phys.Lett. **B65**, 369 (1976)
A.M. Polyakov: Phys.Lett. **B103**, 207 (1981)

8 L. Baulieu, C. Becchi and R. Stora: Phys.Lett. **B180**, 55 (1986)

9 D.W. Düsedau: Phys.Lett. **B188**, 51 (1987)

10 A. Rebhan and U. Kraemmer: Phys.Lett. **B196**, 477 (1987)

11 U. Kraemmer and A. Rebhan: Nucl.Phys. **B315**, 717 (1989)

12 N.K. Nielsen and R. Schroer: Nucl.Phys. **B127**, 493 (1977)
T. Eguchi, P.B. Gilkey and A.J. Hanson: Phys.Rev. **66**, 213 (1980)
B. Zumino: In *Relativity, Groups and topology II*, ed. by B.S. DeWitt and R. Stora (Les Houches 1983, North Holland 1984) p. 1291

13 E. Kraus and K. Sibold: *Yang Mills theory in a nonlinear gauge: quantization and gauge-independence*, preprint MPI-PAE/PTh 31/89, May 1989

14 Note that an even infinite number of counter-terms for unphysical parameters (like the gauge parameters occurring in SUSY YM–theory) is permissible [5].

15 The first derivation of an identity for Γ, albeit still a nonlocal one, goes back to B.W. Lee: Phys.Lett. **B46**, 214 (1973)

16 Y.M.P. Lam: Phys.Rev. **D6**, 2154 (1972); **D7**, 2943 (1973)
P. Breitenlohner, D. Maison: Comm.Math.Phys. **52**, 11 (1977)

17 Actually from $\mathcal{B}(L) = 0$ one should determine the most general solution representing the full action, rather than starting, as we did, from an explicit special solution of this equation [18].

18 A. Becchi, A. Rouet, A. Stora: Annals of Physics (NY) **98**, 287 (1976)

19 Cf. [18] and the contribution of C. Becchi to these Proceedings

20 'Gauge-independence' means that the gauge-dependent part of $\mathcal{A}^{(0)}$ may be removed by counter-terms $\Delta = \mathcal{A}^{(1)}$

21 Note that for the internal anomaly with nonlinear gauges, although $\partial\mathcal{A}^0/\partial p$ was also inequivalent to zero, the disturbing term still remained a local one.

22 L. Baulieu, C. Becchi and R. Stora: Phys.Lett. **B180**, 55 (1986)

23 L. Baulieu, W. Siegel and B. Zwiebach: Nucl.Phys. **B287**, 93 (1987)
D.W. Düsedau: Phys.Lett. **B188**, 51 (1987)

24 A. Rebhan and U. Kraemmer: Phys.Lett. **B196**, 477 (1987)
U. Kraemmer and A. Rebhan: Nucl.Phys. **B315**, 717 (1989)

25 In cases like the bosonic string, where the FP-ghosts are affected by the external symmetry, it is necessary to gaugify $s \to \tilde{s}$ too (see below).

26 W. Kummer: *Symmetry extended BRS-technique and gauge-independence of anomalies*, prep. TU Wien, July 1989

27 W.Kummer: Phys.Lett. **231**, 53 (1989)
28 H. Balasin, O. Piguet, M. Schweda and M. Stierle: Phys.Lett. **215**, 328 (1988)
29 For notational details we refer to [5] and, e.g. W. Kummer et al.: Fortschr. der Physik **375**, 261 (1989)
30 Note that it is the $\not\partial$ in the first factor of (8.5) which is gaugified appropriately
31 K. Konishi and K. Shizuya: Nuovo Cim. **90A**, 111 (1985)
32 T.E. Clarke, O. Piguet and K. Sibold: Nucl.Phys. **B143**, 445 (1978); **B172**, 201 (1980) O. Piguet and K. Sibold: Nucl.Phys. **B196**, 428, 447 (1982)
33 S.L. Adler, W.A. Bardeen: Phys.Rev. **182**, 1517 (1969)
34 Cf. e.g. V.A. Novikov, M.A. Shifman, A.I. Vainshtein, V.I. Zakharov: Phys.Lett. **157B**, 169 (1985).

LEP: The First Hundred Days

F. Dydak

CERN
CH – 1211 Geneva 23, Switzerland

Abstract: The Large Electron–Positron collider (LEP) at CERN came into operation in August 1989. What has been learned from the first hundred days of data-taking and analysis? In these lectures, first results are presented from the following areas of physics: the Standard Model of the electroweak interactions, QCD, and searches for new particles. A brief outlook on the likely future directions of LEP operation is given.

1. Introduction

On 14 July 1989 (Bastille day), positrons were injected for the first time into the newly completed ring of the Large Electron–Positron collider (LEP), the construction of which commenced in 1981 and terminated in 1989. Within a few hours, the commissioning team succeeded in getting the positrons all the way around the 27 km circumference of LEP. One month later, a 'pilot physics run' was conducted. On the machine side, sufficiently intense positron and electron beams were accelerated to Z resonance energy (then known most precisely from a resonance scan [1] performed at the SLC, the SLAC Linear Collider), producing some 20 Z events at each of the four interaction points equipped with detectors. In no way less impressive were the achievements of the four LEP Collaborations, ALEPH, DELPHI, L3, and OPAL, who were constructing their detectors during the years 1983 to 1989. All four detectors were completed to such an extent that several Z events each could be registered. This exercise proved extremely valuable in that it gave important feedback on the performance of the machine and detectors, and helped in preparing the physics run which took place from 19 September to 19 December 1989.

The integrated luminosity delivered by LEP at each of the interaction points was about 1.8 pb^{-1}, with all four experiments being active with an average efficiency of 70%. The centre-of-mass energy was varied between 88.3 and 95.0 GeV to permit a scan of the Z resonance. About half of the time was spent on the peak, the other half at positions off the resonance. Almost 30,000 hadronic Z decays have been recorded by each of the experiments. The maximum luminosity achieved was $L \approx 0.5\ 10^{31}\mathrm{cm}^{-2}\mathrm{s}^{-1}$, which falls short of the design luminosity by only a factor of 3.

The physics analysis of the data had been well prepared in advance, so there was no, or little, delay in the communication of first results, which were obtained during the 1989 data-taking period. At the time of writing up these lectures, newer and more precise results have become available since the lectures given at Schladming. Wherever possible, they have been incorporated into the written version.

2. Electroweak Physics Results

2.1 Outline of the Programme

The Z resonance has always been regarded as a 'bonanza' for electroweak physics. The measurement of the resonance line-shape and the study of the decay characteristics of the Z decay channels allow both a precision measurement of the Z mass (which is considered a fundamental constant of nature) and a stringent test of the Standard Model of electroweak interactions.

The Z line shape is represented by a Breit–Wigner resonance shape:

$$\sigma_{\mathrm{f}}(\sqrt{s}) = \sigma_{\mathrm{f}}^{0} \frac{s\Gamma_{\mathrm{Z}}^{2}}{(s-m_{\mathrm{Z}}^{2})^{2}+s^{2}\Gamma_{\mathrm{Z}}^{2}/m_{\mathrm{Z}}^{2}} , \tag{1}$$

with

$$\sigma_{\mathrm{f}}^{0} = \frac{12\pi}{m_{\mathrm{Z}}^{2}} \frac{\Gamma_{\mathrm{e}}\Gamma_{\mathrm{f}}}{\Gamma_{\mathrm{Z}}^{2}} , \tag{2}$$

where m_{Z}, Γ_{Z}, Γ_{e}, and Γ_{f} denote the mass and total width of the Z, and the partial widths for the decay into a pair of electrons and fermions, respectively. This line shape is quite significantly altered by higher-order electroweak radiative corrections, the dominant contribution of which arises from initial-state bremsstrahlung. These radiative corrections have been calculated by various groups (for a comprehensive review, see Ref. [2]), in such a way that the resulting Z line shape is predicted to better than half of a percent.

The experimental procedure is in principle simple: at several settings of well-known beam energy, both Z decays and elastic Bhabha scattering events with small scattering angles are recorded. The latter are dominated by t-channel exchange of photons, a QED process that is well understood. Its known cross-section together with the observed number of Bhabha events serves to determine the 'luminosity' of the beam–beam crossings.

The events recorded by a real apparatus have imperfections: the momentum and energy resolution is finite; the acceptance for the detection of particles is less than 4π owing to beam holes, cable ducts, and other dead zones; the efficiency for triggering is less than 100%, etc. All these imperfections are simulated in Monte Carlo programs, and the resulting corrections are applied to the real data.

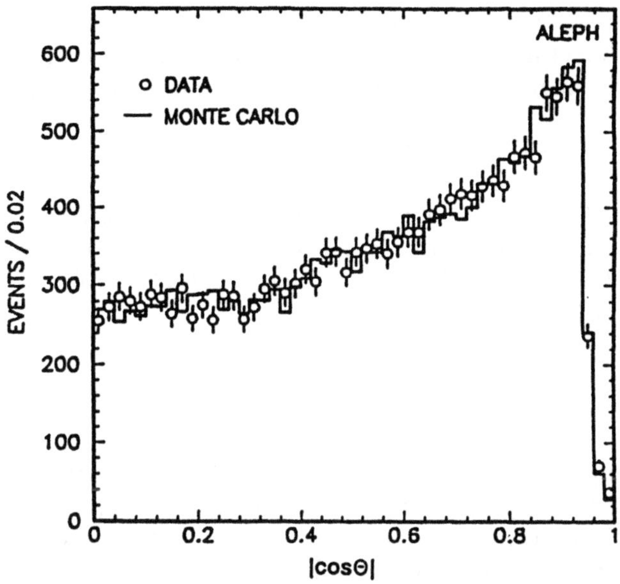

Fig. 1. The $\cos\theta$ of the sphericity axis of hadronic events

All four LEP detectors have been designed in such a way as to come reasonably close to a 4π coverage of the solid angle. As a consequence of this, and of the relatively isotropic distribution of the polar angle of Z decay products, losses due to limited acceptance are small and easily corrected for. As an example, Fig. 1 shows a distribution [3] of the cosine of the polar angle θ of the sphericity axis of hadronic Z decays: the loss of events near $\cos\theta = 1$ is small, and – what is most important – is well reproduced by the Monte Carlo simulation. Fig. 2 shows a distribution [3] of the polar angle θ of small-angle

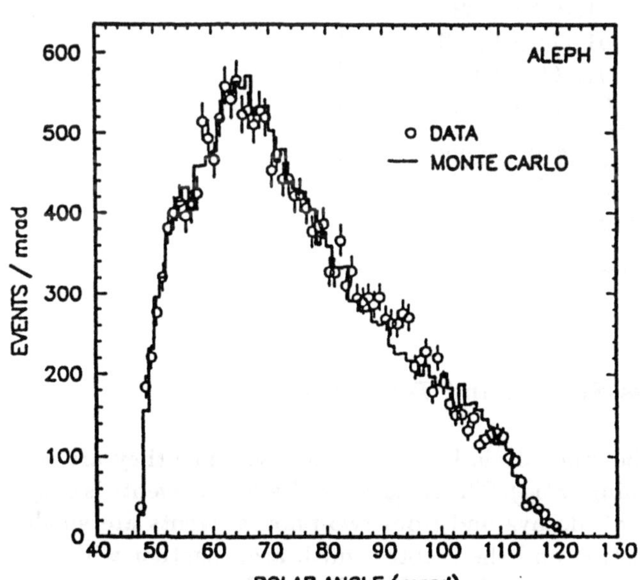

Fig. 2. The θ distribution of small-angle Bhabha events

Bhabha events, which is again well reproduced by the Monte Carlo simulation. Thus the design of the LEP detectors pays dividends: since the corrections for the non-uniformity of the solid-angle coverage are small, the danger of wrong results is greatly diminished.

Eventually, the theoretically predicted distributions (after the implementation of the distortion from radiative processes) are fitted to the experimental distributions, which are corrected for all detector effects. The fits determine the best values of m_Z, Γ_Z, etc., in a way that is *a priori* independent of the Standard Model (even the radiative corrections are to a large extent calculated in a model-independent fashion).

2.2 The Z Mass

The Z mass, as determined by the four LEP Collaborations [4,5,6,7] and by the Mark II Collaboration [8] working at the SLC, is given in Table 1 (earlier measurements of the Z mass in $\bar{p}p$ collision experiments are much less precise, and would have no effect on the average). The overall error is dominated by a 30 MeV uncertainty in the absolute energy of the LEP machine. There are ways to decrease this uncertainty; however, a final overall error of the order of 10 MeV will persist. Still, this ultimate accuracy of $1\ 10^{-4}$ renders m_Z one of the most precise electroweak constants of nature after the fine-structure constant α (known to $5\ 10^{-8}$) and the Fermi coupling constant G_F (known to $2\ 10^{-5}$).

Table 1. Measurements of m_Z at LEP and the SLC

	m_Z (GeV)
ALEPH [4]	91.197 ± 0.021
DELPHI [5]	91.171 ± 0.030
L3 [6]	91.160 ± 0.024
OPAL [7]	91.154 ± 0.021
LEP	$91.171 \pm 0.012 \pm 0.030_{\mathrm{LEP}}$
MARK II [8]	91.14 ± 0.12
m_Z	91.169 ± 0.031

2.3 Hadronic Peak Cross-Section and Total Width

The Z line shape can best be studied via hadronic Z decays since they are the most abundant (70% branching ratio). The selection of hadronic events is easy, since backgrounds from $\tau^+\tau^-$ decays and from two-photon events are small. The hadronic peak cross-section and the Z total width are, together with the Z mass, the three parameters that are determined from a fit to the data.

Table 2. Hadronic peak cross-section and Z total width

	σ_h^0 (nb)	Γ_Z (MeV)
ALEPH [4]	41.30 ± 0.75	2562 ± 47
DELPHI [5]	41.6 ± 1.7	2511 ± 65
L3 [6]	39.8 ± 0.9	2539 ± 54
OPAL [7]	41.4 ± 1.1	2536 ± 45
MARK II [8]	42 ± 4	2420 ± 400
Average	40.92 ± 0.49	2540 ± 26
SM prediction	41.4 ± 0.1	2490 ± 35

The fit results are given in Table 2, and the average of the results is compared with the prediction of the Standard Model. The uncertainty of the latter reflects the uncertainty in m_Z ($\pm$ 31 MeV), in the strong coupling constant ($0.10 \leq \alpha_s \leq 0.13$), in the t-quark mass ($60 < m_t < 220$ GeV), and in the Higgs mass ($25 < m_H < 1000$ GeV). In the averaging of the results on σ_h^0 it has been assumed that a 1% systematic error is common to all experiments because of the theoretical uncertainty of the Bhabha cross-section, which determines the luminosity. For the average of Γ_Z, a common uncertainty of 10 MeV arising from the radiative correction of the line shape has been taken into account.

The measurements of σ_h^0 and Γ_Z are in good agreement with the expectation of the Standard Model.

2.4 Hadronic and Leptonic Partial Widths

In a manner analogous to that described in the preceding subsections, a simultaneous fit of the Breit–Wigner resonance shape to four data sets, namely Z $\to$ hadrons, e^+e^-, $\mu^+\mu^-$, and $\tau^+\tau^-$, enables the determination of six parameters:

$$m_Z,\ \Gamma_Z,\ \sigma_h^0,\ \Gamma_e/\Gamma_h,\ \Gamma_\mu/\Gamma_h,\ \Gamma_\tau/\Gamma_h\ .$$

Since the hadronic data sample is much larger than the leptonic one, the results for m_Z, Γ_Z, and σ_h^0 will improve only marginally compared with a fit of the hadronic data sample alone.

The $Z \to e^+e^-$ decay is different from the other decay modes in that the t-channel exchange of photons contributes, especially at small polar scattering angles. The t-channel contribution is calculated from the known cross-section and the measured luminosity, and is subtracted from the data.

Recalling Eq.2, the partial width Γ_e can be derived from σ_h^0 (which depends on $\Gamma_e\Gamma_h$) and Γ_e/Γ_h, and then, in turn, the partial widths Γ_h, Γ_μ, and Γ_τ. The results quoted by the experiments [4,5,9,10,7,11] are listed in Table 3, and again compared with the predictions of the Standard Model. The partial widths for Γ_e, Γ_μ, and Γ_τ are equal within the experimental errors, as expected from lepton

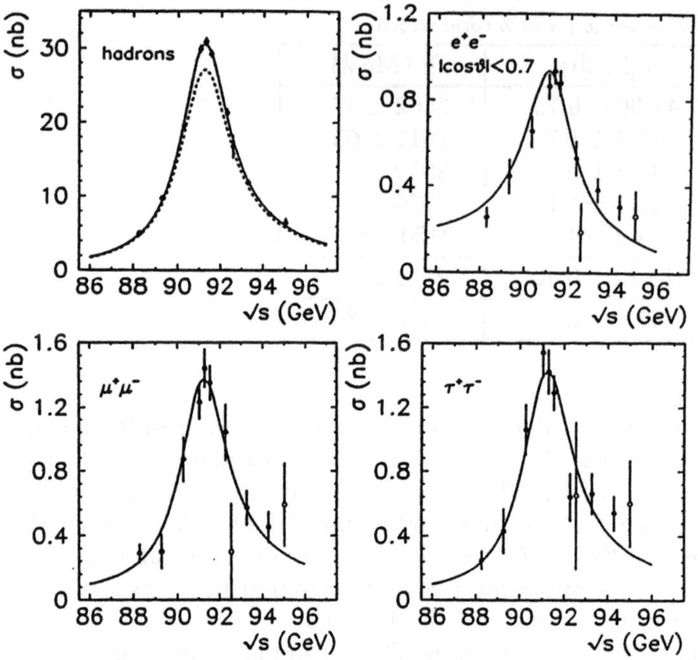

Fig. 3. Hadronic and leptonic cross-sections (OPAL data)

universality. Therefore, the fits were repeated with the constraint $\Gamma_e = \Gamma_\mu = \Gamma_\tau$, resulting in a measurement of the leptonic width Γ_l and hadronic width Γ_h. The measured partial widths are in good agreement with the predictions of the Standard Model. Fig. 3 shows the hadronic and leptonic cross-sections across the Z resonance, as measured by the OPAL Collaboration [7], together with the expected shape from the Standard Model with three families of quarks and leptons. The most marked dependence of the theoretical prediction of Γ_l and Γ_h is the one on m_t: both quantities increase by about 1.5% when m_t changes from 60 GeV to 220 GeV. Hence the ratio Γ_h/Γ_l is quite independent of m_t and constitutes a rather firm prediction (the remaining dependence on α_s and m_H is small). The experimental values of Γ_h/Γ_l are listed in Table 3 together

Table 3. Hadronic and leptonic partial widths (in MeV)

	Γ_h	Γ_e	Γ_μ	Γ_τ	Γ_l	Γ_h/Γ_l
ALEPH [4]	1803 ± 37	85.1 ± 2.4	83.1 ± 3.6	85.1 ± 3.7	85.1 ± 1.7	21.1 ± 0.6
DELPHI [5]	1741 ± 61	83.2 ± 3.8	86.0 ± 7.4	82.0 ± 8.4	85.1 ± 2.9	20.5 ± 1.0
L3 [9,10]	1744 ± 53	79.0 ± 2.8	90.0 ± 5.7	92.3 ± 7.5	82.8 ± 2.4	21.1 ± 0.9
OPAL [7]	1838 ± 46	81.2 ± 2.6	82.6 ± 5.8	85.7 ± 7.1	81.9 ± 2.0	22.4 ± 0.8
MARK II [11]	1560 ± 260					
Average	1790 ± 24	82.2 ± 1.4	84.7 ± 2.5	85.9 ± 2.8	83.8 ± 1.1	21.33 ± 0.37
SM prediction	1739 ± 28	83.7 ± 0.8	83.7 ± 0.8	83.7 ± 0.8	83.7 ± 0.8	20.8 ± 0.2

with the prediction of the Standard Model. Again there is good agreement. However, it would be most interesting to see whether this agreement persists when the experimental error keeps decreasing.

2.5 Constraints on the t-Quark Mass

Before LEP and the SLC came into operation, an upper limit on m_t could be derived from a comparison of the W mass as measured directly in $\bar{p}p$ collider experiments, and as predicted from measurements of $\sin^2\theta_w$ in neutrino–nucleon scattering experiments. The relation between m_W^2 and $\sin^2\theta_w$ is given by

$$m_W^2 = \frac{\pi\alpha}{\sqrt{2}G_F \sin^2\theta_w(1-\Delta r)}\,, \tag{3}$$

where $\sin^2\theta_w \equiv 1 - m_W^2/m_Z^2$, and Δr represents the electroweak radiative correction of m_W. The correction Δr, of a few percent, is strongly dependent on m_t (see Fig. 4), and hence allows a constraint on m_t from independent measurements of m_W and $\sin^2\theta_w$.

With the advent of LEP and the SLC, the constraint on m_t has become more stringent, since the rather imprecise W mass could be replaced by the more precise Z mass:

$$m_Z^2 = \frac{\pi\alpha}{\sqrt{2}G_F \sin^2\theta_w \cos^2\theta_w(1-\Delta r)}\,. \tag{4}$$

The most precise measurements of $\sin^2\theta_w$ are listed in Table 4: direct measurements via the mass ratio m_W/m_Z as measured in the $\bar{p}p$ collider experiments [12,13], and indirect but equivalent measurements [14] from the ratio of neutral-

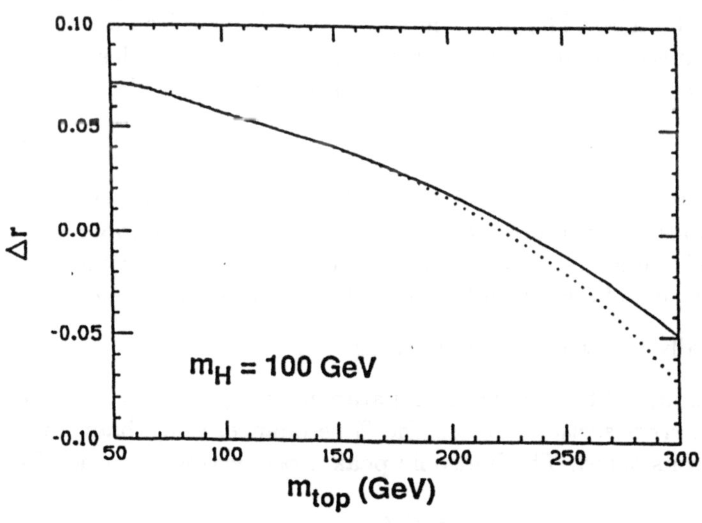

Fig. 4. Δr in $\mathcal{O}(\alpha)$ (dotted line) and $\mathcal{O}(\alpha^2)$ (solid line)

Table 4. Measurements of $\sin^2\theta_w \equiv 1 - m_W^2/m_Z^2$

Direct measurements of m_W/m_Z	
UA2 [12]	0.220 ± 0.010
CDF [13]	0.231 ± 0.009
Average	0.225 ± 0.007
Indirect measurements from R_ν = NC/CC	
CDHS [15]	0.227 ± 0.005(exp.) ± 0.005(th.)
CHARM [16]	0.236 ± 0.005(exp.) ± 0.005(th.)
Average	0.231 ± 0.003(exp.) ± 0.005(th.)
Global average	0.2282 ± 0.0045

to charged-current (NC/CC) scattering in neutrino–nucleon collisions [15,16]. Utilizing this value of $\sin^2\theta_w$ together with m_Z as determined at LEP and the SLC, constrains the t-quark mass to $m_t = 130 \pm 60$ GeV [17]. It should be noted, however, that this constraint is valid only in the context of the Minimal Standard Model, with isospin doublets only in the Higgs sector.

The above bound on m_t is limited by the accuracy of the measurement of $\sin^2\theta_w$, and will therefore improve only slowly with time.

2.6 The Invisible Width and the Number of Neutrino Families

One of the very first measurements at LEP and the SLC was the determination of the Z invisible width. The Standard Model predicts a width of 166.5 MeV for each family of (essentially massless) neutrinos, and hence an invisible width of 500 MeV for three families. The invisible width (which includes also decays into hitherto unknown weakly interacting, stable, and neutral particles) can be measured from the total width by subtracting the hadronic and the leptonic widths:

$$\Gamma_{\rm inv} = \Gamma_Z - \Gamma_h - 3\Gamma_l\ . \tag{5}$$

Dividing the thus determined invisible width by 166.5 MeV gives N_ν, the number of neutrino families. Even more, N_ν is expected to be the number of all families of quarks and leptons, since the neutrinos associated with very massive quarks are still expected to be light and to contribute to the Z invisible width.

There are essentially two ways of determining N_ν:

- In the first method, a fit with two free parameters m_Z and Γ_Z, is made to the measured cross-sections across the Z resonance. Only hadronic Z decays need be considered. The hadronic peak cross-section σ_h^0 is not free but is fixed by

$$\sigma_h^0 = \frac{12\pi}{m_Z^2}\frac{\Gamma_e\Gamma_h}{\Gamma_Z^2}\ ,$$

where Γ_e and Γ_h are taken from the Standard Model. Because of the strong dependence of σ_h^0 on Γ_Z, the precision of Γ_Z is largely determined by the hadronic peak cross-section. The result for Γ_Z is precise, but is dependent on the Standard Model. The method was primarily used in the early days of LEP operation when the statistics were still scarce.

- The second method employs both hadronic and leptonic Z decays, assuming lepton universality. Besides m_Z and Γ_Z, two further parameters, σ_h^0 and $R = \Gamma_h/\Gamma_l$, are introduced, which determine the hadronic and leptonic peak cross-sections, thus breaking the link between the total width and the peak cross-sections. Therefore, Γ_Z is determined in a less precise but model-independent way. From Eq.5, with $\Gamma_{inv} = N_\nu \Gamma_\nu$, one obtains

$$N_\nu = \frac{\Gamma_l}{\Gamma_\nu}\left[\sqrt{\left(\frac{12\pi R}{m_Z^2 \sigma_h^0}\right)} - R - 3\right],$$

where the expression in square brackets is determined by the best fit results for m_Z, σ_h^0, and R, and the ratio Γ_l/Γ_ν is taken from the Standard Model. This procedure reduces the model dependence of N_ν to a minimum since the ratio $\Gamma_l/\Gamma_\tau = 0.5010 \pm 0005$ has a much reduced sensitivity on free parameters of the Standard Model.

Table 5 quotes the results for Γ_{inv} and for N_ν [4,5,18,7], all obtained through the second method (or minor variants thereof). All results agree well with the predictions of the Standard Model.

Table 5. The invisible width and the number of neutrino families

	Γ_{inv} (MeV)	N_ν
ALEPH [4]	503 ± 33	2.97 ± 0.16
DELPHI [5]	515 ± 54	3.05 ± 0.28
L3 [18]	537 ± 48	3.23 ± 0.29
OPAL [7]	453 ± 44	2.73 ± 0.26
Average	499.6 ± 21.7	2.98 ± 0.11
SM prediction	500 ± 5	

2.7 Forward–Backward Asymmetry of Leptons

The partial width for the decay $Z \rightarrow f\bar{f}$ depends on the vector and axial–vector coupling constants of the fermion f:

$$\Gamma_f = \frac{G_F m_Z^3}{6\sqrt{2}\pi}(g_{Vf}^2 + g_{Af}^2).$$

The forward–backward asymmetry in the angular distribution of the final-state fermions is given by

$$A_{\mathrm{f}}^{\mathrm{FB}}(m_{\mathrm{Z}}) = 3\,\frac{g_{\mathrm{Ve}}g_{\mathrm{Ae}}g_{\mathrm{Vf}}g_{\mathrm{Af}}}{(g_{\mathrm{Ve}}^2 + g_{\mathrm{Ae}}^2)(g_{\mathrm{Vf}}^2 + g_{\mathrm{Af}}^2)}\ .$$

Therefore, the fermion coupling constants can be determined from a simultaneous measurement of Γ_{f} and $A^{\mathrm{FB}}(m_{\mathrm{Z}})$.

Since the determination of the primordial quark flavour is rather difficult in hadronic events, the determination of the fermion coupling constants has so far been concentrated on leptons, which are easy to identify experimentally. Since no systematic difference between the three lepton channels has been seen, lepton universality is assumed and a common fit to all leptons is performed. The results are given in Table 6. The agreement with the expectation from the Standard Model, also given in Table 6, is good.

Table 6. Leptonic vector and axial-vector coupling constants

	g_{Vl}^2	g_{Al}^2
L3 [10]	0.0044 ± 0.0048	0.245 ± 0.007
SM prediction	0.0016	0.250

2.8 Summary

On the whole, the agreement between the measurements and the predictions of the Minimal Standard Model of the electroweak interaction, in terms of the three input parameters α, G_{F}, and m_{Z}, is impressive. The room for new, so far unidentified, decay channels of the Z has become small: about 2% of the total Z decay width at most.

However, significant improvements in the experimental accuracy are possible in the near future, and will test the Standard Model quite thoroughly. The interest will focus on 'hard' quantities such as $R = \Gamma_{\mathrm{h}}/\Gamma_{\mathrm{l}}$, where the Standard Model predictions are not very dependent on free parameters.

3. QCD Results

The most prominent Z decay mode is the one into a quark pair, which in turn radiates gluons, which split into quark pairs, etc., thus producing a cascade of quarks and gluons that materializes eventually as a jet of hadrons. The first part of this process, which takes place at the scale $Q^2 = m_{\mathrm{Z}}^2$, is expected to be well described by perturbative QCD. The latter part, which takes place at a scale $Q^2 \approx 1\ \mathrm{GeV}^2$, describes how quarks and gluons get confined in hadrons. In the absence of a theory of confinement, recourse is being made to various 'QCD inspired' models of hadronization.

There exist a number of Monte Carlo programs that simulate the parton cascade and the subsequent hadronization. For the perturbative part, the programs employ either a leading logarithm approximation in the creation of quarks and gluons, or a rigorous (or almost rigorous) QCD calculation of the matrix elements up to $\mathcal{O}(\alpha_s^2)$. In the latter case, the maximum number of partons is limited to four. For the hadronization part, the string fragmentation model of the LUND Group [19] has proved to be particularly successful: hadrons are produced along strings, which are stretched between the outgoing partons. An alternative is cluster fragmentation, proposed by Marchesini and Webber [20]: the partons are combined into colourless clusters, which either decay isotropically, or – if they have a large mass – decay in a string-like fashion into clusters of lower mass.

Naturally, there is quite some freedom in the way the hadronization is modelled, showing up in a number of free parameters that are adjusted in order to achieve close similarity between Monte Carlo and data distributions.

What can one learn from the comparison of the data with such QCD-inspired models? The basic programme of work involves two steps:

- adjust the free parameters of the hadronization model at one given Q^2 (e.g. at the Z pole), so as to reproduce the data optimally;
- evolve the quark–gluon cascade to another Q^2 via the mechanism of perturbative QCD, while retaining the parameters fixed.

Since the hadronization mechanism is expected to be invariant against such a change of scale, the comparison with data constitutes a test of perturbative QCD, and would permit a determination of the strong coupling constant α_s.

Because of the very large statistics that will be accumulated at the Z pole, it seems that such an initial adjustment of parameters should be done there, allowing the evolution down to smaller Q^2 (data from PEP and PETRA), and – some time in the future – the evolution to larger Q^2 (data from LEP 200).

QCD tests of this type are just beginning to come along. The analyses published so far from LEP data tend to focus on the evaluation of the many Monte Carlo programs that are available, and on finding the optimum set of parameters.

A host of variables is being used to compare Monte Carlo predictions with the data. They can be grouped into two categories: global variables, which characterize the event shape; and single-particle inclusive variables, which deal with individual particles of an event.

3.1 Analysis of Global Event Variables

Assuming that an event has n charged tracks with momenta p^i ($1 \le i \le n$), the momentum tensor

$$M_{\alpha\beta} = \sum_{i=1}^{n} p_{\alpha i} p_{\beta i}$$

can be constructed, where α and β refer to the x, y, and z components. Let Q_1, Q_2, and Q_3 be the eigenvalues of the momentum tensor, ordered so that $Q_1 < Q_2 < Q_3$, and normalized so that $Q_1 + Q_2 + Q_3 = 1$. The eigenvector $\hat{n}_3$ defines the 'sphericity axis', and $\hat{n}_3$ together with $\hat{n}_2$ defines the 'event plane'. Two commonly used global variables are then the sphericity

$$S = \frac{3}{2}(Q_1 + Q_2)$$

and the aplanarity

$$A = \frac{3}{2}Q_1 \,.$$

The sphericity ($0 \leq S \leq 1$) is zero for pencil-like two-jet events, and is unity for isotropic events. The aplanarity ($0 \leq A \leq 0.5$) is zero for planar events and 0.5 for isotropic events.

Since the sphericity and aplanarity variables are based on the product of momenta, another variable, which is linear in the momentum (and hence gives relatively more weight to low-momentum particles), is often used: the thrust,

$$T = \max\left(\frac{\sum_{i=1}^{n} |p_{\parallel i}|}{\sum_{i=1}^{n} |p_i|}\right) ,$$

where again the sum runs over all charged particles of the event, and the longitudinal momentum component $p_{\parallel i}$ refers to an axis for which the sum of all longitudinal components assumes a maximum.

Figs. 5, 6, and 7 show the sphericity, aplanarity, and thrust distributions as measured by the DELPHI Collaboration [21] (the other groups draw the

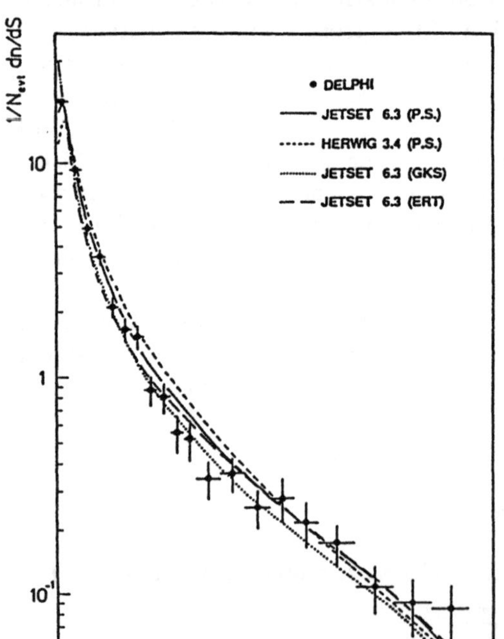

Fig. 5. Data and Monte Carlo predictions of the sphericity distribution

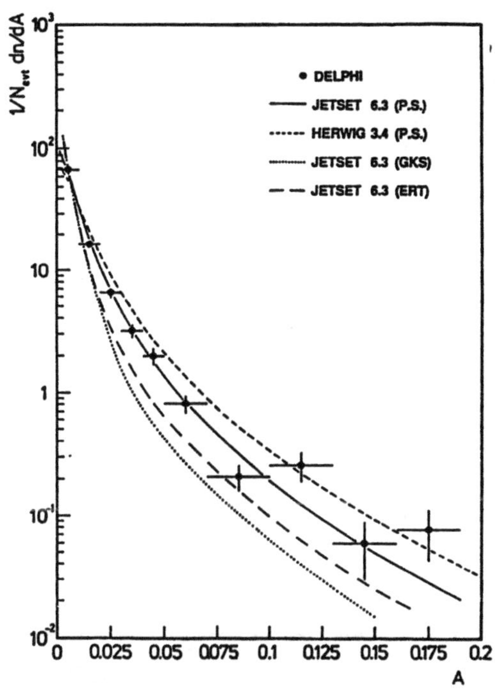

Fig. 6. Data and Monte Carlo predictions of the aplanarity distribution

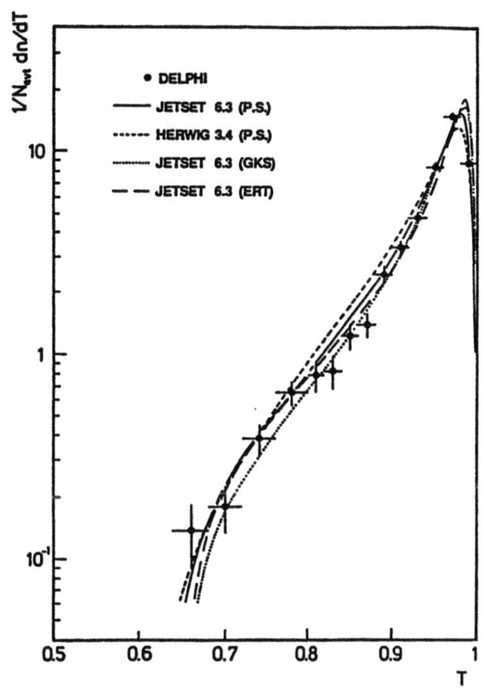

Fig. 7. Data and Monte Carlo predictions of the thrust distribution

same conclusions from their data [22,23]. The data are compared with four Monte Carlo models, of which JETSET 6.3 (P.S.) and HERWIG 3.4 (P.S.) are based on the leading logarithm approximation, whereas JETSET 6.3 (GKS) and JETSET 6.3 (ERT) have a QCD calculation of the matrix elements up to $\mathcal{O}(\alpha_s^2)$ incorporated. The parameters of the hadronization model are, in this particular analysis, optimized with data from PEP (at $\sqrt{s} = 29$ GeV).

First of all, one concludes that all models resemble the data amazingly well. On closer inspection, it turns out that the models with leading logarithm approximation do somewhat better than the $\mathcal{O}(\alpha_s^2)$ models: this is particularly visible in Fig. 6: the $\mathcal{O}(\alpha_s^2)$ model underestimates the occurrence of events with large aplanarity, presumably because multi-gluon emission is not accounted for.

3.2 Analysis of Single-Particle Inclusive Variables

The most commonly used single-particle inclusive variables are the momentum fraction variable

$$x_p = \frac{2p}{\sqrt{s}} = \frac{p}{p_{\text{beam}}} ,$$

where p is the momentum of the particle concerned, and the transverse momentum within and out of the event plane is

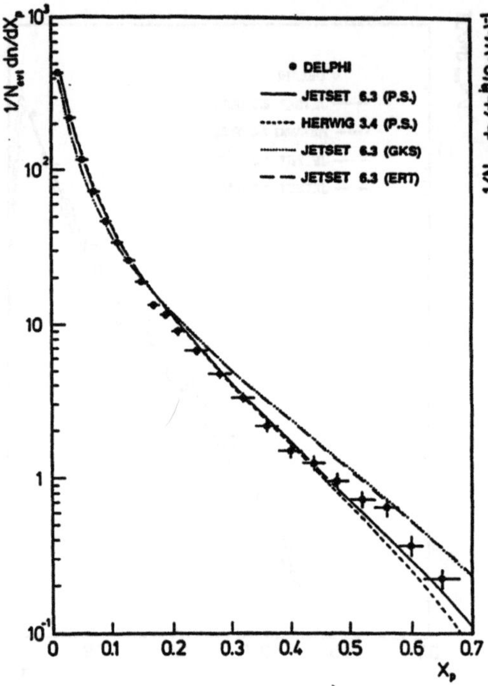

Fig. 8. Data and Monte Carlo predictions of the x_p distribution

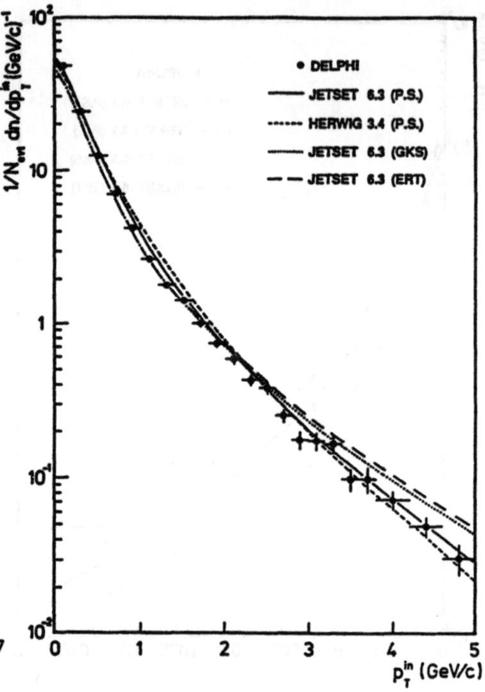

Fig. 9. Data and Monte Carlo predictions of the $p_\perp^{in}$ distribution

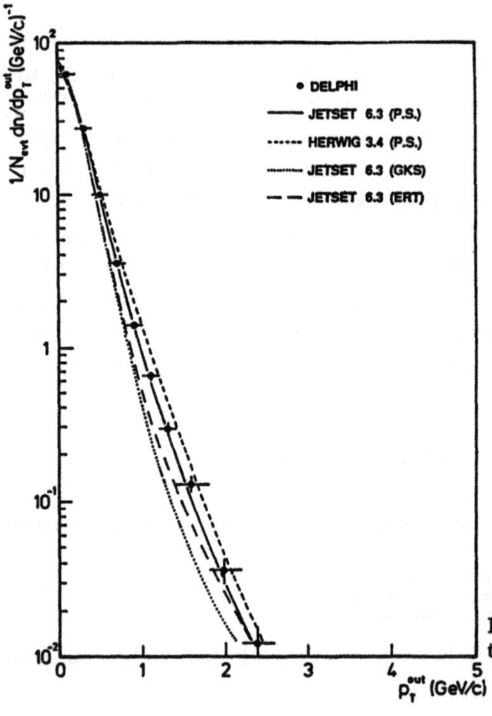

Fig. 10. Data and Monte Carlo predictions of the $p_\perp^{out}$ distribution

$$p_\perp^{\rm in} = |p.\hat{n}_2|$$

and

$$p_\perp^{\rm out} = |p.\hat{n}_3| \; .$$

Figs. 8, 9, and 10 show x_p, $p_\perp^{\rm in}$, and $p_\perp^{\rm out}$ as measured by the DELPHI Collaboration [21], again compared with the predictions of the same four models. The same conclusions can be drawn from the data: models with the leading logarithm approximation reproduce the data somewhat better, especially in $p_\perp^{\rm out}$. It is expected that this qualitative conclusion would not be altered if the parameters of the hadronization model would be optimized at the Z energy rather than taken from low-energy data. However, such a procedure would tend to compensate for the deficiency of the $\mathcal{O}(\alpha_s^2)$ QCD evolution with conveniently adjusted fragmentation parameters.

3.3 Is α_s Running?

The strong coupling constant α_s is expected to decrease logarithmically with increasing Q^2. The Z pole provides data with $Q^2 = 8300$ GeV^2, much higher than for data from PEP and PETRA.

A systematically precise determination of α_s at LEP is a challenging problem because of the uncertainty introduced by the hadronization process. One popular method consists of measuring the relative abundance of 3-jet events with respect to 2-jet events, a ratio which measures α_s. Of course, such an event classification according to the number of jets is anything but trivial. However, one might argue that applying the *same* procedure to data taken at different energies may permit a more reliable determination of the Q^2-dependence of α_s than of the value of α_s itself.

The results of an analysis performed along such lines has been published by the OPAL Collaboration [23]. Fig. 11 shows their result for the relative

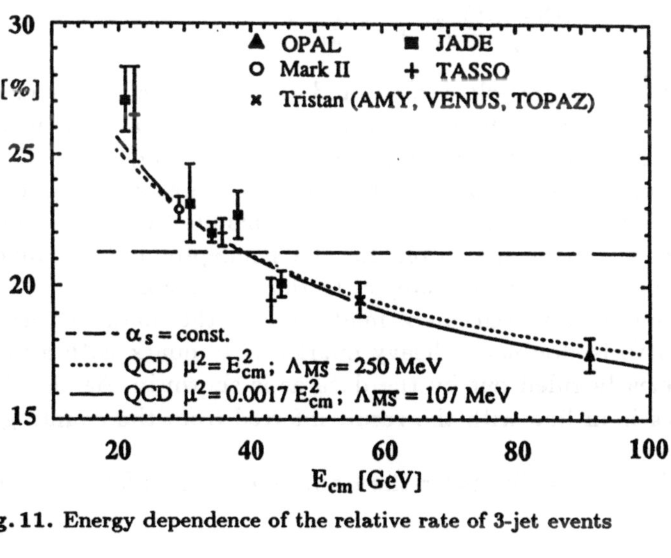

Fig. 11. Energy dependence of the relative rate of 3-jet events

abundance of 3-jet events, as a function of the centre-of-mass energy. Only statistical errors are shown, and no point-to-point systematic errors. However, the data look quite interesting, and are certainly rather in support of a strong coupling constant which is 'running', as expected, by perturbative QCD.

4. Searches and Limits

According to our belief that all constituents of matter – known or unknown – participate in the weak interaction, Z decays are an ideal laboratory for the search for new particles through their weak coupling to the Z. On the basis of their theoretical production cross-sections and the known luminosity, sensitive searches for the existence of such events can be made, in most of the cases the only limit being the beam energy: only particles with mass less than $m_Z/2$ can be pair-produced.

4.1 Heavy Sequential Quarks and Charged Leptons

The t-quark is, besides the ν_τ, the best established unobserved particle. Since the t-quark is heavy, Z decays into a pair of t-quarks would lead to a rather spherical event shape, and event-shape parameters, e.g. sphericity, would constitute a sensitive measure for the existence of t-quarks.

Fig. 12 shows a sphericity distribution of hadronic events as measured by the OPAL Collaboration [24]. The data are well represented by a Monte Carlo simulation involving u, d, c, s, and b-quarks. A t-quark with a mass of 35 GeV is clearly ruled out by the data. If the mass is even heavier, then the production cross-section is reduced by the factor

$$\frac{\beta(3-\beta^2)}{2} g_{\mathrm{Vt}}^2 + \beta^3 g_{\mathrm{At}}^2 \,, \tag{6}$$

where β is the velocity of the t-quark, and g_{Vt} and g_{At} are its vector and axial–vector coupling constants. Since the production cross-section vanishes in the limit $\beta \to 0$, the above test will lose its sensitivity near $m_t \approx m_Z/2$.

The CDF experiment at the Fermilab $\bar{p}p$ Collider has reported a lower limit of 77 GeV [25] for the mass of the t-quark. However, this limit depends on the validity of the Standard Model predictions for the semileptonic decay mode, whereas the limit from Z decays is not subject to this restriction.

In an analogous manner, a search can be made for a b', the charge-$\frac{1}{3}$ partner in a hypothetical fourth isodoublet of heavy quarks. A b' quark with a mass of 35 GeV is again clearly ruled out by the data as is shown in Fig. 12. The absence of a b' quark is in line with the result inferred from the counting of neutrino families.

Table 7 lists the lower mass limits for the t and b' quarks as found at LEP [26,27,24].

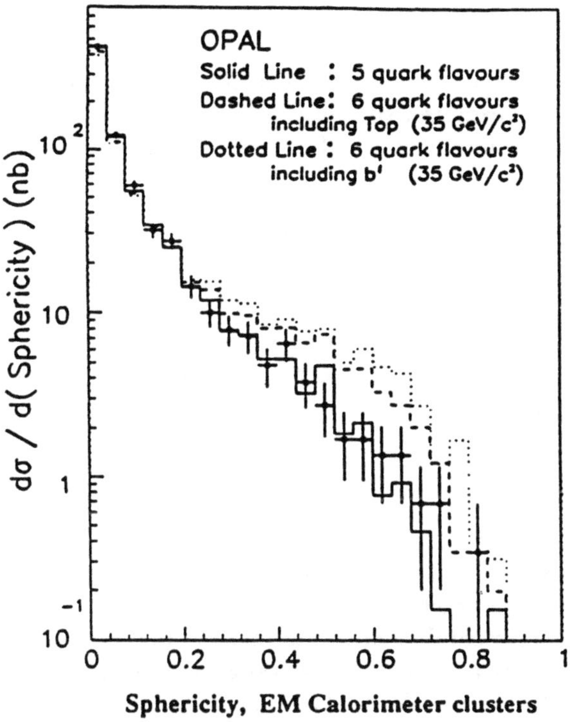

Fig. 12. Sphericity distribution of hadronic events

Table 7. Lower mass limits for the t- and b'-quarks (95% CL)

	m_t (GeV)	$m_{b'}$ (GeV)
ALEPH [26]	45.8	46.0
DELPHI [27]	44.0	44.5
OPAL [24]	44.5	45.2

A sequential charged heavy lepton L is expected to decay into its associated neutrino ν_L and a virtual W boson. The signature would be missing energy and missing $p_\perp$, carried away by the neutrino.

The OPAL Collaboration have carried out a direct search along this line [28], and quote a lower limit of 44.3 GeV (at 95% CL), assuming that the mass of the associated ν_L does not exceed about 10 GeV. In contrast, the ALEPH Collaboration have indirectly concluded from their measurement of the total and hadronic widths that any sequential charged heavy lepton, whose associated neutrino has a mass smaller than 42.7 GeV at 95% CL, is excluded.

4.2 Heavy Neutral Leptons

The heavy neutral lepton in question may be a stable (neutrino-like) lepton, or an unstable lepton, which would be expected to mix with the ordinary neutrinos

ν_e, ν_μ, and ν_τ, and hence to decay into an ordinary charged lepton and a virtual W.

The Z decay into a pair of massive stable neutrino-like leptons is limited by the measured invisible width. In this exercise, the threshold suppression factor of fermions (see Eq.6) has to be taken into account. The ALEPH Collaboration [26] quote a lower limit of 42.7 GeV (at 95% CL) for the mass of a neutral stable heavy lepton.

The Z decay into a pair of massive unstable neutral leptons would modify the observed total and hadronic widths of the Z. From the comparison with the measured values, and from the negative result of a direct search for isolated ordinary leptons from the decay of such a heavy lepton, the mass interval from 25.0 to 42.7 GeV is excluded, independently of the coupling strength of the neutral heavy lepton to the ordinary neutrinos.

4.3 The Neutral Higgs Boson

The neutral Higgs boson H^0, the existence of which is postulated in the Minimal Standard Model of the electroweak interaction, plays an essential role in the theory but has not been seen experimentally. Since its mass is not predicted by the theory, searches for the H^0 over as broad a mass range as possible are a MUST. The Z resonance constitutes an excellent laboratory for such a search, through the radiation of a H^0 off a virtual Z:

$$e^+e^- \to Z \to Z^* H^0 ,$$

where the virtual Z^* decays into a pair of fermions. The Z decay modes into quarks and into all charged leptons, and into neutrinos, have been utilized. As for the H^0, its decay modes are very strongly dependent on its mass. For a very light H^0, only the decays into $\gamma\gamma$ and e^+e^- are possible, and the H^0 is long-lived. The signature is either that of an e^+e^- pair with a vertex that is displaced from the interaction point, or of missing energy and momentum if it leaves the apparatus undetected. Above the muon threshold, the dominant decay mode is $H^0 \to \mu^+\mu^-$. Fig. 13 (taken from Ref. [29]) shows the expected branching ratios of a light H^0. For heavier H^0, the dominant decay mode is into a pair of the heaviest, energetically allowed, fermions.

The decay channel $H^0 \to H^0\bar{\nu}\nu$ allows a particularly sensitive search for the H^0. The signal is large missing energy and momentum. Also useful are the decays of the virtual Z into charged leptons: $Z \to H^0 l^+ l^-$.

The production cross-section of the H^0 falls rapidly with increasing H^0 mass. Therefore, if no candidate event is seen, then the mass limit will be strongly dependent on the amount of integrated luminosity collected by the experiment.

Table 8 lists the mass regions that have so far been excluded by the various LEP experiments [29,30,31,32,33].

The ALEPH Collaboration [31] have extended the search for H^0 down to zero mass by looking for events with just a lepton pair and large missing energy

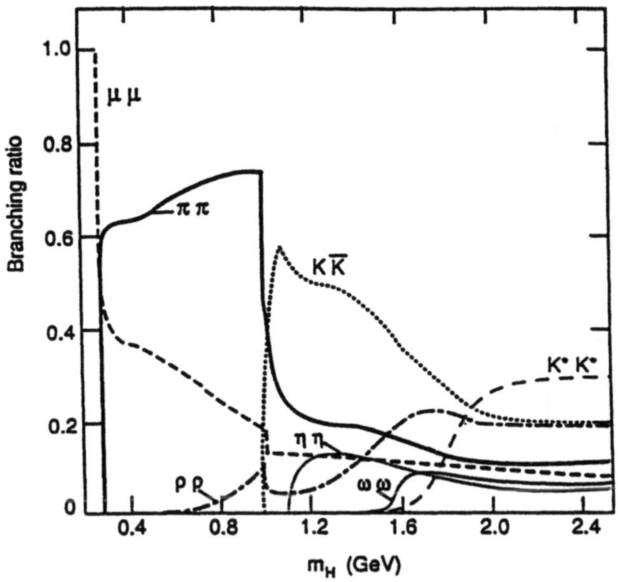

Fig. 13. Expected branching ratios of a light Higgs boson

Table 8. H^0 mass regions excluded by the LEP experiments (95% CL)

ALEPH [29,30,31]	0 – 24.0 GeV
DELPHI [32]	0.21 – 14.0 GeV
OPAL [33]	3.0 – 19.3 GeV

and momentum, supposedly carried away by the undetected, very light H^0. No signal has been found above the background from radiative lepton pair production where, for one reason or another, the photon(s) remain undetected.

4.4 Charged Higgs Bosons

Although simplicity is the only reason in favour of the minimal structure of the Higgs sector of the Standard Model of the electroweak interaction, there may well be a richer structure. The simplest extension would be the addition of another doublet of complex scalar fields. This would give rise to a total of five physical Higgs bosons, usually denoted by H^0, h^0, A^0, H^+, and H^-. The latter two charged Higgs particles would be pair-produced in Z decays, with a partial width of

$$\Gamma(Z \rightarrow H^+H^-) = \frac{G_F m_Z^3}{6\sqrt{2}\pi}\left(\frac{1}{2} - \sin^2\theta_w\right)^2 \beta^3 ,$$

where β is the velocity of the charged Higgs particles, and β^3 is the suppression factor for the p-wave production of scalar particles. The charged Higgs bosons decay into the heaviest fermions that are kinematically allowed; however, the branching ratios of the various channels are not fixed by the theory.

In the mass range accessible at LEP, the decay modes

$$H^+H^- \rightarrow \nu\bar{\tau}\bar{\nu}\tau, \nu\bar{\tau}\bar{c}s, c\bar{s}\bar{c}s$$

are favoured. The signature would be ordinary leptons from τ decay together with missing energy and momentum, or four acoplanar jets.

Table 9 gives the mass regions excluded by the LEP experiments [34,35,36] assuming a 50% branching ratio of the decay mode $H^\pm \rightarrow \nu\tau$.

Table 9. Lower $H^\pm$ mass limits from LEP (95% CL)

ALEPH [34]	40.6 GeV
DELPHI [35]	36.0 GeV
OPAL [36]	42.0 GeV

4.5 Supersymmetric Particles

The supersymmetry (SUSY) model is considered a serious contender for a theory beyond the Standard Model. It predicts a plethora of new particles, differing by half a unit of spin from the known ones. All these particles would be produced in Z decays provided they are low enough in mass. The production cross-sections would be comparable to the ones of the known fermions. Because of the conservation of the supersymmetric R-parity, SUSY particles will be produced in pairs. For the same reason of R-parity conservation, the lightest SUSY particle must be a stable and neutral particle, which will interact only weakly with ordinary fermions. Hence the signature of SUSY particles is missing energy and momentum, which is carried away by the unobserved lightest SUSY particle, in general believed to be the photino.

Sleptons are expected to decay into an ordinary lepton and a photino: $\tilde{l} \rightarrow l\tilde{\gamma}$, the signature being two acoplanar leptons with large missing energy and momentum. As an example, Fig. 14 shows the transverse momentum distribution of $\mu^+\mu^-$ pairs as measured by the L3 Collaboration [37]. The predicted accumulation of events at large $p_\perp$ (for the case $m_{\tilde{\mu}} = 41$ GeV, $m_{\tilde{\gamma}} = 10$ GeV) is clearly ruled out by the data. Table 10 lists the lower mass limits of sleptons determined by the LEP Collaborations [38,37,39] (the limits quoted assume a degeneracy of the mass of left-handed and right-handed sleptons, and are valid for $m_{\tilde{\gamma}} \leq 20$ GeV).

Table 10. Lower mass limits of sleptons (95% CL)

	$m_{\tilde{e}}$ (GeV)	$m_{\tilde{\mu}}$ (GeV)	$m_{\tilde{\tau}}$ (GeV)
ALEPH [38]	43.5	42.6	40.4
L3 [37]	41.0	41.0	
OPAL [39]	43.4	43.0	43.0

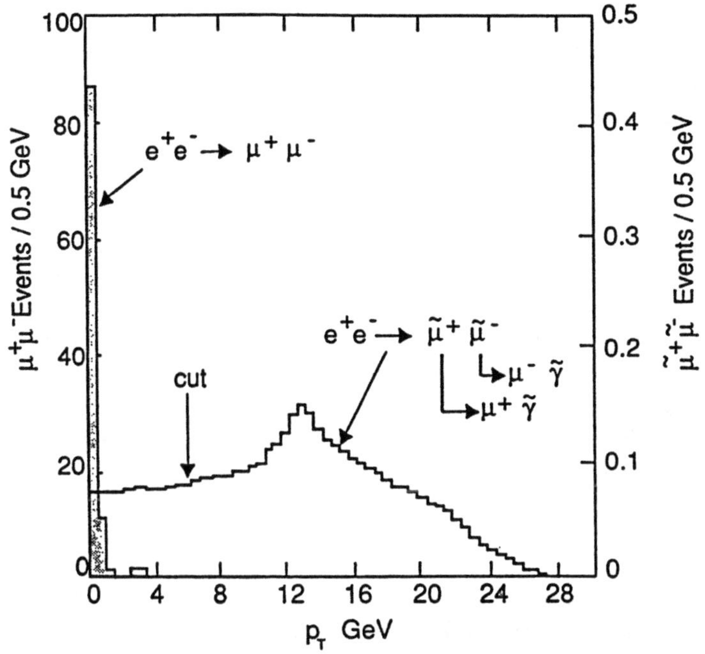

Fig. 14. Transverse momentum distribution of $\mu^+\mu^-$ pairs (L3 data)

The supersymmetric partners of the charged bosons ($W^\pm$, $H^\pm$) are called charginos ($\tilde{\chi}^\pm$), a state which is presumably a mixture of Winos and Higgsinos:

$$\tilde{\chi}^\pm = \alpha \tilde{\mathrm{W}}^\pm + \beta \tilde{\mathrm{H}}^\pm .$$

The chargino is expected to decay into leptons ($\tilde{\chi}^\pm \to l\nu\tilde{\gamma}$) and into quarks ($\tilde{\chi}^\pm \to q'\bar{q}\tilde{\gamma}$). The signature is again missing energy and momentum. Table 11 lists the lower mass limits obtained for charginos [38,37,39] (with the constraint of $m_{\tilde{\gamma}} \leq 20$ GeV).

Table 11. Lower mass limits for charginos (95% CL)

	$\mathrm{m}_{\tilde{\chi}^\pm}$ (GeV)
ALEPH [38]	45
L3 [37]	44
OPAL [39]	45

5. LEP: What Next?

5.1 A Short-Term Perspective

During 1990, one hundred days of LEP running-time are scheduled. Realistically, one could expect a tenfold increase of the statistics from this run, by having a combination of a higher average luminosity ($\approx 1\ 10^{31}\text{cm}^{-2}\text{s}^{-1}$), longer running-time than in 1989, and better overall reliability of the LEP machine and its injector complex. Based on the statistics accumulated in 1989, some 300,000 hadronic Z decays per experiment were expected for 1990.

Meanwhile, with the 1990 running of LEP already in full swing, it was discovered that a substantial increase of the LEP beam current does not produce an equivalent increase in luminosity, because the colliding beams tend to blow each other up ('beam–beam effect'). Studies are being made to find a way of alleviating this problem by changing the optics of the machine. Also, the running efficiency did not improve as much as was hoped. Overall, the 1990 sample of hadronic Z decays may fall short by a factor of 2 with respect to the goal of 300,000 events. Still, it will be an impressive sample for the physics analysis.

It is expected that 1991 will be the year when LEP will be running with the highest luminosity at the Z peak, allowing each of the four experiments to log a million Z decays on tape.

5.2 Plans for the Future

LEP is a machine with quite interesting options for its future running: energy increase, luminosity increase, and longitudinal beam polarization.

- At present, warm radio-frequency (RF) cavities made of copper are installed in LEP, for beam acceleration and for compensation of the continuous energy loss due to synchrotron radiation. For the future, superconducting RF cavities made of copper with a thin surface layer of niobium are considered to be much more economic. By adding such superconducting cavities, and by eventually replacing the warm copper cavities, the LEP energy can be increased from the present 55 GeV to about 100 GeV per beam. Data-taking well above the W pair threshold is scheduled to commence in 1994.

- The higher accelerating power that will be available after the installation of superconducting RF cavities can alternatively be used to compensate the synchrotron energy loss of many more bunches of electrons and positrons: perhaps as many as 36, compared with the 4 bunches that are used at present. This would permit an increase in the luminosity by a factor of 10. Except in the interaction regions that are equipped with detectors, the beam orbits have to be separated in order to minimize beam–beam effects. The proposed geometrical pattern of this separation has given the name 'Pretzel-LEP' to this upgrade option.

- If transverse beam polarization is observed during 1990–91, originating from the emission of synchrotron quanta (a highly non-trivial observation because of the existence of many depolarizing resonances in the beam parameter space), then consideration will be given to the installation of a pair of spin rotators around each of the four interaction points. They would permit the study of the collisions of longitudinally polarized electrons and positrons.

All in all, LEP carries the promise of exciting physics opportunities until the turn of the century: it will permit thorough tests of the Standard Model with unprecedented precision, searches for new particles with mass up to about 100 GeV, and searches for very rare phenomena in Z decays.

Acknowledgement: I wish to thank Professor H Mitter for his kind invitation to give these lectures, and for the warm hospitality that was extended to me in the pleasant atmosphere of Schladming.

References

1 G.S.Abrams et al. (Mark II Collaboration): Phys. Rev. Lett. **63**, 724 (1989)
2 G.Altarelli, R.Kleiss, and C.Verzegnassi (eds.): *Z Physics at LEP* (CERN 89–08, Geneva 1989)
3 D.Decamp et al. (ALEPH Collaboration): Phys. Lett. **B235**, 399 (1990)
4 L.Rolandi: preprint CERN–EP/90–64 (1990)
5 P.Abreu et al. (DELPHI Collaboration): preprint CERN–EP/90–32 (1990)
6 B.Adeva et al. (L3 Collaboration): Phys. Lett. **B237**, 136 (1990)
7 M.Z.Akrawy et al. (OPAL Collaboration): preprint CERN–EP/90–27 (1990)
8 G.S.Abrams et al. (MARK II Collaboration): Phys. Rev. Lett. **63**, 2173 (1989)
9 B.Adeva et al. (L3 Collaboration): Phys. Lett. **B236**, 109 (1990)
10 B.Adeva et al. (L3 Collaboration): L3 preprint L3–005 (1990)
11 G.S.Abrams et al. (MARK II Collaboration): Phys. Rev. Lett. **63**, 2780 (1989)
12 J.Alitti et al. (UA2 Collaboration): Phys. Lett. **B241**, 150 (1990)
13 P.Shalbach et al. (CDF Collaboration): presented at the APS Conference, Washington DC, April 1990
14 A.Blondel: preprint CERN–EP/89–84 (1989)
15 H.Abramowicz et al. (CDHS Collaboration): Phys. Rev. Lett. **57**, 298 (1986); A.Blondel et al. (CDHS Collaboration): Z. Phys. **C45**, 361 (1990)
16 J.V.Allaby et al. (CHARM Collaboration): Phys. Lett. **B177**, 446 (1986); Z. Phys. **C36**, 611 (1987)
17 A.Blondel: preprint CERN–EP/90–10 (1990)
18 V.Plyaskin: presented at the 25th Rencontre de Moriond, Les Arcs, France, March 1990
19 See, for example, B.Andersson et al.: Phys. Rep. **97**, 33 (1983)
20 G.Marchesini and B.R.Webber: Nucl. Phys. **B238**, 1 (1984)
21 P.Aarnio et al. (DELPHI Collaboration): Phys. Lett. **B240**, 271 (1990)
22 D.Decamp et al. (ALEPH Collaboration): Phys. Lett. **B234**, 209 (1990)
23 M.Z.Akrawy et al. (OPAL Collaboration): Phys. Lett. **B235**, 389 (1990); preprint CERN–EP/90–48 (1990)
24 M.Z.Akrawy et al. (OPAL Collaboration): Phys. Lett. **B236**, 364 (1990)
25 F.Abe et al. (CDF Collaboration): University of Pennsylvania preprint UPR-0172E (1989)
26 D.Decamp et al. (ALEPH Collaboration): Phys. Lett. **B236**, 511 (1990)
27 P.Abreu et al. (DELPHI Collaboration): preprint CERN–EP/90–46 (1990)
28 M.Z.Akrawy et al. (OPAL Collaboration): Phys. Lett. **B240**, 250 (1990)
29 D.Decamp et al. (ALEPH Collaboration): Phys. Lett. **B236**,233 (1990)
30 D.Decamp et al. (ALEPH Collaboration): Phys. Lett. **B241**, 141 (1990)
31 D.Decamp et al. (ALEPH Collaboration): preprint CERN–EP/90–70 (1990)
32 P.Abreu et al. (DELPHI Collaboration): preprint CERN–EP/90–44 (1990)
33 M.Z.Akrawy et al. (OPAL Collaboration): Phys. Lett. **B236**, 224 (1990)
34 D.Decamp et al. (ALEPH Collaboration): preprint CERN–EP/90–34 (1990)
35 P.Abreu et al. (DELPHI Collaboration): preprint CERN–EP/90–33 (1990)
36 M.Z.Akrawy et al. (OPAL Collaboration): Phys. Lett. **B242**, 299 (1990)
37 B.Adeva et al. (L3 Collaboration): Phys. Lett. **B233**, 530 (1989)
38 D.Decamp et al. (ALEPH Collaboration): Phys. Lett. **B236**, 86 (1990)
39 M.Z.Akrawy et al. (OPAL Collaboration): Phys. Lett. **B240**, 261 (1990)

QCD and Nuclear Structure

Konrad Bleuler

University of Bonn
Nussalle 12, D – 5300 Bonn, Fed. Rep. of Germany

Abstract Conventional nuclear theory — i.e. visionalizing the nucleus as a bound system of nucleons — is hampered by severe difficulties. They are mainly due to the unavoidable phenomenological character of the expressions for nuclear forces. Even within the well known boson theoretical attempt (so called boson exchange) which so far represents the most common approach, phenomenological parameters represent a basic input (e.g. various coupling constants and a most important cut-off or formfactor). In view of this unpleasant situation the transition from this nucleonic level to an over-all nuclear quark structure represents, in fact, a real revelation: For the first time within the long and tedious history of nuclear science the expression for the basic interaction between the elementary constituents, i.e. the quarks, appears completely determined by a (geometrically) most natural and fundamental theory namely QCD based, in turn, on the all-embracing *local* gauge principle. In other words: Half phenomenological boson exchange appears by now replaced by gluon exchange, i.e. the quanta of the SU(3) gauge field, which, in turn, is uniquely determined by the basic local invariance principle of the theory. In this extremely short and informal report it will be shown within the framework of two different examples, how conventional models may be reinterpreted, or, replaced by using this more basic viewpoint: We first consider the few nucleon systems to be followed — as the main point — by a quark and gauge theoretical reformulation of conventional nuclear shell structure.

1. Introduction

In the course of these last years physics underwent a far reaching and breath taking development, which might best be characterized by a two-dimensional geometrical plot: The horizontal line represents the empirical data with their enormous extension covering by now a domain (delimited on one side by the largest accelerators and the modern radio telescopes on the other) of 40 – 50 decimals if data were compared in typical length scales. In vertical direction one may indicate the developments of basic theories which step by step lead to a characteristic concentration to more general viewpoints (embracing larger domains of empirical data) and ending up (at present!) with the 3 well-known fundamental theories, i.e. general relativity, electroweak and strong interaction.

Fields and Particles Editors: H. Mitter · W. Schweiger

In spite of enormous differences in dimensions and physical meanings, all three are based on the all embracing gauge principle. While the first two cases exhibit additional structures (i.e. the metric structure on one side and the Higgs field on the other which hamper a full application), the 3rd case represents an 'ideal' gauge theory, allowing a.o. (according to general principles) perfect renormalization. The main aim of this paper is to describe a first attempt of englobing nuclear theory (after a long period of separation from the main stream of developments!) into this really beautiful and far reaching picture. As seen from the empirical side it might be remarked in this connection that the periodic table of nuclei represents the best measurable *and* best measured system there is in physics. On the other side, it appears to be an enormous temptation to realize that it leads also to an 'ideal', i.e. non-abelian, gauge theory (e.g. no 'Landau-pole' in contrast to QED!). In other words: A nearly uncountable amount of empirical data (of very different type, e.g. extended level scheme with corresponding assignments; charge distributions, several types of transitions, the enormous realm of reactions, and so on) is to be interpreted by a most natural and instructive geometrical principle, which was first introduced by H. Weyl in 1929 and later enlarged by Yang and Mills in 1963! It might be emphasized that the *local* gauge principle *enforces* the existence of an interaction between the (fermionic) constituents through a bosonic field, whereby the different cases are characterized (or better generated) through the different internal variables, i.e. spin, isospin, flavour and (in our case) colour. These inner variables represent, so to speak, the empirical input for the (automatic) creation of this field action with the help of the local gauge (or invariance) principle. The free choice of the so-called gauge group (acting within the 'fibers' generated in a natural way through the inner variables) leads to a unique determination of the theory. From the empirical side —including the enormous realm of high energy physics— there exists, so far, not the slightest hint of a failure (e.g. a lack of additional terms) of QCD, whereas from the theoretical, or better, mathematical side an enormous effort and a far reaching extension of usual methods are definitely needed. We are a.o. led into the extended world of non-linearity, i.e. a domain which calls for an enormous extension of usual mathematical methods and, at the same time, for a world-wide collaboration including an exchange with recent developments within the corresponding mathematical domains.

As far as the practical applications are concerned it should, first of all, be realized that QCD represents the very basis of the enormous realm of high energy hadron physics: The infinite number of different hadrons with their excited states and their mutual interactions is, by now, to be interpreted as a series of well-defined systems of quarks bound by the (uniquely defined) QCD-gauge field! This means that a practically infinite system of empirical data (i.e. masses, exited states, mutual interactions, or, in other words, reaction processes) appears — in principle — reduced to this geometric idea (thus saving us from introducing an infinite number of elementary constants otherwise needed for describing the world of hadronic states with the enormous realm of interactions and processes).

If we review the present-day situation in conventional nuclear theory under this aspect, we are, in fact, at the very borderline of a similar desperate situation, i.e. introducing a nearly infinite number of parameters: In fact, the enormous domain of empirical data ranging over the whole table of isotopes, their inner excitations and their mutual interaction and various properties as described above, appears, so far, patched by a large number of subdomains corresponding to various, well known nuclear models with their restricted domain of applicability and their particular choice of characterizing parameters (e.g. the shell-model, the theory of deformations, interacting bosons and so on). On one side, we thus have various parameter systems valid in special domains, which in overlapping regions are far from being consistent (e.g. boson exchange and conventional shell-structure); on the other side, it is clear that the various models exhibit each a (restricted) success, in as much as a certain domain of empirical data appears interpreted in an interesting and non trivial way. A well known example represents the conventional shell-structure, which englobes a large number of empirical data with a most restricted number of parameters.

Under these circumstances we are, in a way, forced to ask the following question: Is there a common, deeper theoretical basis behind these special and partly valuable attempts? In other words: Is there hope of reaching a deeper understanding of empirical facts in view of this partly desperate situation, or, might we have reached the very end of understanding of nature? The answer: Definitely not! QCD offers, in fact, a most natural way out of these problems and difficulties, which are due to the fact that empirical data were interpreted on an artificial and unsafe basis, i.e. the nucleonic level of our approach. We realized — and that was a real revelation for those who had suffered for years working within this conventional framework — that there exists a deeper, at first sight unexpected deeper level of a quark structure, in which, for the first time within history of nuclear research, the binding forces between the elementary constituents (the quarks) are due to a most natural and uniquely defined interaction, namely the gluon exchange replacing the half phenomenological boson exchange on the nucleonic level. In a more concise form: For the first time in the long history of nuclear science, calculations are to be based on a well defined, i.e. consistent and even most natural basic theory. Hereby the decisive point is not the existence of quarks as basic constituents, but the fact that the local gauge principle yields on this level a practically parameterfree expression for the binding of the corresponding systems. Under this new and basic aspect the various models discussed above may, by now, be considered to represent more or less valuable approximation schemes, or better, intermediate steps on the way to this basic insight. Certain inconsistencies between these approaches thus become understandable and were even to be expected. Our task, therefor, consists in reinterpreting (by enlarging and partly by replacing) various conventional models, or better, in proving (in spite of their formal differences) their common origin. We will try this here in two characteristic cases, namely in the few nucleon problem, i.e. on the lowest part of the periodic table on one side, and in reinterpreting conventional shell structure in the realm of heavy nuclei on the other side. In this connection it should be emphasized that we are *not* aiming to obtain 'better' results (such claims, which are unfortunately often heard, are not always really meaningful in view of the situation described above), but we rather look for a deeper and more consistent understanding of nuclear properties.

2. The Realm of Light Nuclei

Let me start with the role of QCD within the 'Few-Body-Problem'.

1. The two-body problem below the pion production threshold is — as well known — reasonably well described on the basis of boson exchange [1], e.g. the so-called Bonn potential. This result is based on the following scheme:
   (a) One considers the exchange of 3 kinds of bosons, i.e. π, ω, ρ.
   (b) The corresponding coupling constants as well as the formfactors or cut-off parameters Λ are to be determined by 'fitting' the experimental scattering data.
   (c) It is of decisive importance to include the virtual Δ-excitations by including different types of so called block diagrams, which a.o. also contain a direct $\pi - \pi$ interaction, that is to be determined empirically with the help of the (measured) $\pi - 2\pi$ scattering process. The global effect of these tedious and far reaching calculations is usually replaced by the simplifying — but by no means rigorous — σ-exchange containing only 2 new parameters.

   As a non-trivial result, one then realizes that practically *all* experimental scattering phases (i.e. about 12 empirical functions in the energy interval 0 – 300 MeV) are reasonably well reproduced by adapting a relatively small number (about 8) of parameters mentioned above, whereby the Λ -value (about 1.3 GeV) plays a decisive role.

2. A most interesting — in fact to the founder of this method unexpected — result is by now the following: Inserting these — by far overdetermined — parameter-values into enormously enlarged calculations of the inelastic domain between 300 MeV and 1 GeV, one again obtains a reasonable representation of the empirical phases. This includes in particular the empirically measured inelasticity (i.e. π-emission above the threshold) as well as the broad resonance corresponding to the (real!) Δ-excitation. The corresponding calculations had to take into account (in order to maintain unitarity!) the pionic selfenergies (about 1 GeV) including the corresponding mass *and* coupling constant renormalization, i.e. a 'new dimension' of numerical calculations. Apart from the decisive practical importance of this result in connection with so-called medium-energy accelerator-experiments, it was realized that these calculations lead us close to the 'abyss' of the appearance

of 'negative probabilities', or very close to the so-called Landau-singularity, as suggested by unpublished results by Wilson based on the 'renormalization group'! In this connection it appears interesting to observe, that empirical results above these critical energies exhibit the appearance of 'resonances' due to 'non-perturbative' quasi-bound states.

3. These non-trivial results are based on the assumption of fixed empirical values for the various vertices, i.e. coupling constants and formfactors. Therefore the natural role of QCD is a theoretical determination of these parameters through the uniquely determined gauge theoretical interaction between the elementary constituents of the various hadrons, the quarks, i.e. a program partly started in the case of the pion-interactions. In addition it should be checked that — as to be expected! — a direct gauge-theoretical interaction between the various hadrons (i.e. without exchanging bosons which by now play the role of an enlarged 'instanton-exchange') plays only a minor role in this realm of relatively low energies.

4. With the very same basic assumption, i.e. fixed vertices, an enormous amount of work has been done within the 3-body problem. Although the binding energy of He^3 and H^3 has eventually been reached, severe problems remain: the theoretical understanding of the experimental charge distribution and the low energy 3-body scattering results. In both cases the difficulties appear related to the small-distances where the value of Λ (i.e. the inner structure of the various hadrons) becomes essential. One therefor is under the impression that a more detailed quark-theoretical description of the processes appears — at least to some extent — already needed.
In this way success *and* limitations of the boson exchange in the realm of light nuclei become apparent. At the same time one realizes the important role of QCD, that is definitely needed for a complete picture of this domain.

3. A New Interpretation of Conventional Shell Structure

The main point of this short contribution is a definitely needed *reinterpretation* of nuclear shell within the framework of QCD, i.e. a quark structure with interaction determined by the gauge principle [2].

1. First of all it should be emphasized that boson exchange (on the nuclear level) leeds — in contrast to a widespread belief — to severe and even unavoidable difficulties:

   (a) A discussion of the binding energy based on the complete nuclear state functional containing simultaneously fermionic *and* bosonic variables (i.e. occupation numbers of the bosonic states) tells us that the (simplifying) σ-exchange plays the dominating role (about 90%). Due to its scalar properties, this term immediately leads to an average nuclear potential which, however, appears (in view of the most complicated block diagrams which generate the σ-interaction) as a rather artificial construction and thus calls for a more natural interpretation.

(b) The main difficulty arises in particular from the shell closedness, i.e. the empirically (sharp) magic numbers. The σ-terms, i.e. the block diagrams, contain, however, a relatively strong Δ-admixture (about 5%), a fact which forbids automatically sharp fermionic numbers of identical fermions.

(c) A more general problem arises from the automatic contribution due to bosonic self-energies and the corresponding definitely needed renormalization, i.e. a program which — as far as I can see— lies beyond practical feasibility.

2. In view of this rather desperate situation, a quark- (i.e. a gauge theoretical) reinterpretation of nuclear shell structure appears to be a real revelation. The nucleon is by now visualized from the outset as a system of quarks bound through characteristic (non-perturbative!) gauge-theoretical expressions. In concentrated and summarizing form the proposal reads as follows:

(a) Start with independent single quark states in a spherical enclosure. Its radius follows — in agreement with a well-known empirical law — from the equilibrium condition between the vacuum pressure (taken from high energy QCD calculations) and the inner pressure due to the occupied single particle states. A spin-orbit term with the characteristic (opposite) sign follows naturally from the QCD-surface conditions. (Within the conventional, i.e. nucleonic, theory the experimental value of this term was never reproduced (factor about 1/2) in a satisfactory way).

(b) The basic condition of colourfree bound states yields to 3-quark colourfree substructures (i.e. a preformation of embedded nucleons, to be enlarged for the introduction of antiquarks leading to boson-like structures) by using a simple group-theoretical argument.

(c) The degeneracy of these substructures, which contain so far 3 different individual quark spins, is split into a ground state with nucleon-like quantum numbers (i.e. a diquark and a valence quark [4]) and an excited state corresponding to a nucleonic Δ-excitation with the help of a generalized quark-quark pairing force due to a *non*-perturbative instanton exchange (generalizing a former result of 't Hooft).

(d) A still needed correlation in space between the 3 quarks of an embedded nucleon may be obtained by using a Kogut term or so-called Mercedes star interaction characteristic for QCD. As a result of this construction (starting from independent quarks and building up the needed correlations for the nucleonic structure step by step) one obtains the well-known shell-structure by now (having all the way used non-perturbative

QCD, i.e. vacuum-pressure, colour-condition, Instanton-exchange and so on). Although this work constitutes just a very first attempt, which calls for a far reaching enlargement, it appears to be a kind of a revelation that QCD provides strangely enough just the needed and most characteristic terms for an understanding of conventional shell structure. At the same time this construction overcomes the enormous difficulties (and contradictions!) of the bosonic approach and represents also an enormous simplification!

4. Conclusion

In two different domains of the periodic table QCD may lead to a natural interpretation of empirical data practically without the use of phenomenological parameters. In spite of the apparent simplicity of these very first steps, it should be emphasized that a more complete nuclear theory based on QCD represents a great challenge for a full generation. It would lead, however, to a central domain of applications of recent and far reaching methods of modern mathematical physics, very much in contrast to the conventional half-phenomenological approaches that lead to practically unlimited enlargements of the corresponding computer programs.

References

1 Charlotte Elster: *Nucleon-Nucleon Interactions at Intermediate Energies and Related Nuclear Processes*, 12th International Conference on 4-Body Problems, Vancouver, July 1989 (and literature mentioned there!)

2 K. Bleuler et. al.: Zeitschrift Naturforschung **38a**, 705 (1983);
H. R. Petry et. al.: Physics Letters **159B**, 93 (1985);
K. Bleuler in *Fundamental Aspects of Quantum Theory*, V. Gorini, A. Frigenio editors (Plenum Publ. Corp. 1986) p. 279;
K. Bleuler in *Symmetries in Science II*, B. Gruker, R. Lenezewski eds. (Plenum Publ. Corp. 1986) p. 61;
K. Bleuler: *QCD and Nuclear Structure* in *Quarks and Nuclei*, Proceedings of the 10th Kikuchi Spring School, O. Hashimoto and F. Sakata ed., Shimoda, April 1987, p. 311

3 K. Bleuler et. al.: *Pairing Approximation in Spherical Nuclei I and II*, Nouvo Cim. **52B**,45 (1967); **52B**, 149 (1967);
K. Bleuler et. al.: *Nuclear Separation Energies ...*, Arkiv for Fysik, Vol. 36, p. 385

4 K. Bleuler et. al.: *Pairing Theory and Diquarks* in *Workshop on Diquarks*, M. Anselmino and E. Predazzi eds. (World Scientific 1988) p. 58

MIX
Papier aus verantwortungsvollen Quellen
Paper from responsible sources
FSC® C105338

If you have any concerns about our products,
you can contact us on
ProductSafety@springernature.com

In case Publisher is established outside the EU,
the EU authorized representative is:
Springer Nature Customer Service Center GmbH
Europaplatz 3, 69115 Heidelberg, Germany

Printed by Libri Plureos GmbH
in Hamburg, Germany